Lecture Notes in Computational Science and Engineering

19

Springer-Verlag Berlin Heidelberg GmbH

Ivo Babuška
Philippe G. Ciarlet
Tetsuhiko Miyoshi
Editors

Mathematical Modeling and Numerical Simulation in Continuum Mechanics

Proceedings of the International Symposium on Mathematical Modeling and Numerical Simulation in Continuum Mechanics, September 29 - October 3, 2000 Yamaguchi, Japan

With 83 Figures

Springer

Editors

Ivo Babuška

Department of Aerospace Engineering
& Engineering Mechanics
The University of Texas at Austin
WRW 215, C0600
Austin, Texas 78712-1085, USA
e-mail: babuska@ticam.utexas.edu

Tetsuhiko Miyoshi

Department of Mathematical Sciences
Faculty of Science
Yamaguchi University
Yoshida 1677-1
753-8512 Yamaguchi, Japan
e-mail: miyoshi@po.cc.yamaguchi-u.ac.jp

Philippe G. Ciarlet

Laboratoire d'Analyse Numérique
Université Pierre et Marie Curie
Boîte courrier 187
75252 Paris cedex 05, France
e-mail: pgc@ann.jussieu.fr

Cataloging-in-Publication Data applied for

Die Deutsche Bibliothek - CIP-Einheitsaufnahme

Mathematical modeling and numerical simulation in continuum mechanics : proceedings of the International Symposium on Mathematical Modeling and Numerical Simulation in Continuum Mechanics, September 29 - October 3, 2000, Yamaguchi, Japan / Ivo Babuška ... ed.. - Berlin ; Heidelberg ; New York ; Barcelona ; Hong Kong ; London ; Milan ; Paris ; Tokyo : Springer, 2002
(Lecture notes in computational science and engineering ; Vol. 19)
ISBN 978-3-540-42399-7 ISBN 978-3-642-56288-4 (eBook)
DOI 10.1007/978-3-642-56288-4

Mathematics Subject Classification (2000):
65-06, 65Mxx, 65Nxx, 74-06, 74Bxx, 74Cxx, 74Kxx, 74Rxx, 76-06, 76Dxx

ISSN 1439-7358
ISBN 978-3-540-42399-7

http://www.springer.de

Originally published by Springer-Verlag Berlin Heidelberg in 2002

Cover Design: Friedhelm Steinen-Broo, Estudio Calamar, Spain
Cover production: *design & production*
Typeset by the authors using a Springer TEX macro package

Printed on acid-free paper SPIN: 10778671 46/3142/LK - 5 4 3 2 1 0

Preface

The first international symposium on mathematical foundations of the finite element method was held at the University of Maryland in 1973. During the last three decades there has been great progress in the theory and practice of solving partial differential equations, and research has extended in various directions. Full-scale nonlinear problems have come within the range of numerical simulation. The importance of mathematical modeling and analysis in science and engineering is steadily increasing. In addition, new possibilities of analysing the reliability of computations have appeared. Many other developments have occurred: these are only the most noteworthy.

This book is the record of the proceedings of the International Symposium on Mathematical Modeling and Numerical Simulation in Continuum Mechanics, held in Yamaguchi, Japan from 29 September to 3 October 2000. The topics covered by the symposium ranged from solids to fluids, and included both mathematical and computational analysis of phenomena and algorithms. Twenty-one invited talks were delivered at the symposium. This volume includes almost all of them, and expresses aspects of the progress mentioned above. All the papers were individually refereed. We hope that this volume will be a stepping-stone for further developments in this field.

The symposium was supported by many people and organizations. Special thanks should go to Professor H. Fujita and Professor H. Hironaka, for their support and advice in planning this symposium, and also to Professor Ohtsuka, Professor Tabata and Mr. Hataya, the members of the local organization committee, for their enthusiastic and continued cooperation. It should be acknowledged that the symposium was sponsored by the Japan Association for Mathematical Sciences, the Inoue Foundation for Science, Mitsui Zosen System Research Inc., the Yamaguchi Tourism and Convention Association, Grants-in-Aid for Scientific Research (A)-10304012, 11304004, (B)-10440035,12440041, and Grant-in-Aid for Exploratory Research 11874022. Finally we wish to express our gratitude to The Japan Society for Industrial and Applied Mathematics, The Japan Society for Computational Engineering and Science, and the Department of Mathematical Science of Yamaguchi University, for their support and cooperation.

March 20, 2001

Ivo Babuška
Philippe G. Ciarlet
Tetsuhiko Miyoshi

Table of Contents

Nonlinear Shell Models of Koiter's Type

Philippe G. Ciarlet

Laboratoire d'Analyse Numérique, Université Pierre et Marie Curie,
4 place Jussieu, 75005 Paris, France

Abstract. We describe, and we discuss the merits of, a two-dimensional nonlinear shell model analogous to a model proposed by W.T. Koiter in 1966, where the exact change of curvature tensor is suitably modified. A first interest of this model, from the computational viewpoint, is that the resulting stored energy function becomes a polynomial with respect to the unknown components of the deformation field and their partial derivatives.

A second interest of this model is its amenability to a formal asymptotic analysis of its solution, with the thickness as the "small" parameter. Such an analysis yields exactly the same conclusions as the formal asymptotic analysis of the solution of the three-dimensional equations, thus providing a justification of the proposed model.

1 A Two-Dimensional Nonlinear Shell Model Proposed by W.T. Koiter

Greek indices and exponents, except ε and ν in ∂_ν, take their values in the set $\{1, 2\}$, Latin indices and exponents take their values in the set $\{1, 2, 3\}$, and the summation convention with respect to repeated indices and exponents is systematically used. The Euclidean inner product and the exterior product of $\mathbf{a}, \mathbf{b} \in \mathbb{R}^3$ are denoted $\mathbf{a} \cdot \mathbf{b}$ and $\mathbf{a} \wedge \mathbf{b}$ and the Euclidean norm of $\mathbf{a} \in \mathbb{R}^3$ is denoted $|\mathbf{a}|$.

Let ω be a bounded, open, connected subset of $\mathbb{R}^2$ with a Lipschitz-continuous boundary γ, the set ω being locally situated on a same side of γ. A generic point in the set $\overline{\omega}$ being denoted $y = (y_\alpha)$, we let $\partial_\alpha := \partial/\partial y_\alpha$ and $\partial_{\alpha\beta} = \partial^2/\partial y_\alpha \partial y_\beta$.

Let there be given an injective mapping $\boldsymbol{\theta} \in \mathcal{C}^2(\overline{\omega}; \mathbb{R}^3)$ such that the two vectors $\mathbf{a}_\alpha(y) := \partial_\alpha \boldsymbol{\theta}(y)$ are linearly independent at all points $y \in \overline{\omega}$. The two vectors $\mathbf{a}_\alpha(y)$ span the tangent plane to the *surface* $S := \boldsymbol{\theta}(\overline{\omega})$ at the point $\boldsymbol{\theta}(y) \in S$ and the unit vector $\mathbf{a}_3(y) := \dfrac{\mathbf{a}_1(y) \wedge \mathbf{a}_2(y)}{|\mathbf{a}_1(y) \wedge \mathbf{a}_2(y)|}$ is normal to S at $\boldsymbol{\theta}(y)$. The three vectors $\mathbf{a}_i(y)$ form the *covariant basis* at $\boldsymbol{\theta}(y)$, while the three vectors $\mathbf{a}^i(y)$ defined by the relations $\mathbf{a}^i(y) \cdot \mathbf{a}_j(y) = \delta^i_j$ form the *contravariant basis* at $\boldsymbol{\theta}(y)$. Note that the vectors $\mathbf{a}^\alpha(y)$ defined in this fashion also span the tangent plane to S at $\boldsymbol{\theta}(y)$ and that $\mathbf{a}^3(y) = \mathbf{a}_3(y)$. The coordinates y_α of the points $y \in \overline{\omega}$ constitute *curvilinear coordinates* for the surface S.

The covariant components $a_{\alpha\beta}$ and the contravariant components $a^{\alpha\beta}$ of the *metric tensor* of S, also called the *first fundamental form* of S, and the covariant components $b_{\alpha\beta}$ of the *curvature tensor* of S, also called the *second fundamental form* of S, are respectively defined by

$$a_{\alpha\beta} := \mathbf{a}_\alpha \cdot \mathbf{a}_\beta, \quad a^{\alpha\beta} := \mathbf{a}^\alpha \cdot \mathbf{a}^\beta, \quad b_{\alpha\beta} := \mathbf{a}^3 \cdot \partial_\alpha \mathbf{a}_\beta \ .$$

The *area element* along the surface S is $\sqrt{a}\, dy$, where $a := \det(a_{\alpha\beta})$. Note that $\sqrt{a} = |\mathbf{a}_1 \wedge \mathbf{a}_2|$, so that we also have

$$b_{\alpha\beta} = \frac{1}{\sqrt{a}} \partial_{\alpha\beta}\boldsymbol{\theta} \cdot \{\partial_1 \boldsymbol{\theta} \wedge \partial_2 \boldsymbol{\theta}\} \ .$$

For more details about the differential geometry of surfaces, see, e.g., do Carmo (1976), Klingenberg (1973), or Ciarlet (2000a, Chap. 2).

Let $\Omega^\varepsilon := \omega \times]-\varepsilon, \varepsilon[$, let $x^\varepsilon = (x_i^\varepsilon)$ denote a generic point in the set $\overline{\Omega}^\varepsilon$, and let $\partial_i^\varepsilon := \partial/\partial x_i^\varepsilon$. Then, for $\varepsilon > 0$ small enough, the mapping $\boldsymbol{\Theta} : \overline{\Omega}^\varepsilon \to \mathbb{R}^3$ defined by

$$\boldsymbol{\Theta}(y, x_3^\varepsilon) := \boldsymbol{\theta}(y) + x_3^\varepsilon \mathbf{a}_3(y) \text{ for all } (y, x_3^\varepsilon) \in \overline{\omega} \times [-\varepsilon, \varepsilon] = \overline{\Omega}^\varepsilon \ ,$$

is injective and the three vectors $\mathbf{g}_i^\varepsilon(x^\varepsilon) := \partial_i^\varepsilon \boldsymbol{\Theta}(x^\varepsilon)$ are linearly independent at all points $x^\varepsilon \in \overline{\Omega}^\varepsilon$ (see Ciarlet (2000a, Thm. 3.1-1)). The vectors $\mathbf{g}_i^\varepsilon(x^\varepsilon)$ then form the *covariant basis* at the point $\boldsymbol{\Theta}(x^\varepsilon)$.

Consider then a *shell* with *middle surface S and constant thickness* $2\varepsilon > 0$, i.e., a body whose *reference configuration* is the set $\boldsymbol{\Theta}(\overline{\Omega}^\varepsilon)$.

The shell is subjected to a *homogeneous boundary condition of place* on the portion $\boldsymbol{\Theta}(\gamma_0 \times [-\varepsilon, \varepsilon])$ of its lateral face, where γ_0 is a subset of γ satisfying *length* $\gamma_0 > 0$. This means that the displacement field vanishes on this portion.

The shell is subjected to *applied body forces* in its interior $\boldsymbol{\Theta}(\Omega^\varepsilon)$ and to *applied surface forces* on its "upper" and "lower" faces $\boldsymbol{\Theta}(\omega \times \{\varepsilon\})$ and $\boldsymbol{\Theta}(\omega \times \{-\varepsilon\})$, given by their contravariant components $f^{i,\varepsilon} \in L^2(\Omega^\varepsilon)$ and $h^{i,\varepsilon} \in L^2(\Gamma_+^\varepsilon \cup \Gamma_-^\varepsilon)$, i.e., their components over the vectors $\mathbf{g}_i^\varepsilon$ of the covariant bases. We then define functions $p^{i,\varepsilon} \in L^2(\omega)$ by letting

$$p^{i,\varepsilon} := \int_{-\varepsilon}^{\varepsilon} f^{i,\varepsilon}\, dx_3^\varepsilon + h^{i,\varepsilon}(\cdot, \varepsilon) + h^{i,\varepsilon}(\cdot, -\varepsilon) \ .$$

Note that the remaining portion $\boldsymbol{\Theta}((\gamma - \gamma_0) \times [-\varepsilon, \varepsilon])$ of the lateral face of the shell is free.

Finally, it is assumed that the shell is constituted by a *nonlinearly elastic, homogeneous*, and *isotropic*, material and that its reference configuration is a *natural state*. Hence (cf., e.g., Ciarlet (1988, Sect. 3.8)) the behavior of the constituting material is governed by its two *Lamé constants* $\lambda > 0$ and $\mu > 0$ (that they are > 0 follows from experimental evidence). The functions

$$a^{\alpha\beta\sigma\tau} := \frac{4\lambda\mu}{\lambda + 2\mu} a^{\alpha\beta} a^{\sigma\tau} + 2\mu(a^{\alpha\sigma} a^{\beta\tau} + a^{\alpha\tau} a^{\beta\sigma})$$

denote the contravariant components of the *two-dimensional elasticity tensor of the shell*, the form of which can be fully justified by an asymptotic analysis with the thickness as the "small" parameter (see, e.g., Ciarlet (2000a)).

The unknown in the nonlinear model proposed by Koiter (1966) is the vector field $\boldsymbol{\zeta}^\varepsilon = (\zeta_i^\varepsilon) : \overline{\omega} \to \mathbb{R}^3$, where the functions $\zeta_i^\varepsilon : \overline{\omega} \to \mathbb{R}$ are the covariant components, i.e., the components over the vectors $\mathbf{a}^i$ of the contravariant bases, of the *displacement* vector field of the points of the middle surface S; in other words, the displacement vector of each point $\boldsymbol{\theta}(y) \in S$, $y \in \overline{\omega}$, is the vector $\zeta_i^\varepsilon(y)\mathbf{a}^i(y)$.

Remark. Koiter's model, as well as the model proposed in Sect. 2, can be re-formulated in terms of *deformation* vector fields: This means that the unknown then becomes the *deformation field of the surface* S, i.e., the vector field $\boldsymbol{\phi}^\varepsilon : \overline{\omega} \to \mathbb{R}^3$ defined by

$$\boldsymbol{\phi}^\varepsilon(y) = \boldsymbol{\theta}(y) + \zeta_i^\varepsilon(y)\mathbf{a}^i(y) \ .$$

In other words, $\boldsymbol{\phi}^\varepsilon(y)$ is the new position occupied by the point $\boldsymbol{\theta}(y) \in S$, under the influence of the applied forces.

Not only does this re-formulation avoid the introduction of covariant derivatives, but it also shows that the two-dimensional shell models discussed here can be equivalently expressed in terms of *Cartesian* components of the unknown. Note that the *variables* in both the "displacement" and the "deformation" approaches are the *curvilinear coordinates* y_α of the surface S. □

Given a sufficiently smooth, but otherwise arbitrary, field $\boldsymbol{\eta} = (\eta_i) : \overline{\omega} \to \mathbb{R}^3$ and its associated displacement field $\eta_i\mathbf{a}^i$ of the surface S, let

$$a_{\alpha\beta}(\boldsymbol{\eta}) := \mathbf{a}_\alpha(\boldsymbol{\eta}) \cdot \mathbf{a}_\beta(\boldsymbol{\eta}), \text{ where } \mathbf{a}_\alpha(\boldsymbol{\eta}) := \partial_\alpha(\boldsymbol{\theta} + \eta_i\mathbf{a}^i) \ ,$$

denote the covariant components of the *metric tensor of the "deformed" surface* $(\boldsymbol{\theta} + \eta_i\mathbf{a}^i)(\overline{\omega})$. The functions

$$G_{\alpha\beta}(\boldsymbol{\eta}) := \frac{1}{2}(a_{\alpha\beta}(\boldsymbol{\eta}) - a_{\alpha\beta})$$

then designate the covariant components of the *change of metric tensor* associated with the displacement field $\eta_i\mathbf{a}^i$ of S. If, *in addition*, the two vectors $\mathbf{a}_\alpha(\boldsymbol{\eta})$ are linearly independent at all points $y \in \overline{\omega}$, the functions

$$b_{\alpha\beta}(\eta) := \frac{1}{\sqrt{a(\boldsymbol{\eta})}}\partial_{\alpha\beta}(\boldsymbol{\theta} + \eta_i\mathbf{a}^i) \cdot \{\mathbf{a}_1(\boldsymbol{\eta}) \wedge \mathbf{a}_2(\boldsymbol{\eta})\} \ ,$$

where

$$a(\boldsymbol{\eta}) := \det(a_{\alpha\beta}(\boldsymbol{\eta})) = |\mathbf{a}_1(\boldsymbol{\eta}) \wedge \mathbf{a}_2(\boldsymbol{\eta})|^2 \ ,$$

are well defined in $\overline{\omega}$. They denote the covariant components of the *curvature tensor of the "deformed" surface* $(\boldsymbol{\theta} + \eta_i\mathbf{a}^i)(\overline{\omega})$. The functions

$$R_{\alpha\beta}(\boldsymbol{\eta}) := b_{\alpha\beta}(\boldsymbol{\eta}) - b_{\alpha\beta}$$

then designate the covariant components of the *change of curvature tensor* associated with the displacement field $\eta_i \mathbf{a}^i$ of S.

Founding his approach on various *a priori* assumptions, of a geometrical and mechanical nature, Koiter (1966) has proposed the following *two-dimensional minimization problem* for modeling the shell problem described above: The unknown displacement field $\zeta_i^\varepsilon \mathbf{a}^i$ of the middle surface S should be such that $\boldsymbol{\zeta}^\varepsilon := (\zeta_i^\varepsilon)$ is a minimizer, or more generally a stationary point, of the *energy* j_K^ε defined by (see *ibid.*, eqs. (4.2), (8.1), and (8.3)):

$$j_K^\varepsilon(\boldsymbol{\eta}) := \frac{\varepsilon}{2} \int_\omega a^{\alpha\beta\sigma\tau} G_{\sigma\tau}(\boldsymbol{\eta}) G_{\alpha\beta}(\boldsymbol{\eta}) \sqrt{a}\, \mathrm{d}y$$
$$+ \frac{\varepsilon^3}{6} \int_\omega a^{\alpha\beta\sigma\tau} R_{\sigma\tau}(\boldsymbol{\eta}) R_{\alpha\beta}(\boldsymbol{\eta}) \sqrt{a}\, \mathrm{d}y - \int_\omega p^{i,\varepsilon} \eta_i \sqrt{a}\, \mathrm{d}y$$

for smooth enough fields $\boldsymbol{\eta} = (\eta_i)$ satisfying *ad hoc* boundary conditions on γ_0. Note that this model is indeed *nonlinear*, in that the functional j_K^ε is not a quadratic function of $\boldsymbol{\eta}$.

However, the functions $b_{\alpha\beta}(\boldsymbol{\eta})$, whence the functions $R_{\alpha\beta}(\boldsymbol{\eta})$, are *not* defined at those points of $\overline{\omega}$ where the vectors $\mathbf{a}_\alpha(\boldsymbol{\eta}) = \partial_\alpha(\boldsymbol{\theta} + \eta_i \mathbf{a}^i)$ are linearly dependent. Hence *this minimization problem is not well posed.*

Our objective consists in showing, first, how the above model of W.T. Koiter can be modified so as to avoid this difficulty; then, how the modified model can be fully justified. The results described here were first announced in Ciarlet (2000b) and Ciarlet & Roquefort (2000). Full details are found in Ciarlet (2001) and Ciarlet & Roquefort (2001).

2 A Two-Dimensional Nonlinear Shell Model "of Koiter's Type"

The *strain energy in Koiter's model* (i.e., the part of the expression $j_K^\varepsilon(\boldsymbol{\eta})$ that does not involve the applied forces) is exactly the *sum* of the *strain energy of a nonlinearly elastic "membrane" shell*, i.e., the part in $j_K^\varepsilon(\boldsymbol{\eta})$ with $\frac{\varepsilon}{2}$ as a factor, and of the *strain energy of a nonlinearly elastic "flexural" shell*, i.e., the part in $j_K^\varepsilon(\boldsymbol{\eta})$ with $\frac{\varepsilon^3}{6}$ as a factor, as they have been recently identified and justified by Miara (1998) and Lods & Miara (1998) by means of a formal asymptotic analysis of the appropriately "scaled" three-dimensional displacement field, with the thickness 2ε as the "small" parameter.

We recall that, in this approach, a nonlinearly elastic shell is deemed either a *"membrane"* or a *"flexural"* one, according to whether a certain manifold of *"inextensional" displacement fields* $\eta_i \mathbf{a}^i$, i.e., that satisfy the relations $G_{\alpha\beta}(\boldsymbol{\eta}) = 0$ in ω, together with *ad hoc* boundary conditions on γ_0, either contains only $\eta_i \mathbf{a}^i = \mathbf{0}$, or contains nonzero displacement fields, in which case the tangent space at each point of the manifold must also contain nonzero elements (for a discussion about the difficulties inherent to a satisfactory classification of nonlinearly elastic shells, see Ciarlet (2000a, Sects. 9.1 and 10.2)).

Remark. Another "membrane" strain energy for a nonlinearly elastic shell has been identified and justified by Le Dret & Raoult (1996). Using *Γ-convergence theory*, they have established the *weak convergence*, in an *ad hoc* space $\mathbf{W}^{1,p}(\Omega)$, of a subsequence of the minimizers of the three-dimensional energy, appropriately scaled over the fixed domain $\Omega = \omega \times]-1, 1[$, toward a minimizer of a "limit energy" as the thickness of the shell approaches zero.

The *Le Dret-Raoult strain energy* appearing in this limit energy is indeed that of a "membrane" shell, in the sense that it is again only a function of the *change of metric tensor* $G_{\alpha\beta}(\boldsymbol{\eta})$. However, this strain energy does not coincide with the "membrane" strain energy found *via* the formal approach by Miara (1998), save for particular deformations identified by Genevey (1997) (for further comments about the comparison between these "membrane" theories, see Ciarlet (2000a, Sect. 9.5)).

A likely explanation for this difference may lie in that the formal approach corresponds to "rigid" nonlinearly elastic materials, while the Γ-convergence approach corresponds to "soft" nonlinearly elastic materials. This assertion remains yet to be mathematically substantiated, however. □

A careful scrutiny of the formal asymptotic approach of Lods & Miara (1998) reveals that, instead of the expected components $R_{\alpha\beta}(\boldsymbol{\eta})$ of the "exact" change of curvature tensor, the components that naturally appear in the course of the asymptotic analysis are different, though closely related. These functions, originally denoted $\widehat{\mathcal{E}}^1_{\alpha\|\beta}(\boldsymbol{\eta})$ by Lods & Miara (1998, Lemma 3), have later been given a remarkably simple expression by Roquefort (2001), viz.,

$$R^\sharp_{\alpha\beta}(\boldsymbol{\eta}) := \frac{1}{\sqrt{a}} \partial_{\alpha\beta}(\boldsymbol{\theta} + \eta_i \mathbf{a}^i) \cdot \{\mathbf{a}_1(\boldsymbol{\eta}) \wedge \mathbf{a}_2(\boldsymbol{\eta})\} - b_{\alpha\beta} \ .$$

Remarks. (1) Clearly, $R^\sharp_{\alpha\beta}(\boldsymbol{\eta}) = R_{\alpha\beta}(\boldsymbol{\eta})$ if the displacement field $\eta_i \mathbf{a}^i$ is *inextensional*, since $a(\boldsymbol{\eta}) = a$ in this case. This explains why the strain energy of a nonlinearly elastic "flexural" shell can be as well expressed in terms of the functions $R_{\alpha\beta}(\boldsymbol{\eta})$, since the energy of such a shell is to be minimized over a *manifold of inextensional displacements.* This is also a first indication that the associated minimization problem could be well posed, since there is no longer a possibly vanishing denominator in the stored energy function. Indeed, one can establish the *existence* of a least one minimizer; cf. Ciarlet & Coutand (1998).

(2) Interestingly, exactly the same functions $R^\sharp_{\alpha\beta}(\boldsymbol{\eta})$ are also mentioned by Koiter (1966, eq. (4.11)), who calls them the covariant components of a *"modified" change of curvature tensor.* However, W.T. Koiter does not provide any hint about the *raison d'être* of these functions, which he most likely found by means of an entirely different approach.

(3) The above considerations suggest that the functions $R^{\sharp}_{\alpha\beta}(\boldsymbol{\eta})$ could be aptly called the covariant components of the *"modified change of curvature tensor of Koiter-Lods-Miara-Roquefort"*. □

On the basis of the aforementioned observations, Ciarlet (2000b, 2001) has proposed the following *two-dimensional nonlinear shell model "of Koiter's type"* for modeling the same shell problem as in Sect. 1: The unknown displacement field $\zeta_i^\varepsilon \mathbf{a}^i$ of the middle surface should be such that $\boldsymbol{\zeta}^\varepsilon := (\zeta_i^\varepsilon)$ is a minimizer, or more generally a stationary point, of the *energy* j^ε defined by

$$j^\varepsilon(\boldsymbol{\eta}) := \frac{\varepsilon}{2}\int_\omega a^{\alpha\beta\sigma\tau} G_{\sigma\tau}(\boldsymbol{\eta}) G_{\alpha\beta}(\boldsymbol{\eta})\sqrt{a}\,\mathrm{d}y$$
$$+\frac{\varepsilon^3}{6}\int_\omega a^{\alpha\beta\sigma\tau} R^{\sharp}_{\sigma\tau}(\boldsymbol{\eta}) R^{\sharp}_{\alpha\beta}(\boldsymbol{\eta})\sqrt{a}\,\mathrm{d}y - \int_\omega p^{i,\varepsilon}\eta_i\sqrt{a}\,\mathrm{d}y\ ,$$

over an *affine space* of sufficiently smooth vector fields $\boldsymbol{\eta}$ (e.g., $\boldsymbol{\eta} \in \mathbf{W}^{2,p}(\omega)$ for some $p > 2$) satisfying *ad hoc* boundary conditions (e.g., the boundary conditions "of strong clamping" $\boldsymbol{\eta} = \mathbf{0}$ and $\partial_\nu \boldsymbol{\eta} = \mathbf{0}$ on γ_0, where ∂_ν denotes the outer normal derivative along γ; cf. Ciarlet (2000a, Sect. 10.5).

A *first interest* of this model is that, contrary to Koiter's model described in Sect. 1, *its stored energy function no longer possesses a possibly vanishing denominator* (viz., $\sqrt{a(\boldsymbol{\eta})}$, which has been "replaced" by $\sqrt{a}$), so that the corresponding minimization problem can be posed over an "entire" vector space.

Its *second interest* is its (relative) simplicity from a computational viewpoint, since its stored energy function is a polynomial (of degree $\leqq 6$) with respect to the unknown covariant components of the displacement field and their partial derivatives.

Its *third interest* is its amenability to a justification, by means of a formal asymptotic analysis of its solution, described in the next section.

3 Justification of the Two-Dimensional Shell Model "of Koiter's Type"

With the same notations as in Sect. 1, we now consider a *family* of nonlinearly elastic shells with *thickness* $2\varepsilon > 0$ *approaching zero*, with each having the *same* middle surface $S = \boldsymbol{\theta}(\overline{\omega})$. Each shell is subjected to a *homogeneous boundary condition of place* on a portion $\boldsymbol{\Theta}(\gamma_0 \times [-\varepsilon, \varepsilon])$ of its lateral face, i.e., each having the *same* set $\boldsymbol{\theta}(\gamma_0)$ as its middle curve, where $\gamma_0 \subset \gamma$ and *length* $\gamma_0 > 0$. Each shell is subjected to body forces in its interior and to surface forces on its upper and lower faces, given by their contravariant components $p^{i,\varepsilon} \in L^2(\omega)$. All the shells in the family are made with the *same nonlinearly elastic, homogeneous*, and *isotropic material*, and their reference configurations are *natural states*. Hence the material constituting the shells is characterized by two Lamé constants $\lambda > 0$ and $\mu > 0$ that are *independent* of ε.

Each shell in the family is modeled by the *nonlinear shell model of Koiter's type* proposed in Sect. 2. This means that, for each $\varepsilon > 0$, the field $\boldsymbol{\zeta}^\varepsilon = (\zeta_i^\varepsilon) : \overline{\omega} \to \mathbb{R}^3$, where $\zeta^\varepsilon \mathbf{a}^i$ is the displacement field of the middle surface S, is a *stationary point* of the *energy* j^ε defined by

$$\begin{aligned} j^\varepsilon(\boldsymbol{\eta}) := & \frac{\varepsilon}{2} \int_\omega a^{\alpha\beta\sigma\tau} G_{\sigma\tau}(\boldsymbol{\eta}) G_{\alpha\beta}(\boldsymbol{\eta}) \sqrt{a}\, \mathrm{d}y \\ & + \frac{\varepsilon^3}{6} \int_\omega a^{\alpha\beta\sigma\tau} R^\sharp_{\sigma\tau}(\boldsymbol{\eta}) R^\sharp_{\alpha\beta}(\boldsymbol{\eta}) \sqrt{a}\, \mathrm{d}y - \int_\omega p^{i,\varepsilon} \eta_i \sqrt{a}\, \mathrm{d}y \ , \end{aligned}$$

where

$$\begin{aligned} G_{\alpha\beta}(\boldsymbol{\eta}) &:= \frac{1}{2}(a_{\alpha\beta}(\boldsymbol{\eta}) - a_{\alpha\beta}) \ , \\ R^\sharp_{\alpha\beta}(\boldsymbol{\eta}) &:= \frac{1}{\sqrt{a}} \partial_{\alpha\beta}(\boldsymbol{\theta} + \eta_i \mathbf{a}^i) \cdot \{\mathbf{a}_1(\boldsymbol{\eta}) \wedge \mathbf{a}_2(\boldsymbol{\eta})\} - b_{\alpha\beta} \ , \\ a^{\alpha\beta\sigma\tau} &:= \frac{4\lambda\mu}{\lambda + 2\mu} a^{\alpha\beta} a^{\sigma\tau} + 2\mu(a^{\alpha\sigma} a^{\beta\tau} + a^{\alpha\tau} a^{\beta\sigma}) \ , \\ p^{i,\varepsilon} &:= \int_{-\varepsilon}^{\varepsilon} f^{i,\varepsilon}\, \mathrm{d}x_3^\varepsilon + h^{i,\varepsilon}(\cdot, \varepsilon) + h^{i,\varepsilon}(\cdot, -\varepsilon) \ , \end{aligned}$$

$\lambda > 0$ and $\mu > 0$ being the two *Lamé constants* of the material constituting the shells.

The problem of finding a stationary point $\boldsymbol{\zeta}^\varepsilon$ of the energy j^ε is first recast as a set of variational equations posed over the space

$$\mathbf{W}(\omega) := \{\boldsymbol{\eta} = (\eta_i) \in \mathbf{W}^{2,p}(\omega);\ \boldsymbol{\eta} = \partial_\nu \boldsymbol{\eta} = \mathbf{0} \text{ on } \gamma_0\} \ ,$$

for some fixed $p > 0$. Then, following a well-established procedure (*see*, e.g., Ciarlet (2000a, Chap. 8)), the unknown and the data are first *"scaled"*, by letting $\boldsymbol{\zeta}(\varepsilon) := \boldsymbol{\zeta}^\varepsilon$ and $p^i(\varepsilon) := \varepsilon^{-1} p^{i,\varepsilon}$. It is then assumed that the field $\boldsymbol{\zeta}(\varepsilon)$ admits a *formal asymptotic expansion* in terms of the thickness as the "small" parameter, viz.,

$$\boldsymbol{\zeta}(\varepsilon) = \boldsymbol{\zeta}^0 + \varepsilon \boldsymbol{\zeta}^1 + \varepsilon^2 \boldsymbol{\zeta}^2 + \cdots \ , \text{ with } \boldsymbol{\zeta}^0 \in \mathbf{W}(\omega) \ .$$

Remark. It can be *demonstrated* that the leading term of this formal asymptotic expansion is indeed of order zero; cf. Ciarlet & Roquefort (2001, Thm. 3). □

The main results are that, according to two mutually exclusive sets of assumptions on an associated *manifold* $\boldsymbol{\mathcal{M}}(\omega)$ of fields $\boldsymbol{\eta} = (\eta_i)$ corresponding to *"inextensional" displacements* (i.e., that satisfy $a_{\alpha\beta}(\boldsymbol{\eta}) - a_{\alpha\beta} = 0$ in ω), the leading term $\boldsymbol{\zeta}^0$ satisfies either the variational problem $\mathcal{P}_M(\omega)$ of a *nonlinearly elastic "membrane" shell* (Theorem 1) or the variational problem $\mathcal{P}_F(\omega)$ of a *nonlinearly elastic "flexural" shell* (Theorem 2).

Remark. Naturally, both problems $\mathcal{P}_M(\omega)$ and $\mathcal{P}_F(\omega)$ should be "de-scaled", in order to acquire physical significance. In particular, such a de-scaling introduces the expected factors ε and ε^3 in the left-hand sides of their respective variational equations. □

Theorem 1. *Assume that the manifold*

$$\boldsymbol{\mathcal{M}}(\omega) := \{\boldsymbol{\eta} \in \mathbf{W}^{2,p}(\omega);\ \boldsymbol{\eta} = \mathbf{0} \text{ on } \gamma_0,\ a_{\alpha\beta}(\boldsymbol{\eta}) - a_{\alpha\beta} = 0 \text{ in } \omega\}$$

contains only $\boldsymbol{\eta} = \mathbf{0}$ and that the applied forces are "of order ε^0 with respect to ε", in the sense that $p^i(\varepsilon) = p^{i,0}$ for all $\varepsilon > 0$, where the functions $p^{i,0} \in L^2(\omega)$ are independent of ε. Then $\boldsymbol{\zeta}^0$ satisfies the following variational problem $\mathcal{P}_M(\omega)$:

$$\boldsymbol{\zeta}^0 \in \mathbf{W}_M(\omega) := \{\boldsymbol{\eta} \in \mathbf{W}^{1,4}(\omega),\ \boldsymbol{\eta} = \mathbf{0} \text{ on } \gamma_0\}\ ,$$
$$\int_\omega a^{\alpha\beta\sigma\tau} G_{\sigma\tau}(\boldsymbol{\zeta}^0)(G'_{\alpha\beta}(\boldsymbol{\zeta}^0)\boldsymbol{\eta})\sqrt{a}\,\mathrm{d}y = \int_\omega p^{i,0}\eta_i\sqrt{a}\,\mathrm{d}y$$

for all $\boldsymbol{\eta} = (\eta_i) \in \mathbf{W}_M(\omega)$, where $G'_{\alpha\beta}(\boldsymbol{\zeta}^0)$ denotes the Fréchet derivative of the function $G_{\alpha\beta}$ at $\boldsymbol{\zeta}^0$. □

Theorem 2. *Assume that $\boldsymbol{\mathcal{M}}(\omega) \neq \{\mathbf{0}\}$ and that, at each point of $\boldsymbol{\mathcal{M}}(\omega)$, the tangent space to $\boldsymbol{\mathcal{M}}(\omega)$ contains nonzero elements. Also, assume that the applied forces are "of order ε^2 with respect to ε", in the sense that $p^i(\varepsilon) = \varepsilon^2 p^{i,2}$ for all $\varepsilon > 0$, where the functions $p^{i,2} \in L^2(\omega)$ are independent of ε. Then $\boldsymbol{\zeta}^0$ satisfies the following variational problem $\mathcal{P}_F(\omega)$:*

$$\boldsymbol{\zeta}^0 \in \boldsymbol{\mathcal{M}}_F(\omega) := \{\boldsymbol{\eta} = (\eta_i) \in \mathbf{W}^{2,p}(\omega);$$
$$\boldsymbol{\eta} = \partial_\nu\boldsymbol{\eta} = \mathbf{0} \text{ on } \gamma_0,\ G_{\alpha\beta}(\boldsymbol{\eta}) = 0 \text{ in } \omega\}\ ,$$
$$\frac{1}{3}\int_\omega a^{\alpha\beta\sigma\tau} R^\sharp_{\sigma\tau}(\boldsymbol{\zeta}^0)(R^\sharp_{\alpha\beta})'(\boldsymbol{\zeta}^0)\boldsymbol{\eta})\sqrt{a}\,\mathrm{d}y = \int_\omega p^{i,2}\eta_i\sqrt{a}\,\mathrm{d}y$$

for all $\boldsymbol{\eta} = (\eta_i)$ in the tangent space at $\boldsymbol{\zeta}^0$ to the manifold $\boldsymbol{\mathcal{M}}_F(\omega)$, where $(R^\sharp_{\alpha\beta})'(\boldsymbol{\zeta}^0)$ denotes the Fréchet derivative of the function $R^\sharp_{\alpha\beta}$ at ζ_0. □

The assumptions and the conclusions of Thms. 1 and 2 being identical to the assumptions and to the conclusions reached by Miara (1998) and Lods & Miara (1998) about the leading term (shown in particular to be independent of the "transverse" variable, so that it can be identified with a two-dimensional vector field) of a formal asymptotic expansion of the *scaled three-dimensional solution* (the vector field whose components are the scaled covariant components of the three-dimensional displacement field, i.e., the components over the contravariant bases $\mathbf{g}^{i,\varepsilon}$ defined by $\mathbf{g}^{j,\varepsilon}\cdot\mathbf{g}_i^\varepsilon = \delta_i^j$), again with the thickness as the "small" parameter, the nonlinear shell model of Koiter's type proposed in Sect. 2 is thus justified, at least formally.

Remarks. (1) The assumption in Thm. 2 about the tangent space to the manifold $\mathcal{M}(\omega)$, which first appeared in this form in Ciarlet (2000a, Thm. 10.1-1), was implicit in Lods & Miara (1998).

(2) The function spaces in Miara (1998) and Lods & Miara (1998) are not the same as in Thms. 1 and 2. This observation bears no consequence, however, inasmuch as only *formal* methods are compared.

(3) This justification of a nonlinear shell model of Koiter's type is analogous in its *principle* to the justification of the *linear* model proposed by Koiter (1970). But while the former justification is only *formal*, the latter is substantiated by *convergence theorems* as ε approaches zero (see Ciarlet (2000a, Sect. 7) and the references therein). □

References

do Carmo, M.P. (1976): *Differential Geometry of Curves and Surfaces*, Prentice-Hall, Englewood Cliffs.

Ciarlet, P.G. (1988): *Mathematical Elasticity, Volume I: Three-Dimensional Elasticity*, North-Holland, Amsterdam.

Ciarlet, P.G. (2000a): *Mathematical Elasticity, Volume III: Theory of Shells*, North-Holland, Amsterdam.

Ciarlet, P.G. (2000b): Un modèle bi-dimensionnel non linéaire de coque analogue à celui de W.T. Koiter, *C.R. Adad. Sci. Paris, Sér. I*, **331**, 405–410.

Ciarlet, P.G. (2001): A two-dimensional nonlinear shell model of Koiter's type, in *Proceedings, Conference in the Honor of the Memory of Jean Leray*, to appear.

Ciarlet, P.G., Coutand, D. (1998): An existence theorem for nonlinearly elastic "flexural" shells, *J. Elasticity* **50**, 261–277.

Ciarlet, P.G., Roquefort, A. (2000): Justification d'un modèle bi-dimensionnel non linéaire de coque analogue à celui de W.T. Koiter, *C.R. Acad. Sci. Paris, Sér. I*, **331**, 441–416.

Ciarlet, P.G., Roquefort, A. (2001): Justification of a two-dimensional nonlinear shell model of Koiter's type, *Chinese Annals Math.*, to appear.

Genevey, K. (1997): Remarks on nonlinear membrane shell problems, *Math. Mech. Solids* **2**, 215–237.

Klingenberg, W. (1973): *Eine Vorlesung über Differentialgeometrie*, Springer-Verlag, Berlin (English translation: *A Course in Differential Geometry*, Springer-Verlag, Berlin, 1978).

Koiter, W.T. (1966): On the nonlinear theory of thin elastic shells, *Proc. Kon. Ned. Akad. Wetensch.* **B69**, 1–54.

Koiter, W.T. (1970): On the foundations of the linear theory of thin elastic shells, *Proc. Kon. Ned. Akad. Wetensch.* **B73**, 169–195.

Le Dret, H., Raoult, A. (1996): The membrane shell model in nonlinear elasticity: A variational asymptotic derivation, *J. Nonlinear Sci.* **6**, 59–84.

Lods, V., Miara, B. (1998): Nonlinearly elastic shell models. II. The flexural model, *Arch. Rational Mech. Anal.* **142**, 355–374.

Miara, B. (1998): Nonlinearly elastic shell models. I. The membrane model, *Arch. Rational Mech. Anal.* **142**, 331–353.

Roquefort, A. (2001): *Sur Quelques Questions Liées aux Modèles Non Linéaires de Coques Minces*, Doctoral Dissertation, Université Pierre et Marie Curie, Paris.

A Survey of Stabilized Plate Elements

Mikko Lyly[1] and Rolf Stenberg[2]

[1] Center for Scientific Computing
P.O. Box 405
FIN-02101 Espoo, Finland
[2] Department of Mathematics
Tampere University of Technology
P.O. Box 692
FIN-33101 Tampere, Finland

Abstract. We present two families of finite element methods for the Reissner-Mindlin plate model. The families are based on a stabilized formulation which circumvents the requirement that the finite element spaces should satisfy the Babuška-Brezzi conditions. In the first family the polynomial order of the basis functions for the deflection is one higher than that for the rotation. In the second family the stabilization is combined with the MITC interpolation technique, which enables equal order basis functions. We review the stability and error estimates which show that the methods are "locking-free" and optimally convergent.

1 Introduction

In the last decade great progress has been achieved in the understanding of the "locking" phenomena connected with the finite element solution of the Reissner-Mindlin plate model. Based on the Babuška-Brezzi theory [6,7,11] for saddle point problems it has been possible to design optimally convergent methods. Among of the most successful are the so-called MITC families of Bathe, Brezzi and Fortin [9]. The performance of these is not only documented by numerous benchmark computations (cf. e.g. [8]) but also by a rigorous mathematical analysis, cf. [9,13,18,10]. When designing an element based on the traditional energy formulation the necessary Babuška-Brezzi conditions are quite restrictive, and hence none of the elements employ standard basis functions.

In our work we have followed another approach that has its origin in fluid mechanics where the class of methods are known as "stabilized" or "Galerkin Least-Squares" methods. In this, properly weighted least-squares terms of the strong form of the differential equations is added to the saddle point functional. This has the consequence that the finite element spaces do not have to satisfy the "inf-sup" condition and hence much more freedom is possible when choosing the finite elements.

For plates this technique was first used by Franca and Hughes [15] and in our work we have continued in this direction. In [17] we have shown that the stabilization be done directly in the displacement variables avoiding the intermediate step of stabilizing the saddle point functional obtained by choosing the shear force as an independent unknown. This is the case both for the error analysis and for the implementation. A consequence is that the method leads to a positively definite stiffness matrix which is better conditioned than those obtained from traditional methods.

This first stabilized formulation suffers from the drawback that the displacement has to be chosen as polynomial of one degree higher than that for the rotation in order to have a right balance in the consistency error. It turns out that equal order polynomials can be chosen if the stabilization is combined with the MITC interpolation technique. These stabilized MITC elements have the following favorable properties:

- They employ standard basis functions, equal for the rotation and the deflection. No special "bubble" degrees of freedom are needed.
- They contain stable and optimally convergent methods with linear (or bilinear) basis functions. (These elements were already introduced in [13].)
- They give rise to well conditioned stiffness methods which is an advantage for iterative solvers.

The purpose of this paper is to give a short review of these elements.

2 The Plate Model of Mindlin and Reissner

Let $\Omega \subset \mathbb{R}^2$ be the midsurface of the plate and suppose that the plate is clamped along the boundary Γ. The variational formulation of the Reissner-Mindlin model (appropriately scaled, cf. e.g. [9]) is: find the deflection $w \in W = H_0^1(\Omega)$ and the rotation vector $\boldsymbol{\theta} = (\theta_x, \theta_y) \in \boldsymbol{V} = [H_0^1(\Omega)]^2$ such that

$$a(\boldsymbol{\theta}, \boldsymbol{\eta}) + t^{-2}(\nabla w - \boldsymbol{\theta}, \nabla v - \boldsymbol{\eta}) = (g, v) \quad \forall (v, \boldsymbol{\eta}) \in W \times \boldsymbol{V}, \tag{1}$$

with the bilinear form a representing bending energy

$$a(\boldsymbol{\theta}, \boldsymbol{\eta}) = \frac{1}{6}\Big\{(\boldsymbol{\varepsilon}(\boldsymbol{\theta}), \boldsymbol{\varepsilon}(\boldsymbol{\eta})) + \frac{\nu}{1-\nu}(\operatorname{div}\boldsymbol{\theta}, \operatorname{div}\boldsymbol{\eta})\Big\}, \tag{2}$$

where t is the thickness of the plate, G is the shear modulus, ν the Poisson ratio and g is the (scaled) transverse load. $\boldsymbol{\varepsilon}(\cdot)$ is the small strain tensor and "div" stands for the divergence:

$$\boldsymbol{\varepsilon}(\boldsymbol{\theta}) = \frac{1}{2}\Big\{\nabla\boldsymbol{\theta} + (\nabla\boldsymbol{\theta})^T\Big\}, \tag{3}$$

$$\operatorname{div}\boldsymbol{\theta} = \frac{\partial\theta_x}{\partial x} + \frac{\partial\theta_y}{\partial y}. \tag{4}$$

By taking the (scaled) shear force

$$\boldsymbol{q} = t^{-2}(\nabla w - \boldsymbol{\theta}) \tag{5}$$

as an independent unknown in $\boldsymbol{S} = [L_2(\Omega)]^2$ one gets the following mixed formulation: find $(w, \boldsymbol{\theta}, \boldsymbol{q}) \in W \times \boldsymbol{V} \times \boldsymbol{S}$ such that

$$\begin{aligned} a(\boldsymbol{\theta}, \boldsymbol{\eta}) + (\boldsymbol{q}, \nabla v - \boldsymbol{\eta}) &= (g, v) \quad \forall (v, \boldsymbol{\eta}) \in W \times \boldsymbol{V}, \\ (\nabla w - \boldsymbol{\theta}, \boldsymbol{s}) - t^2(\boldsymbol{q}, \boldsymbol{s}) &= 0 \qquad\quad \forall \boldsymbol{s} \in \boldsymbol{S}. \end{aligned} \tag{6}$$

We here remark that the problem is singularly perturbated; in the limit obtained when $t \to 0$ the shear force is not longer in $\boldsymbol{S}$, but in the space $H^{-1}(\text{div} : \Omega)$, cf. [12]. On consequence of this is that there for small values of the thickness t there is a boundary layer in the solution [1]. The differential equations of this system are obtained by integrating by parts:

$$\begin{aligned} \boldsymbol{L\theta} + \boldsymbol{q} &= \boldsymbol{0} \quad &&\text{in } \Omega, \\ -\text{div}\, \boldsymbol{q} &= g \quad &&\text{in } \Omega, \\ -t^2 \boldsymbol{q} + \nabla w - \boldsymbol{\theta} &= \boldsymbol{0} \quad &&\text{in } \Omega, \\ w = 0, \ \boldsymbol{\theta} &= \boldsymbol{0} \quad &&\text{on } \Gamma. \end{aligned} \tag{7}$$

Here the differential operator $\boldsymbol{L}$ is defined from

$$\boldsymbol{L}\eta = \frac{1}{6} \mathbf{div} \left\{ \boldsymbol{\varepsilon}(\boldsymbol{\eta}) + \frac{\nu}{1-\nu} \text{div}\, \boldsymbol{\eta} \boldsymbol{I} \right\} \tag{8}$$

and $\boldsymbol{m}$ is the moment tensor

$$\boldsymbol{m} = \frac{1}{6} \left\{ \boldsymbol{\varepsilon}(\boldsymbol{\theta}) + \frac{\nu}{1-\nu} \text{div}\, \boldsymbol{\theta} \boldsymbol{I} \right\}. \tag{9}$$

The notation **div** stands for the divergence of second order tensors:

$$\mathbf{div}\, \boldsymbol{r} = \left(\frac{\partial r_{xx}}{\partial x} + \frac{\partial r_{xy}}{\partial y}, \frac{\partial r_{yx}}{\partial x} + \frac{\partial r_{yy}}{\partial y} \right). \tag{10}$$

The first two equations in (7) are the local equilibrium equations between the moment, shear force and load. The third equation is the constitutive relation between the shear strain and shear force.

3 The Stabilized Finite Element Methods

Let $\mathcal{C}_h$ be a partitioning of $\bar{\Omega}$ into triangular or quadrilateral finite elements. The elements $K \in \mathcal{C}_h$ are images of the reference element $\hat{K}$ under the (bi)linear mapping $\boldsymbol{F}_K : \hat{K} \to K$. The diameter of an element $K \in \mathcal{C}_h$ is denoted by h_K. In the mesh we allow both triangles and quadrilaterals and hence we will use the notation

$$R_s(K) = \begin{cases} P_s(K) & \text{when } K \text{ is a triangle,} \\ Q_s(K) & \text{when } K \text{ is a quadrilateral.} \end{cases} \tag{11}$$

The finite element subspaces for the deflection and rotation are denoted by $W_h \subset W$ and $\boldsymbol{V}_h \subset \boldsymbol{V}$, respectively.

In the next two sections we will present the two families of elements.

3.1 A Consistent Formulation

The spaces are specified as:

$$W_h = \{v \in W \mid v_{|K} \in R_{k+1}(K),\ \forall K \in \mathcal{C}_h\}, \tag{12}$$
$$\boldsymbol{V}_h = \{\boldsymbol{\eta} \in \boldsymbol{V} \mid \boldsymbol{\eta}_{|K} \in [R_k(K)]^2,\ \forall K \in \mathcal{C}_h\}, \tag{13}$$

where $k \geq 1$ is the polynomial index. The method is then defined as follows.

Method 1. *Find* $(w_h, \boldsymbol{\theta}_h) \in W_h \times \boldsymbol{V}_h$ *such that*

$$\mathcal{B}_h(w_h, \boldsymbol{\theta}_h; v, \boldsymbol{\eta}) = (g, v) \quad \forall (v, \boldsymbol{\eta}) \in W_h \times \boldsymbol{V}_h, \tag{14}$$

with the bilinear form

$$\mathcal{B}_h(z, \boldsymbol{\phi}; v, \boldsymbol{\eta}) = a(\boldsymbol{\phi}, \boldsymbol{\eta}) - \sum_{K \in \mathcal{C}_h} \alpha h_K^2 (\boldsymbol{L}\,\boldsymbol{\phi}, \boldsymbol{L}\,\boldsymbol{\eta})_K \tag{15}$$
$$+ \sum_{K \in \mathcal{C}_h} (t^2 + \alpha h_K^2)^{-1} (\nabla z - \boldsymbol{\phi} - \alpha h_K^2 \boldsymbol{L}\,\boldsymbol{\phi}, \nabla v - \boldsymbol{\eta} - \alpha h_K^2 \boldsymbol{L}\,\boldsymbol{\eta})_K.$$

¿From the solution $(w_h, \boldsymbol{\theta}_h)$ *we then calculate the approximation for the shear by*

$$\boldsymbol{q}_{h|K} = (t^2 + \alpha h_K^2)^{-1} (\nabla w_h - \boldsymbol{\theta}_h - \alpha h_K^2 \boldsymbol{L}\,\boldsymbol{\theta}_h)_{|K} \qquad \forall K \in \mathcal{C}_h. \tag{16}$$

Here α *is a positive parameter lying in a fixed range which will be specified in Theorem 2 below.* □

Note that (5) and the first equation of (7) give

$$\boldsymbol{q}_{|K} = (t^2 + \alpha h_K^2)^{-1} (\nabla w - \boldsymbol{\theta} - \alpha h_K^2 \boldsymbol{L}\,\boldsymbol{\theta})_{|K} \qquad \forall K \in \mathcal{C}_h. \tag{17}$$

Hence, we see that the approximation (16) is consistent with the exact shear.

The formulation, although nonstandard, is easily seen to be consistent.

Theorem 1. *The solution* $(w, \boldsymbol{\theta})$ *to* (7) *satisfies the equation*

$$\mathcal{B}_h(w, \boldsymbol{\theta}; v, \boldsymbol{\eta}) = (g, v) \quad \forall (v, \boldsymbol{\eta}) \in W \times \boldsymbol{V}.$$

Proof: Recalling the first equation in (7), the expression (17), and the variational form (6), we get

$$\begin{aligned}
&\mathcal{B}_h(w, \boldsymbol{\theta}; v, \boldsymbol{\eta}) \\
&= a(\boldsymbol{\theta}, \boldsymbol{\eta}) - \sum_{K \in \mathcal{C}_h} \alpha h_K^2 (\boldsymbol{L}\,\boldsymbol{\theta}, \boldsymbol{L}\,\boldsymbol{\eta})_K \\
&\quad + \sum_{K \in \mathcal{C}_h} (t^2 + \alpha h_K^2)^{-1} (\nabla w - \boldsymbol{\theta} - \alpha h_K^2 \boldsymbol{L}\,\boldsymbol{\theta}, \nabla v - \boldsymbol{\eta} - \alpha h_K^2 \boldsymbol{L}\,\boldsymbol{\eta})_K \\
&= a(\boldsymbol{\theta}, \boldsymbol{\eta}) + \sum_{K \in \mathcal{C}_h} \alpha h_K^2 (\boldsymbol{q}, \boldsymbol{L}\,\boldsymbol{\eta})_K
\end{aligned}$$

$$
\begin{aligned}
&+ \sum_{K\in\mathcal{C}_h} (\boldsymbol{q}, \nabla v - \boldsymbol{\eta} - \alpha h_K^2 \boldsymbol{L}\boldsymbol{\eta})_K \\
&= a(\boldsymbol{\theta},\boldsymbol{\eta}) + \sum_{K\in\mathcal{C}_h} \alpha h_K^2 (\boldsymbol{q}, \boldsymbol{L}\boldsymbol{\eta})_K + (\boldsymbol{q}, \nabla v - \boldsymbol{\eta}) - \sum_{K\in\mathcal{C}_h} \alpha h_K^2 (\boldsymbol{q}, \boldsymbol{L}\boldsymbol{\eta})_K \\
&= a(\boldsymbol{\theta},\boldsymbol{\eta}) + (\boldsymbol{q}, \nabla v - \boldsymbol{\eta}) \\
&= (g, v). \qquad \square
\end{aligned}
$$

Let us next outline the steps needed in the error analysis of the method, i.e. the stability and estimation of the interpolation error. (The details of the analysis are given in [17].)

First, let us denote by $C_I > 0$ the biggest constant in the inverse inequality

$$C_I \sum_{K\in\mathcal{C}_h} h_K^2 \|\boldsymbol{L}\boldsymbol{\eta}\|_{0,K}^2 \le a(\boldsymbol{\eta},\boldsymbol{\eta}), \qquad \forall \boldsymbol{\eta} \in \boldsymbol{V}_h, \tag{18}$$

which is valid as $\boldsymbol{V}_h$ consists of continuous piecewise polynomial functions (cf. e.g. [14]).

For the discrete solution space $W_h \times \boldsymbol{V}_h$ we then define the mesh dependent norm

$$|||(v,\boldsymbol{\eta})|||_h = \|v\|_1 + \|\boldsymbol{\eta}\|_1 + \Big(\sum_{K\in\mathcal{C}_h} (t^2 + h_K^2)^{-1} \|\nabla v - \boldsymbol{\eta}\|_{0,K}^2 \Big)^{1/2}. \tag{19}$$

The stability with respect to this norm now follows then from the Poincaré and Korn inequalities.

Theorem 2. *Suppose that* $0 < \alpha < C_I$. *Then there is a positive constant* C *such that*

$$\mathcal{B}_h(v,\boldsymbol{\eta}; v,\boldsymbol{\eta}) \ge C |||(v,\boldsymbol{\eta})|||_h^2 \qquad \forall (v,\boldsymbol{\eta}) \in W_h \times \boldsymbol{V}_h.$$

Proof: Using the inverse estimate (18) and Korn's inequality we get

$$
\begin{aligned}
&\mathcal{B}_h(v,\boldsymbol{\eta}; v,\boldsymbol{\eta}) \\
&= a(\boldsymbol{\eta},\boldsymbol{\eta}) - \sum_{K\in\mathcal{C}_h} \alpha h_K^2 \|\boldsymbol{L}\boldsymbol{\eta}\|_{0,K}^2 + \sum_{K\in\mathcal{C}_h} (t^2 + \alpha h_K^2)^{-1} \|\nabla v - \boldsymbol{\eta} - \alpha h_K^2 \boldsymbol{L}\boldsymbol{\eta}\|_{0,K}^2 \\
&\ge \sum_{K\in\mathcal{C}_h} (1 - \alpha C_I^{-1})\, a_K(\boldsymbol{\eta},\boldsymbol{\eta}) + \sum_{K\in\mathcal{C}_h} (t^2 + \alpha h_K^2)^{-1} \|\nabla v - \boldsymbol{\eta} - \alpha h_K^2 \boldsymbol{L}\boldsymbol{\eta}\|_{0,K}^2 \\
&\ge C\Big(a(\boldsymbol{\eta},\boldsymbol{\eta}) + \sum_{K\in\mathcal{C}_h} (t^2 + \alpha h_K^2)^{-1} \|\nabla v - \boldsymbol{\eta} - \alpha h_K^2 \boldsymbol{L}\boldsymbol{\eta}\|_{0,K}^2\Big) && (20) \\
&\ge C\Big(\|\boldsymbol{\eta}\|_1^2 + \sum_{K\in\mathcal{C}_h} (t^2 + \alpha h_K^2)^{-1} \|\nabla v - \boldsymbol{\eta} - \alpha h_K^2 \boldsymbol{L}\boldsymbol{\eta}\|_{0,K}^2\Big).
\end{aligned}
$$

Using the triangle inequality, the same inverse estimate and the boundedness of the bilinear form a one obtains

$$\sum_{K\in\mathcal{C}_h} (t^2 + \alpha h_K^2)^{-1} \|\nabla v - \boldsymbol{\eta}\|_{0,K}^2 \tag{21}$$

$$\leq C\big(\sum_{K\in\mathcal{C}_h}(t^2+\alpha h_K^2)^{-1}\|\nabla v-\boldsymbol{\eta}-\alpha h_K^2\boldsymbol{L}\,\boldsymbol{\eta}\|_{0,K}^2+\|\boldsymbol{\eta}\|_1^2\big).$$

Combining (20) and (21) gives

$$\mathcal{B}_h(v,\boldsymbol{\eta};v,\boldsymbol{\eta})\geq C(\|\boldsymbol{\eta}\|_1^2+|(v,\boldsymbol{\eta})|_h^2)\geq C|||(v,\boldsymbol{\eta})|||_h^2,$$

where we used an equivalence of norms, which is easily proved by scaling. □

Remark 1. For triangular elements with $k=1$ it holds

$$\boldsymbol{L}\,\boldsymbol{\phi}_{|K}=\boldsymbol{0},\quad \forall K\in\mathcal{C}_h,\quad \forall\boldsymbol{\phi}\in\boldsymbol{V}_h,$$

and hence the bilinear form $\mathcal{B}_h$ reduces to

$$\mathcal{B}_h(w,\boldsymbol{\theta};v,\boldsymbol{\eta})=a(\boldsymbol{\theta},\boldsymbol{\eta})+\sum_{K\in\mathcal{C}_h}(t^2+\alpha h_K^2)^{-1}(\nabla w-\boldsymbol{\theta},\nabla v-\boldsymbol{\eta})_K,$$

and we obtain a formulation proposed by Pitkäranta [19]. Now, the method is stable for all positive values of α. It is easily seen that the above bilinear form can also be used in the quadrilateral case for $k=1$ without a decrease in accuracy. □

In the convergence analysis we take the effect of the boundary layer into account. (We refer to [1–3] for a detailed analysis of the boundary layers.) In the limit $t\to 0$ the solution $(w,\boldsymbol{\theta})=(w_t,\boldsymbol{\theta}_t)$ of the Reissner–Mindlin equations converges to the Kirchhoff solution for which it holds

$$\boldsymbol{\theta}_0=\nabla w_0. \tag{22}$$

The limit solution w_0 satisfies the biharmonic equation in the domain Ω.

The following theorem (that can be deduced [17] from results by Arnold and Liu [4]) gives the interior and global regularity for the "Kirchhoff" component and the "residual" component of the solution.

Theorem 3. *Let Ω be a convex polygonal domain and let Ω_i be a domain compactly embedded in Ω. Denote by $(w,\boldsymbol{\theta},\boldsymbol{q})$ the Reissner–Mindlin solution for the clamped plate and let $w=w_0+w_r$, where w_0 is the deflection obtained from the Kirchhoff model. With $g\in H^{s-2}(\Omega)$ and $tg\in H^{s-1}(\Omega)$, $s\geq 1$, it then holds*

$$\|w_0\|_3+t^{-1}\|w_r\|_2+\|\boldsymbol{\theta}\|_2+\|\boldsymbol{q}\|_0+t\|\boldsymbol{q}\|_1\leq C(\|g\|_{-1}+t\|g\|_0) \tag{23}$$

and

$$\begin{aligned}\|w_0\|_{s+2,\Omega_i}+t^{-1}\|w_r\|_{s+1,\Omega_i}+\|\boldsymbol{\theta}\|_{s+1,\Omega_i}+\|\boldsymbol{q}\|_{s-1,\Omega_i}+t\|\boldsymbol{q}\|_{s,\Omega_i}&\\ \leq C(\|g\|_{s-2}+t\|g\|_{s-1}).&\end{aligned} \tag{24}$$

When estimating the approximation error we consider separately both components of the solution. Furthermore, we consider the case of a finer mesh along the boundary. Hence, we measure the size of the elements in the interior region Ω_i and the boundary region $\Omega_b = \Omega \setminus \Omega_i$ by the mesh parameters

$$h_i = \max_{K \subset \Omega_i} h_K, \quad h_b = \max_{K \not\subset \Omega_i} h_K. \tag{25}$$

The estimate so obtained is [17]:

Theorem 4. *Suppose that* $0 < \alpha < C_I$. *For the solution* $(w_h, \boldsymbol{\theta}_h, \boldsymbol{q}_h)$ *of* (14) *it then holds*

$$\begin{aligned} &|||(w - w_h, \boldsymbol{\theta} - \boldsymbol{\theta}_h)|||_h + \|\boldsymbol{q} - \boldsymbol{q}_h\|_{-1,h} + t\,\|\boldsymbol{q} - \boldsymbol{q}_h\|_0 \\ &\quad \le C\Big\{h_i^k(\|g\|_{k-2} + t\|g\|_{k-1}) + h_b(\|g\|_{-1} + t\|g\|_0)\Big\}. \end{aligned}$$ □

3.2 The Stabilized MITC Elements

These elements use identical basis functions for the deflection and both components of the rotation. For the index $k \ge 1$ they are defined as

$$W_h = \{v \in W \mid v_{|K} \in R_k(K),\ \forall K \in \mathcal{C}_h\}, \tag{26}$$

$$\boldsymbol{V}_h = \{\boldsymbol{\eta} \in \boldsymbol{V} \mid \boldsymbol{\eta}_{|K} \in [R_k(K)]^2,\ \forall K \in \mathcal{C}_h\}. \tag{27}$$

The use of equal basis functions is enabled by modifying the shear energy term. The shear force will be interpolated in the space

$$\boldsymbol{S}_h = \{\boldsymbol{s} \in \boldsymbol{S} \mid \boldsymbol{s}_{|K} \in \boldsymbol{S}_k(K),\ \forall K \in \mathcal{C}_h\}, \tag{28}$$

with

$$\boldsymbol{S}_k(K) = \{\boldsymbol{\eta} = \boldsymbol{J}_K^{-T}\hat{\boldsymbol{\eta}} \circ \boldsymbol{F}_K^{-1} \mid \hat{\boldsymbol{\eta}} \in \boldsymbol{S}_k(\hat{K})\}, \tag{29}$$

where $\boldsymbol{J}_K$ is the Jacobian matrix of $\boldsymbol{F}_K$ and $\boldsymbol{J}_K^{-T}$ is the transpose of $\boldsymbol{J}_K^{-1}$. The spaces on the reference element $\boldsymbol{S}_k(\hat{K})$ will be defined separately for triangles and quadrilaterals. In addition, we will define the the MITC reduction operator $\boldsymbol{R}_K : [H^1(K)]^2 \to \boldsymbol{S}_k(K)$. It is defined from the operator $\boldsymbol{R}_{\hat{K}} : [H^1(\hat{K})]^2 \to \boldsymbol{S}_k(\hat{K})$ on the reference element by a covariant transformation through the equation

$$\boldsymbol{R}_K\boldsymbol{\eta} = \boldsymbol{J}_K^{-T}\boldsymbol{R}_{\hat{K}}\boldsymbol{J}_K^T\boldsymbol{\eta}. \tag{30}$$

The shear spaces and reduction operators are now the following.

Triangular elements. *For a triangle* K *we choose*

$$\boldsymbol{S}_k(\hat{K}) = [P_{k-1}(\hat{K})]^2 \oplus (\eta, -\xi)\widetilde{P}_{k-1}(\hat{K}), \tag{31}$$

where $\widetilde{P}_{k-1}(\hat{K})$ *is space of homogeneous polynomials of degree* $k-1$*.* ξ *and* η *are the coordinates of* $\hat{K}$ *(i.e. the natural coordinates of* K*). This is the rotated Raviart-Thomas space [20].*

The reduction operator $\boldsymbol{R}_{\hat{K}}$ *is defined through the conditions*

$$\int_{\hat{E}} [(\boldsymbol{R}_{\hat{K}}\hat{\boldsymbol{\eta}} - \hat{\boldsymbol{\eta}}) \cdot \hat{\boldsymbol{\tau}}]\hat{v} \, d\hat{s} = 0, \ \forall \hat{v} \in P_{k-1}(\hat{E}), \text{ for every edge } \hat{E} \text{ of } \hat{K}, \tag{32}$$

and

$$\int_{\hat{K}} (\boldsymbol{R}_{\hat{K}}\hat{\boldsymbol{\eta}} - \hat{\boldsymbol{\eta}}) \cdot \hat{\boldsymbol{s}} \, d\xi d\eta = 0, \ \forall \hat{\boldsymbol{s}} \in [P_{k-2}(\hat{K})]^2, \tag{33}$$

where $\hat{\boldsymbol{\tau}}$ *is the unit tangent to the edge.*

Quadrilateral elements. *For a quadrilateral* K *we choose*

$$\boldsymbol{S}_k(\hat{K}) = P_{k-1,k}(\hat{K}) \times P_{k,k-1}(\hat{K}), \tag{34}$$

which is the rotated rectangular Raviart-Thomas space [20]. *The reduction operator is defined through the conditions*

$$\int_{\hat{E}} [(\boldsymbol{R}_{\hat{K}}\hat{\boldsymbol{\eta}} - \hat{\boldsymbol{\eta}}) \cdot \hat{\boldsymbol{\tau}}]\hat{v} \, d\hat{s} = 0, \ \forall \hat{v} \in P_{k-1}(\hat{E}), \text{ for every edge } \hat{E} \text{ of } \hat{K}, \tag{35}$$

and

$$\int_{\hat{K}} (\boldsymbol{R}_{\hat{K}}\hat{\boldsymbol{\eta}} - \hat{\boldsymbol{\eta}}) \cdot \hat{\boldsymbol{s}} \, d\xi d\eta = 0, \ \forall \hat{\boldsymbol{s}} \in P_{k-1,k-2}(\hat{K}) \times P_{k-2,k-1}(\hat{K}). \tag{36}$$

We are now ready to define the method.

Method 2. *Find the approximate deflection* $w_h \in W_h$ *and the rotation vector* $\boldsymbol{\theta}_h \in \boldsymbol{V}_h$ *such that*

$$\mathcal{B}_h(w_h, \boldsymbol{\theta}_h; v, \boldsymbol{\eta}) = (g, v) \quad \forall (v, \boldsymbol{\eta}) \in W_h \times \boldsymbol{V}_h, \tag{37}$$

with

$$\begin{aligned} \mathcal{B}_h(z, \boldsymbol{\phi}; v, \boldsymbol{\eta}) &= a(\boldsymbol{\phi}, \boldsymbol{\eta}) - \sum_{K \in \mathcal{C}_h} \alpha h_K^2 (\boldsymbol{L\phi}, \boldsymbol{L\eta})_K \\ &+ \sum_{K \in \mathcal{C}_h} (t^2 + \alpha h_K^2)^{-1} (\boldsymbol{R}_K(\nabla z - \boldsymbol{\phi} - \alpha h_K^2 \boldsymbol{L\phi}), \boldsymbol{R}_K(\nabla v - \boldsymbol{\eta} - \alpha h_K^2 \boldsymbol{L\eta}))_K. \end{aligned} \tag{38}$$

The approximate shear is computed from

$$\boldsymbol{q}_{h|K} = (t^2 + \alpha h_K^2)^{-1} \boldsymbol{R}_K (\nabla w_h - \boldsymbol{\theta}_h - \alpha h_K^2 \boldsymbol{L\theta}_h)_{|K}, \ \forall K \in \mathcal{C}_h. \tag{39}$$

As before α *is a numerical parameter satisfying* $0 < \alpha < C_I$*.*

Remark 2. For the triangular linear elements with $k = 1$, it holds

$$L\boldsymbol{\eta}_{|K} = \mathbf{0}, \quad \forall K \in \mathcal{C}_h, \ \forall \boldsymbol{\eta} \in \boldsymbol{V}_h, \tag{40}$$

and we have

$$\mathcal{B}_h(z, \boldsymbol{\phi}; v, \boldsymbol{\eta}) = a(\boldsymbol{\phi}, \boldsymbol{\eta}) + \sum_{K \in \mathcal{C}_h} (t^2 + \alpha h_K^2)^{-1} (\boldsymbol{R}_K(\nabla z - \boldsymbol{\phi}), \boldsymbol{R}_K(\nabla v - \boldsymbol{\eta}))_K. \tag{41}$$

This gives the linear element introduced and analyzed in [13]. In [16] it has been shown that it is essentially equivalent to an element introduced by Hughes and Tessler [22]. Later, it has been rediscovered in [23,5,21]. □

The analysis of the method requires a careful study of the properties of the reduction operator. First, it has to be included in meshdependent norm.

$$|||(v, \boldsymbol{\eta})|||_h = \|v\|_1 + \|\boldsymbol{\eta}\|_1 + \Big(\sum_{K \in \mathcal{C}_h} (t^2 + h_K^2)^{-1} \|\boldsymbol{R}_K(\nabla v - \boldsymbol{\eta})\|_{0,K}^2 \Big)^{1/2}. \tag{42}$$

Similarly as for the first method we automatically have a stable formulation.

Theorem 5. *There exists a constant $C > 0$ such that for $0 < \alpha < C_I$ it holds*

$$\mathcal{B}_h(v, \boldsymbol{\eta}; v, \boldsymbol{\eta}) \geq C|||(v, \boldsymbol{\eta})|||_h^2 \quad \forall (v, \boldsymbol{\eta}) \in W_h \times \boldsymbol{V}_h. \quad \square$$

Second, in contrast to the first method, this formulation is not consistent. The error introduced is however of the right order due to the orthogonality properties (35) and (35) of the reduction operators.

Third, it can be shown that for there is an interpolation operator I_h for the deflection such that

$$\boldsymbol{R}_K \nabla(w - I_h w) = \mathbf{0} \quad \forall K \in \mathcal{C}_h. \tag{43}$$

Without this crucial property we would have an error term of order $\mathcal{O}(h^{k-1})$ which would exclude the use of equal order interpolation.

We refer to the original article [17] for all the details in this error analysis. The final error estimate obtained is:

Theorem 6. *Suppose that $0 < \alpha < C_I$. Then it holds*

$$\|w - w_h\|_1 + \|\boldsymbol{\theta} - \boldsymbol{\theta}_h\|_1 + \Big(\sum_{K \in \mathcal{C}_h} (t^2 + h_K^2) \|\boldsymbol{q} - \boldsymbol{q}_h\|_{0,K}^2 \Big)^{1/2}$$
$$\leq C\Big\{ h_i^k (\|g\|_{k-2} + t\|g\|_{k-1}) + h_b(\|g\|_{-1} + t\|g\|_0) \Big\}. \quad \square$$

References

1. D.N. Arnold and R.S. Falk. Edge effects in the Reissner-Mindlin plate theory. In A.K. Noor, T. Belytschko, and J.C. Simo, editors, *Analytical and Computational Models of Shells.*, pages 71–89, New York, 1989. ASME.
2. D.N. Arnold and R.S. Falk. The boundary layer for the Reissner-Mindlin plate model. *SIAM J. Math. Anal.*, 21:10–40, 1990.
3. D.N. Arnold and R.S. Falk. Asymptotic analysis of the boundary layer for the Reissner-Mindlin plate model. *SIAM J. Math. Anal.*, 27:486–5140, 1996.
4. D.N. Arnold and X. Liu. Interior estimates for a low order finite element method for the Reissner-Mindlin plate model. *Adv. Comp. Math.*, 7:337–360, 1997.
5. F. Aurichhio and R.L. Taylor. A triangular thick plate finite element with an exact thin limit. *Finite Elements in Analysis and Design*, 19:57–68, 1995.
6. I. Babuška. The finite element method with Lagrangian multipliers. *Numer. Math.*, 20:179–192, 1973.
7. I. Babuška and A. Aziz. Survey lectures on the mathematical foudations of the finite element method. In A. Aziz, editor, *The Mathematical Foundations of the Finite Element Method with Application to Partial Differential Equations*, pages 5–359, New York, 1973. Academic Press.
8. K.J. Bathe. *Finite Element Procedures in Engineering.* Prentice Hall, 1995.
9. K.J. Bathe, F. Brezzi, and M. Fortin. Mixed-interpolated elements for Reissner-Mindlin plates. *Int. J. Num. Meths. Eng.*, 28:1787–1801, 1989.
10. D. Braess. *Finite Elemente.* Springer-Verlag, 1997.
11. F. Brezzi. On the existence, uniqueness and approximation of saddle poit problems arising from lagrangian multipliers. *RAIRO Anal. Num.*, 2:129–151, 1974.
12. F. Brezzi and M. Fortin. *Mixed and Hybrid Finite Element Methods.* Springer-Verlag, 1991.
13. F. Brezzi, M. Fortin, and R. Stenberg. Error analysis of mixed-interpolated elements for Reissner-Mindlin plates. *Mathematical Models and Methods in Applied Sciences*, 1:125–151, 1991.
14. P.G. Ciarlet. *The Finite Element Method for Elliptic Problems.* North - Holland, 1978.
15. T.J.R. Hughes and L.P. Franca. A mixed finite element formulation for Reissner-Mindlin plate theory: Uniform convergence of all higher-order spaces. *Comp. Meths. Appl. Mech. Engrg.*, 67:223–240, 1988.
16. M. Lyly. On linear triangular finite elements for Reissner-Mindlin plates. *Num. Math.*, 85:77–107, 2000.
17. M. Lyly and R. Stenberg. Stabilized finite elements for Reissner-Mindlin plates. Technical Report 4, Universität Innsbruck, Institut für Mathematik und Geometrie, 1999. Availabe from: http://www.tut.fi/~rstenber/publications.
18. P. Peisker and D. Braess. Uniform convergence of mixed interpolated elements for Reissner-Mindlin plates. M^2AN, 26:557–574, 1992.
19. J. Pitkäranta. Analysis of some low-order finite element schemes for Mindlin-Reissner and Kirchhoff plates. *Numer. Math.*, 53:237–254, 1988.
20. P.A. Raviart and J.M. Thomas. A mixed finite element method for second order elliptic problems. In *Mathematical Aspects of the Finite Element Method. Lecture Notes in Math. 606*, pages 292–315. Springer-Verlag, 1977.

21. R.L. Taylor and F. Aurichhio. Linked interpolation for Reissner-Mindlin plate elements. *Int. J. Num. Meths. Engng.*, 36:3057–3066, 1993.
22. A. Tessler and T.J.R. Hughes. A three-node Mindlin plate element with improved transverse shear. *Comp. Meths. Appl. Mech. Engng.*, 50:71–101, 1985.
23. Z. Xu. A thick-thin triangular plate element. *Int. J. Num. Meths. Eng.*, 33:963–973, 1992.

Prediction of the Fatigue Crack Growth Life in Microelectronics Solder Joints

Ken Kaminishi

Department of Mechanical Engineering, Yamaguchi University
2557 Tokiwadai, Ube City, 755-8611, Japan

Abstract. In order to predict the crack growth life in microelectronics solder joints, an FEA(finite element analysis) program employing a new scheme for crack growth analysis is developed. Also some experimental data necessary for the practical application of this program are obtained. Above all, the data related to the crack growth rate play a key role and are obtained in terms of the maximum opening stress range $\Delta\sigma_{\theta max}$ as

$$da/dN = \beta\,[\Delta\sigma_{\theta max} - \gamma]^{\alpha}\,,$$

where $\alpha = 2.0$ and $\beta = 2.5 \times 10^{-9} mm^5/N^2$ are independent of the test conditions, and γ is dependent on the solder material. The calculated values of the crack growth life by the FEA are in good agreement with the experimental ones. This indicates at the same time that the crack growth rate and path are certainly controlled, through the above equation, by $\Delta\sigma_{\theta max}$ measured at a certain radial distance from the crack tip.

1 Introduction

Microelectronics solder joints are constantly subjected to a fatigue induced by a thermal expansion mismatch between IC-package and substrate. From a viewpoint of reliability assessment of the microelectronics solder joints, it is important to predict the fatigue life of the joints, which should help in improving the package design accuracy and efficiency. The fatigue life of solder joints may be divided into the crack initiation and growth life. As to crack initiation life, many formulae based on the Coffin-Manson's law or the modified Coffin-Manson's law have been proposed [1]-[4]. By applying the strain range partitioning approach and a linear cumulative damage concept, and using material-dependent parameters, the authors also have proposed a fatigue life prediction formula on eutectic (63Sn - 37Pb) and low melting-point (37Sn - 45Pb - 18Bi) solders, expressing the number of cycles to failure as a function of cyclic frequency and equivalent inelastic strain range, which can be understood to successfully incorporate a creep damage effect [5]. However, little is known about the crack growth behavior of the solder joints used in surface mount technology[6]. In the meantime, the toxicity of lead, Pb, included in the conventional solder material is also a problem. Therefore the development of Pb-free solder joints with high reliability is hurried up and considerable attention has been paid to the fatigue strength of Pb-free solder [7]-[10].

In this work, fatigue tests are carried out on surface-mounted solder joints in product size electronic package models using conventional Sn-37Pb eutectic solder and two kinds of Pb-free solder (Sn-2.8Ag-15Bi-0.5Cu and Sn-3.5Ag-0.7Cu) in order to examine the crack initiation and growth behavior. Torsion fatigue tests are performed in parallel on bulk specimen of solder used in this work to obtain basic fatigue data of the solder materials, such as mechanical properties, cyclic stress-strain curves and prediction formulae used in numerical simulation. Further, a finite element program employing a new scheme for crack growth analysis is developed, and finite element analyses (FEA) of fatigue crack growth in microelectronics solder joints are carried out by this program. The calculated results are compared with the experimental ones to examine the feasibility of crack growth life prediction by the FEA.

2 Testing procedures

The geometry of the surface mount type specimen is shown in Fig.1. The specimen was manufactured by cutting the quad flat type LSI-package, and consisted of LSI-package, lead wire, solder joints and substrate. The configuration and the dimension of the specimen are shown magnified in Fig.2. The lead wire, being of a Gullwing type shown in the figure, is made of 42 Alloy and the substrate of glass epoxy resin. The dimensions shown in this figure are averages of several measured values in specimens. Conventional Sn-37Pb eutectic solder and two kinds of Pb-free solder (Sn-2.8Ag-15Bi-0.5Cu and Sn-3.5Ag -0.7Cu) are used.

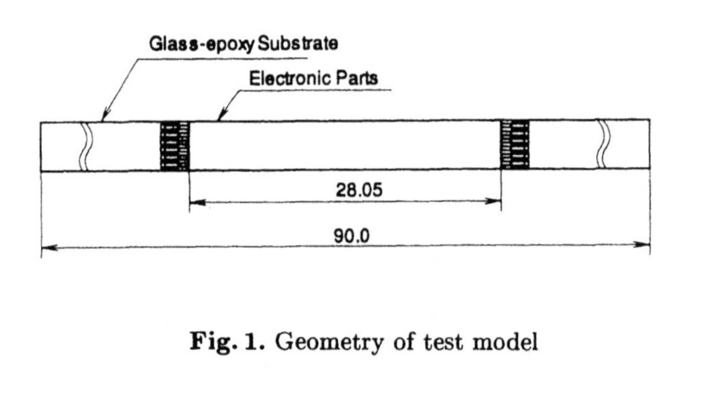

Fig. 1. Geometry of test model

The fitting of the specimen to fatigue testing system is as shown in Fig.3. The top surface of the test model was clamped to a displacement-controlled reciprocating pulse-stage, as shown in the figure. The fatigue tests on solder joints were carried out at temperatures controlled to 303±2K and 333±2K, the reciprocating frequencies 0.1 and 1.0Hz, and the displacement amplitudes being 25 and 50μm. The crack initiation and growth loci were followed under stereo-microscope. In addition, torsion fatigue tests on bulk specimen of Sn-2.8Ag-15Bi-0.5Cu and Sn-3.5Ag -0.7Cu solder were performed in the same way as shown in the previous paper [11] in order to obtain the material constant using FEA.

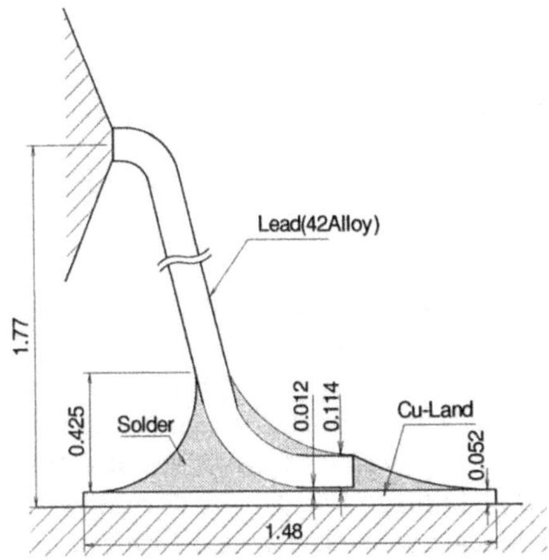

Fig. 2. Test section construction details

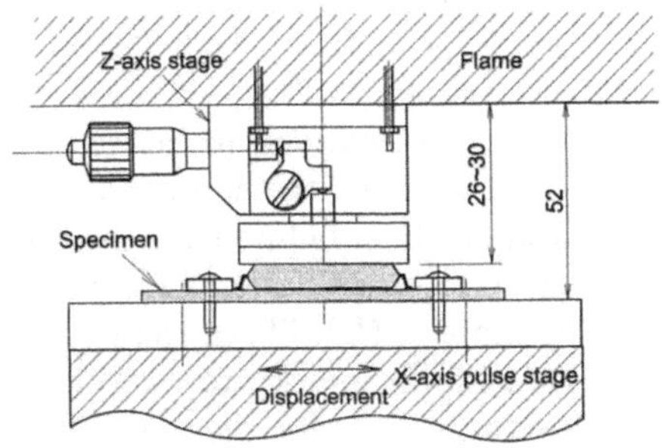

Fig. 3. Fitting of testing model to fatigue testing system

3 Experimental results

With increase in number of cycles, a surface of the solder fillet loses metallic luster, followed by the appearance of a fissured pattern, which grows into cracks in the comparatively upper position of the fillet edge. Fatigue crack initiation life, NC, is defined by the value of N, number of cycles, at which a microcrack has grown to approximately 30m, long enough to be detectable as a surface crack. After crack initiation, the crack grew along the above-mentioned fissured pattern bi-directionally, i.e., both to the fillet center and to the longitudinal direction. The cross-sectional views which were cut from solder fillet along the center line were examined by microscope to determine the longitudinal crack growth locus in the fillet. An example is shown in Fig.4.

The crack grew initially downwards in direction. When the crack approached the interface with lead wire, its direction changed and began to grow along the vicinity of the solder-lead interface thereafter.

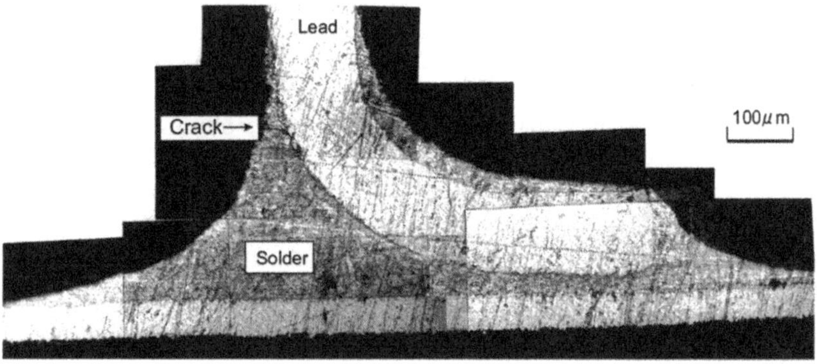

Fig. 4. Cracked solder joint (Sn-2.8Ag-15Bi-0.5Cu, $\delta = \pm 50\mu m$ 0.1Hz, 303K)

A typical relationship between the crack length, a, along the interface with lead wire and the number of cycles, N, is shown in Fig. 5. The process of initiation to final failure can reasonably be divided into four phases, namely, life up to crack initiation (T), life from crack initiation to crack growth up to the plateau (U), life of the plateau, i.e., the period of crack stabilization (V) and life for crack propagation to final failure (W). Figure 6 represents the averaged lives for the above four phases under each test condition. We see from this figure that the rate of domain U to the total life is very large for all test conditions. Consequently, for the prediction of the total fatigue life, crack growth analyses including the plateau region are indispensable. As for the influence of the test condition on fatigue life, it should be noted that an increase in applied displacement from ± 25 to $\pm 50\mu m$, abruptly reduced the domains U and V for Sn-2.8Ag-15Bi-0.5Cu solder joints, while it reduced the life to a quarter only for 63Sn-37Pb solder joints. As to Sn-3.5Ag -0.7Cu solder, the crack did not grow to final failure until 30,000 cycles for all test conditions.

4 Prediction of Crack Initiation Locus and Life

4.1 Crack initiation life prediction formula of bulk solder

In a previous paper, we proposed a crack initiation life, Nc, prediction formula based on the strain range partitioning method and linear cumulative damage concept for Sn-37Pb solder material [5]. In the present work, the material constants of the following prediction formula for Sn-2.8Ag-15Bi-0.5Cu and

Sn-3.5Ag-0.7Cu were obtained in the same manner as for Sn-37Pb solder material, and list them in Table 1.

$$N_C^{-1} = A_T \left\{ f_T(\nu)\, \Delta\varepsilon_{in} \right\}^{B_T} + C_T \left\{ (1 - f_T(\nu))\, \Delta\varepsilon_{in} \right\}^{D_T} \tag{1}$$

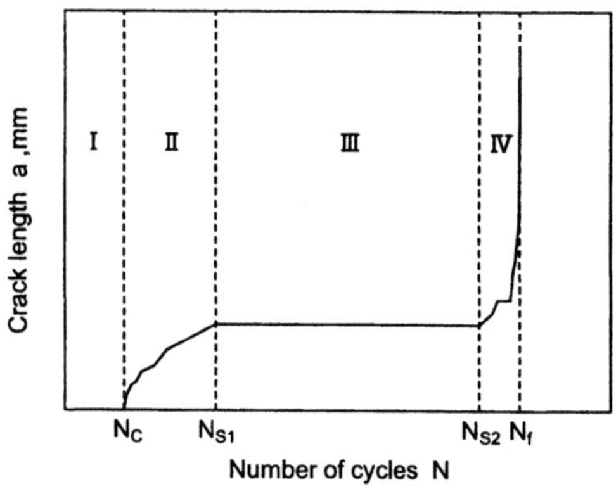

Fig. 5. Illustration of crack against number of cycles

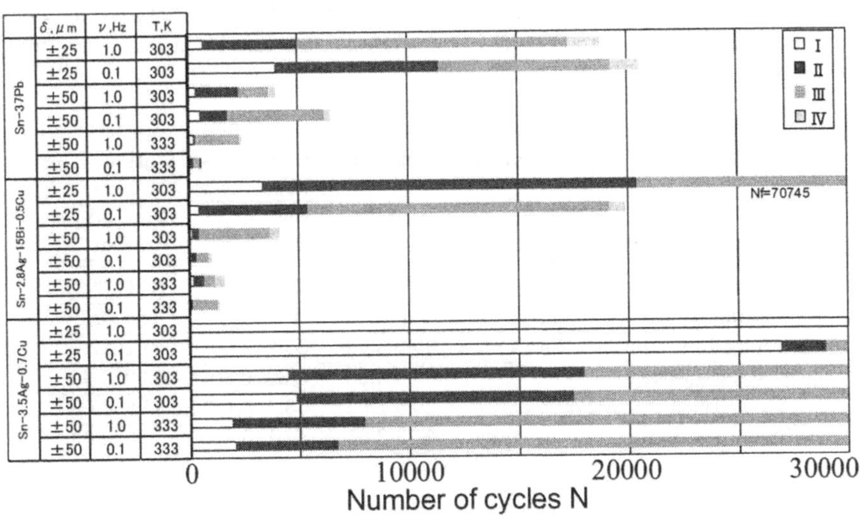

Fig. 6. Fatigue lives divided into four phases

Table 1. Material constants used in the fatigue life prediction formula

(a) $T = 303$K

	A_{303}	B_{303}	C_{303}	D_{303}	$f_{303}(\nu)$
Sn-37Pb	7.68×10^{-3}	1.74	5.44×10^{-3}	1.19	$1.56\nu^{0.724}$
Sn-2.8Ag-15Bi-0.7Cu	1.79×10^{-2}	1.66	9.22×10^{-2}	2.09	$1.84\nu^{0.437}$
Sn-3.5Ag-0.7Cu	1.19×10^{-3}	1.74	1.79×10^{-3}	1.33	$1.004\nu^{0.358}$

(b) $T = 333$K

	A_{333}	B_{333}	C_{333}	D_{333}	$f_{333}(\nu)$
Sn-37Pb	3.68×10^{-3}	1.74	4.98×10^{-3}	1.31	$0.402\nu^{0.704}$
Sn-2.8Ag-15Bi-0.7Cu	5.48×10^{-2}	2.38	2.67×10^{-2}	2.21	$1.790\nu^{0.705}$
Sn-3.5Ag-0.7Cu	4.90×10^{-3}	2.30	1.95×10^{-3}	1.84	$0.978\nu^{0.650}$

4.2 Stress-strain analysis of solder joints

The stress-strain analyses of the solder joints were performed by three dimensional elasto-inelastic FEM in order to predict the crack initiation locus and life in the solder joint. In our strain analyses, an experimentally determined cyclic stress range versus inelastic strain range relation was incorporated into the constitutive equation to consider the creep effect. First, the whole finite element model as shown in Fig.7 was calculated, and then a detailed analysis of the solder joint as shown in Fig.8 was carried out by using a zooming technique.

Table 2. Mechanical properties used for numerical simulation

	Sn-37Pb		Sn-2.8Ag-15Bi-0.5Cu		Sn-3.5Ag-0.7Cu	
Temperature T, K	303	303	303	303	303	303
Cycling frequency ν, Hz	1.0	0.1	1.0	0.1	1.0	0.1
Young's module E, GPa	35.0	32.8	33.6	29.4	19.1	22.7
Poisson's ratio	0.30	0.30	0.30	0.30	0.30	0.30
Yield stress σ_Y, MPa	26.5	23.3	63.1	60.0	32.5	28.0
Strain hardening rate H', GPa	2.78	2.37	4.21	3.40	0.63	0.30

	Lead	Package	Substrate	Cu-land
Young's module E, GPa	147.1	23.5	18.6	117
Poisson's ratio	0.3	0.25	0.16	0.3

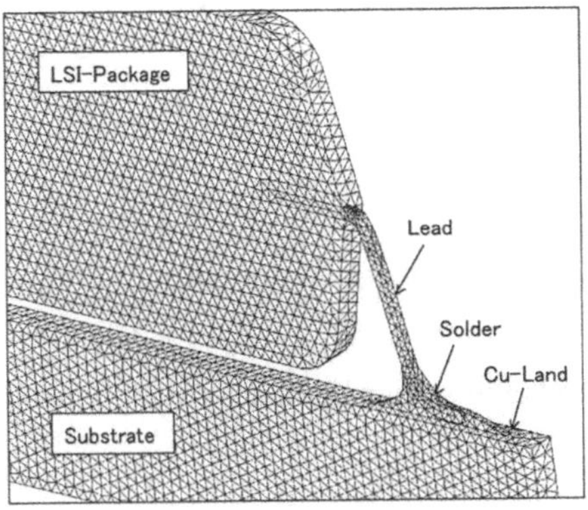

Fig. 7. Finite element meshes of the test model

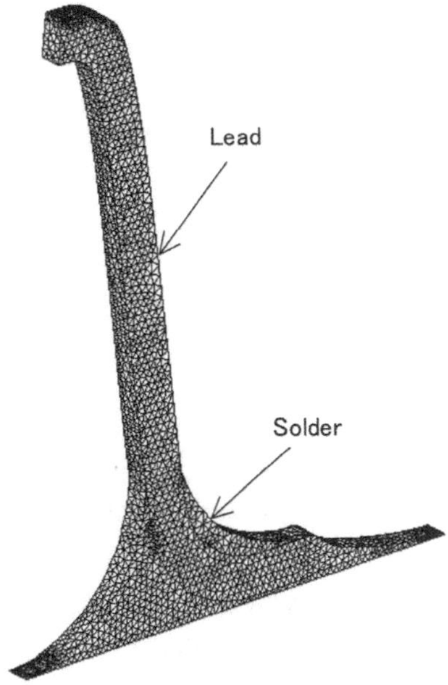

Fig. 8. Finite element meshes of solder joint

The cyclic equivalent stress, σ_{eq}, - inelastic strain, ε_{eq}^{in}, relations used in this calculation were obtained from torsion fatigue test data on the bulk solder. Mechanical properties used for the numerical simulation are shown in Table 2. The modulus of longitudinal elasticity and yield stress was obtained from the above-mentioned fatigue test data on bulk solder, the Poisson's ratio being assumed to be 0.3. Perfect elasticity of the lead wire, package, substrate and Cu-land were assumed. The sum of the equivalent inelastic strain ε_{eq}^{in} for downward displacement,δ, and that for backward displacement, $-\delta$, was given at the upper surface of the LSI-package, was regarded as equivalent inelastic strain range, $\Delta\varepsilon_{eq}^{in}$. Figure 9 shows the distribution of the equivalent inelastic strain range $\Delta\varepsilon_{eq}^{in}$. The highest zone of $\Delta\varepsilon_{eq}^{in}$ is in good agreement with the experimentally observed crack initiation site.

4.3 Crack initiation life of solder joints

By substituting equivalent inelastic strain range, $\Delta\varepsilon_{eq}^{in}$, calculated by three-dimensional inelastic stress analysis into eq.(1), the calculated crack initiation life, $[N_C]_{cal}$ is obtained. Table 3 represents several examples of $[N_C]_{cal}$ and compares with experimental results Nc; the ratios $N_C/[N_C]_{cal}$ are found to be 1.1 to 5.37. Thus it would be reasonable to assume that fatigue crack

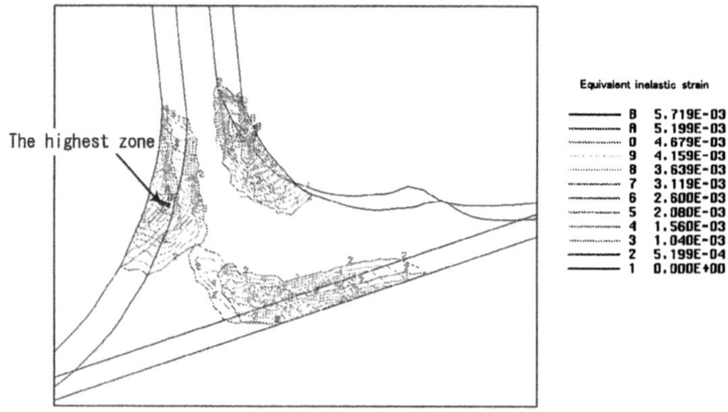

Fig. 9. Distribution of the equivalent inelastic strain range

initiation life of solder joints can be predicted somewhat conservatively using the proposed formulae.

Table 3. Results of experiments and numerical analyses for crack initiation life

Material	Disp. $\delta, \mu m$	frequency ν, Hz	Temp. T, K	N_C	$\Delta\varepsilon_{eq}^{in}$ %	$[N_C]_{cal}$	$N_C/[N_C]_{cal}$
Sn-37Pb	±25	1.0	303	592	0.453	519	1.1
	±50	0.1	303	618	1.643	115	5.4
Sn-2.8Ag-	±50	1.0	303	133	1.255	36	3.7
15Bi-0.7Cu	±50	0.1	303	161	1.107	47	3.4
Sn-3.5Ag-	±25	1.0	303	34200	0.906	15907	2.1
0.7Cu	±50	0.1	303	4880	2.24	4468	1.1

5 Prediction of the crack growth path and life

5.1 Crack growth path

Because the inner part of solder fillet in Gullwing type joints can be regarded as in an approximate plane strain condition, two-dimensional FEM program for crack growth analyses was developed in order to predict the fatigue crack growth life in solder joints. In this work, the super-element shown in Fig.10, which can be rotated in conformity with the direction of crack growth, was embedded in a crack tip region; triangular elements were regenerated by Delaunay Triangulation technique[12] in the outer region for arbitrary boundary geometry including crack faces. A special attention was paid in this work regarding the crack surface deformation problem as a dynamic contact problem by the use of penalty function method[13]. Cyclic stress-strain curves obtained in torsion fatigue tests of the solder material were used to consider both plasticity and creep effect.

As for the prediction of crack growth path, some criterions based on maximum tangential stress (MTS), maximum energy release rate (ERR), minimum strain energy density (SED) and modified SED have been proposed. In this work it is assumed that the crack extends in the direction of maximum $\Delta\sigma_\theta$ at a small radial distance of $r = d$, where d is chosen to be a grain diameter's distance, 1.0m, in solder material. See reference[14] for general discussions including MTS and the present assumption. Concretely, enforcement displacements ($+\delta-\delta+\delta-\delta$cc) are given cyclically at the certain crack length stage. Then, the relationship between σ_θ and angle θ in the last cycle ($+\delta-\delta$) is interpolated by the spline function of degree 3, and the maximum value of σ_θ was made to be $\sigma_{\theta max}$. In the meantime, the minimum value of σ_θ becomes a negative value. However, since the negative part is not concerned in crack growth, a maximum value of $\Delta\sigma_\theta$ is regarded as $\sigma_{\theta max}$. Therefore it is defined as $(\Delta\sigma_\theta)_{CT}$ in respect of the maximum value of $\Delta\sigma_\theta$ as shown in Fig.11, and a line crack of $10\mu m$ length is made to grow in the direction θ_1. By the way it is well known that the fatigue resistance in intermetallic compound formed at the solder-lead interface is improved. Therefore we assumed that the thickness of the compound is $7\mu m$ and the next crack tip never enter this zone.

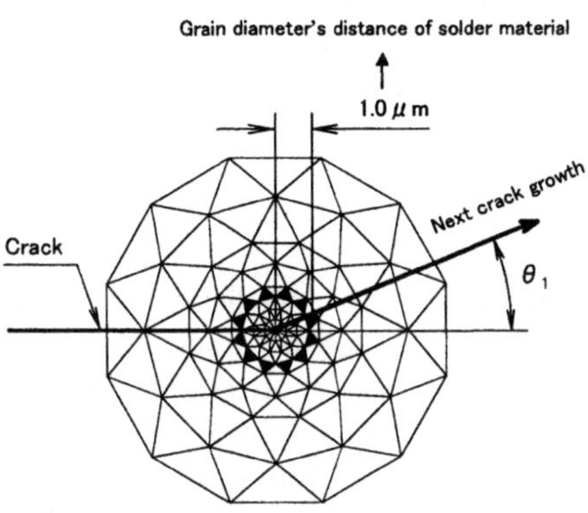

Fig. 10. Super-element embedded in the crack-tip region

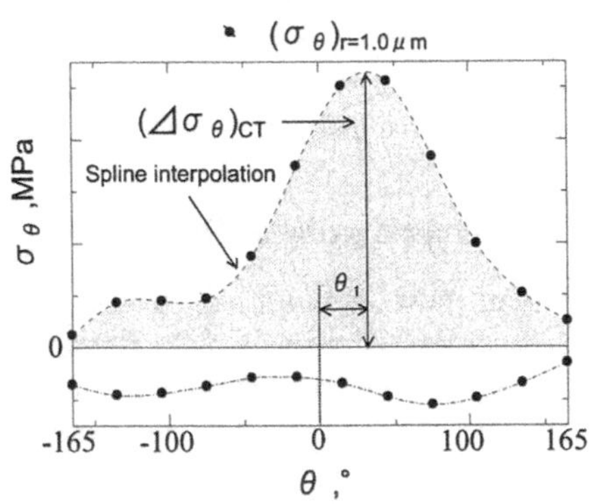

Fig. 11. Determination of crack growth path

Figure 12 illustrates an initial finite element meshing which embeds the super-element. A crack 10m long is introduced as an initial crack at the location of numerically obtained maximum equivalent inelastic strain range described in section 4.2.

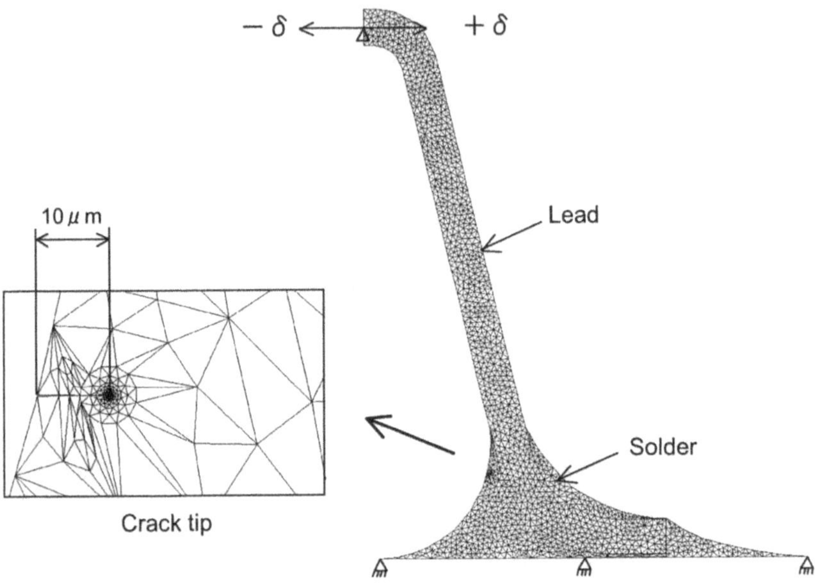

Fig. 12. Initial finite element meshes for crack growth analysis

A typical example of the crack growth path calculated by FEA is demonstrated in Fig.13. The crack grew along the vicinity of the solder-lead interface after it approached the interface. The simulated crack growth path is in good agreement with the experimentally observed one. It would be reasonable to suppose that the crack growth path in microelectronics solder joint can be predicted by using the present FEA program.

5.2 Crack growth life

Let us now attempt to extend this simulation to the prediction of crack growth life. Figure 14 plots the computed $(\Delta\sigma_\theta)_{CT}$ against crack length for test conditions indicated in the figure, where $(\Delta\sigma_\theta)_{CT}$ denotes the maximum $\Delta\sigma_\theta$ at $r = 1.0\mu m$. It is characteristic of this plot that $(\Delta\sigma_\theta)_{CT}$ decreases monotonically as crack length exceeds $0.03mm$ until $0.18mm$ and increases again to final failure due to the contact effect of the crack surface. This tendency is similar to the experimentally obtained relationship between crack growth rate and crack length. In Fig. 15 crack growth rate, da/dN, is plotted against $(\Delta\sigma_\theta)_{CT}$ for the conditions indicated in the figure, which describes crack growth rate as a function of $(\Delta\sigma_\theta)_{CT}$ of the form;

$$da/dN = \beta\,[(\Delta\sigma_\theta)_{CT} - \gamma]^\alpha \tag{2}$$

where $\alpha = 2.0$ and $\beta = 2.5 \times 10^{-9} mm^5/N^2$ are determined independently of test conditions and γ is a kind of material constant as follows; In case

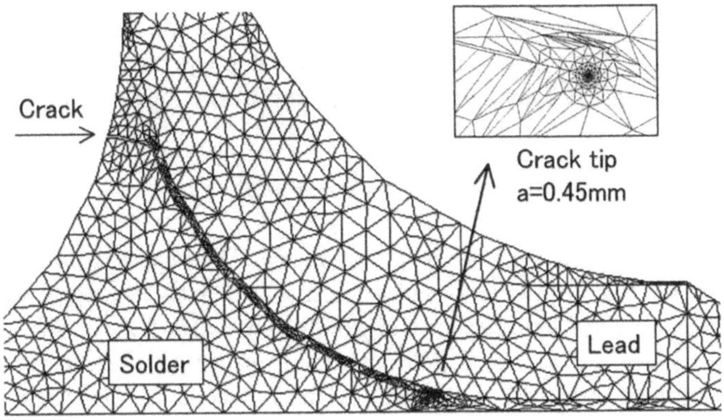

Fig. 13. Simulated crack growth path (Sn-2.5Ag-15Bi-0.5Cu, $\delta = \pm 50\mu m$, 0.1Hz, 303K)

Sn-37Pb :γ = 20MPa, in case Sn-2.8Ag-15Bi-0.5Cu :γ = 5 MPa, in case Sn-3.5Ag-0.7Cu : γ = 220MPa.

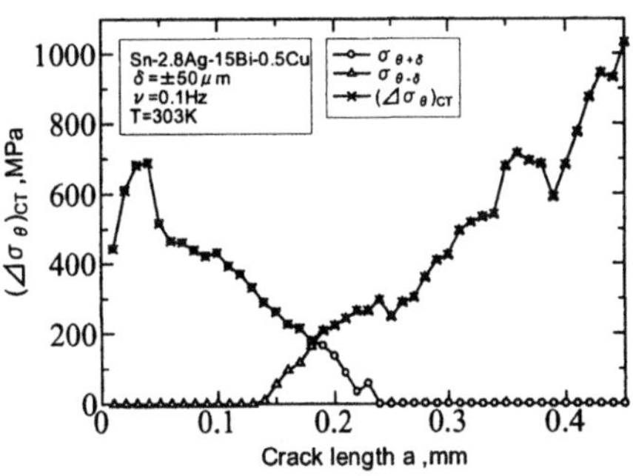

Fig. 14. Plot of computed maximum tangential stress range against observed crack length

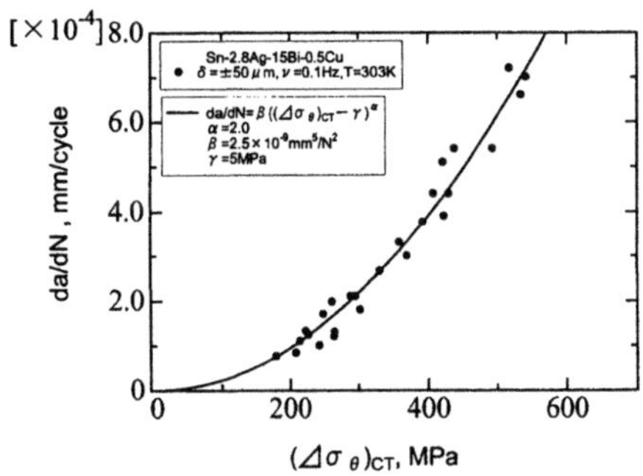

Fig. 15. Plot of experimentally determined crack growth rate against maximum tangential stress range

In the FEA the same super-element is embedded in the near crack-tip region to avoid the effect of mesh density on the results. It is considered that α, β and γ in the above equation would be regarded as material constants being independent of the applied displacement amplitude, δ, and joint type, provided that the same super-element is used in the FEA. For this reason, fatigue crack growth life can be predicted by integrating Eq.(2).

Crack length-to-number of cycles relations for $\pm 25\mu m$ and $\pm 50\mu m$ were calculated by using the above-mentioned values of α, β and γ, and compared with the experimental ones. Several examples of the results for each materials are shown in Fig.16(a) to (f). Agreement between calculation and experiment would be satisfactory for all experimental conditions in Gullwing type joints.

6 Application to a Butt type joint

The FEA of the Butt type joint shown in Fig.17 is carried out in the same manner as for the Gullwing type and compared with the experiment in order to examine the feasibility of the prediction method for crack growth life proposed in this work. See Fig.18, which compare the crack growth paths between simulation and experiment. Although the fatigue crack growth in butt joint takes place in different fashion compared with the case of Gullwing type, crack growth path is controlled by maximum stress range as well as Gullwing type. An example of the prediction of fatigue crack growth life in Butt type by integrating Eq.(2) using above-mentioned values of α, β and γ, which are obtained from Gullwing type model is shown in Fig.19. In spite of the quite different joint geometry, the numerical result is in a fair agreement with the experimental one. These comparison would suggest the validity of the present FEA for the prediction of fatigue crack growth life of the microelectronics solder joint independently of the joint geometry.

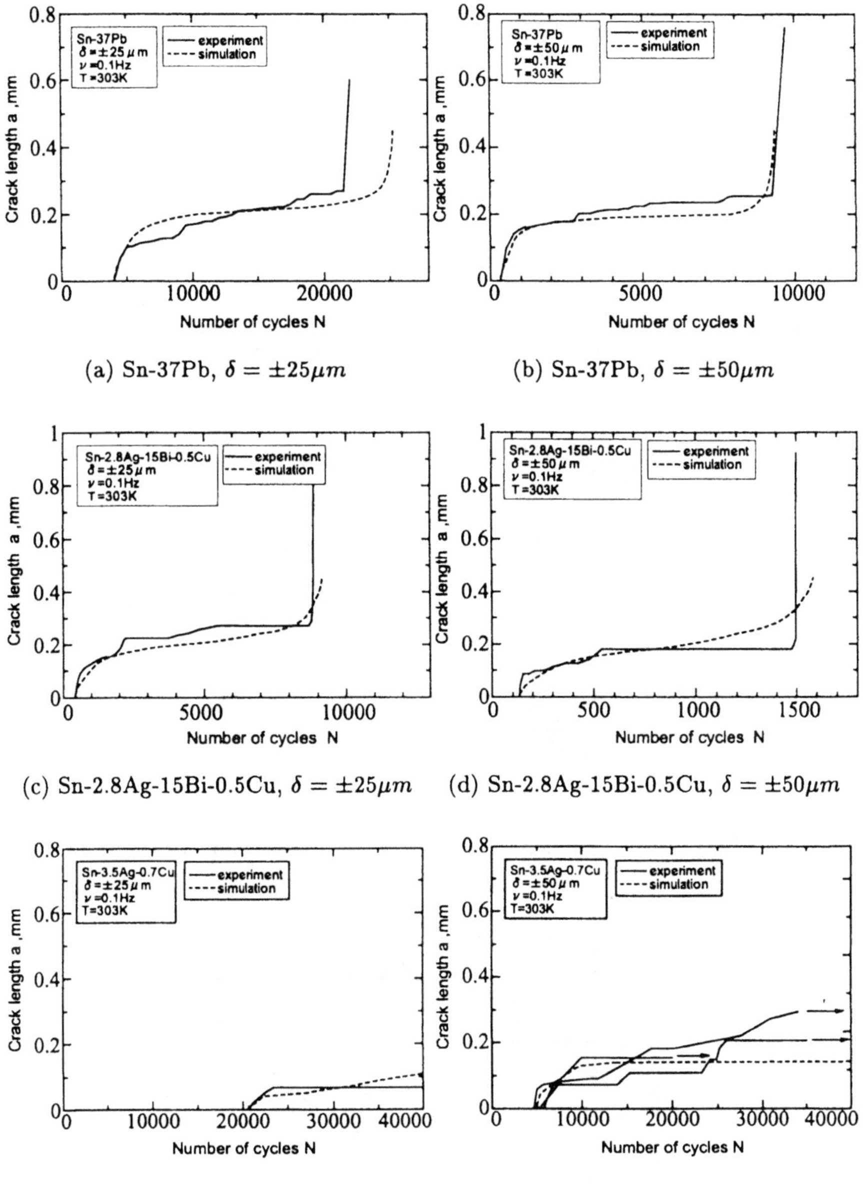

(a) Sn-37Pb, $\delta = \pm 25\mu m$

(b) Sn-37Pb, $\delta = \pm 50\mu m$

(c) Sn-2.8Ag-15Bi-0.5Cu, $\delta = \pm 25\mu m$

(d) Sn-2.8Ag-15Bi-0.5Cu, $\delta = \pm 50\mu m$

(e) Sn-3.5Ag-0.7Cu, $\delta = \pm 25\mu m$

(f) Sn-3.5Ag-0.7Cu, $\delta = \pm 50\mu m$

Fig. 16. Relationship between crack length and number of cycles in Gullwing type joint

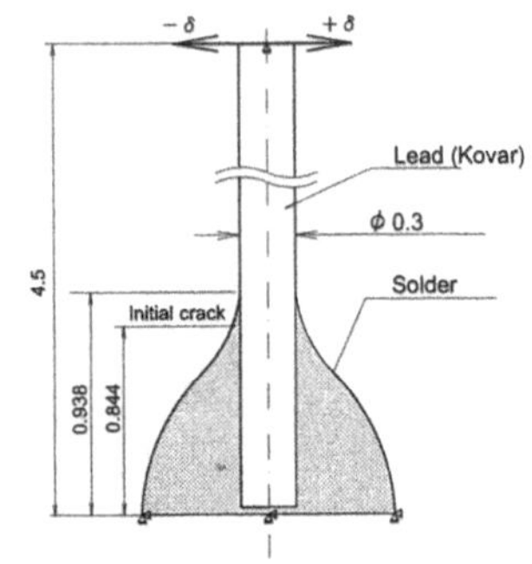

Fig. 17. Geometry of Butt type model

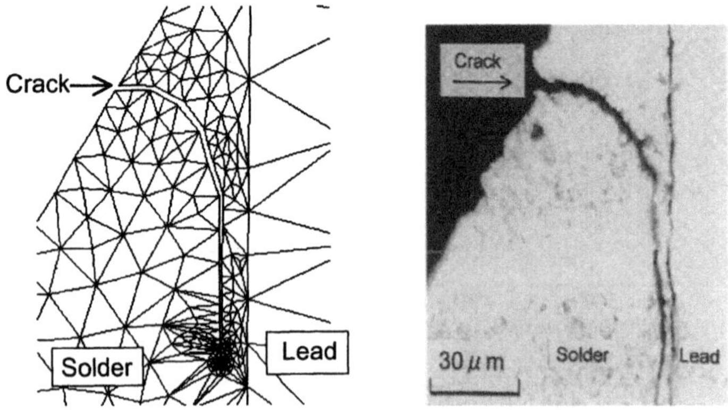

Fig. 18. Comparison of simulated crack growth with experiment in Butt type joint

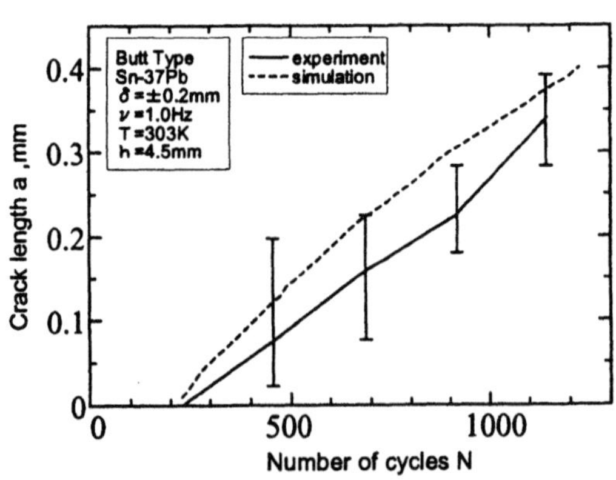

Fig. 19. Relationship between crack length and number of cycles in Butt type joint

7 Conclusions

An FEA program employing a new scheme for crack growth analysis was developed, by which fatigue crack growth in microelectronics solder joints were numerically analyzed. The FEA results show that crack growth rate and path are controlled by a maximum opening stress range, $\Delta\sigma_{\theta max}$, at a small radial distance of $r = d$, where d is chosen to be a grain diameter's distance, 1.0m, in solder material. Experimentally obtained crack growth rate is found to be related to $\Delta\sigma_{\theta max}$ by

$$da/dN = \beta\left[\Delta\sigma_{\theta max} - \gamma\right]^{\alpha},$$

where $\alpha = 2.0$ and $\beta = 2.5 \times 10^{-9} mm^5/N^2$ are determined independently of test conditions and γ is a kind of material constant. Conclusively it is shown that experimentally observed fatigue crack growth path and life can be predicted by the maximum opening stress range criterion and numerically integrating the above equation.

References

[1] Mukai, M., Kawakami, T., Endo, T. and Takahashi, K., " Elastic-creep behavior and fatigue life in an IC package solder joint", 4th Annual Meeting on Computational Mechanics, JSME, No.910- 79, (1991), pp.223-224.

[2] Lau, J.H., Rice, D.W. and Erasmus, S., "Thermal fatigue life of 256-Pin, 0.4mm pitch plastic quad flat pack(QFP) solder joint", ASME Advances in Electronic Packaging 1992, EEP-Vol- 2, (1992), pp.855-865.

[3] Busso, E.P., Kitano, M. and Kumazawa, T., "Modeling complex inelastic deformation processes in IC packages solder joints", ASME Journal of Electronic Packaging, Vol.116, (1994), pp.6-15.

[4] Shiratori, M. and Qiang, Y., "Fatigue-strength prediction of microelectronics solder joints under thermal cyclic loading", Fifth Intersociety Conference on Thermal Phenomena in Electronic Systems, (1996), pp.151-157.

[5] Taneda, M. and Kaminishi, K., "Effect of cycling frequency on fatigue life of solder", ASME Advances in Electronic Packaging 1992, EEP-Vol-1, (1992), pp.337-342.

[6] Kaminishi, K., Iino, M. and Taneda, M.,"Evaluation of fatigue crack initiation and extension life in microelectronics solder joints of surface mount type", JSME Int. J. Vol.42, No.2,(1999), pp.272-279.

[7] Mavoori, H., Chin, J., Vayman, S., Moran, B., Keer, L. and Fine, M., "Creep Stress Relaxation, and Plastic Deformation in Sn-Ag and Sn-Zn Eutectic Solders", Journal of Electronic Materials, Vol.26, No.7, (1997), pp.783-790.

[8] Trumble, B., "Get the Lead Out", IEEE Spectrum, Vol.35, No.5, (1998), pp.55-62.

[9] Takemoto, T., Takahasi, M., Mtsunawa, A., Ninomiya, R. and Tai, H., "Tensile Deformation Behavior of Sn-Ag-Bi System Lead-Free Solders", Journal of Japan Welding Society, Vol.16, No.1, (1998), pp.87-92.

[10] Kariya, Y. and Otsuka, M., "Mechanical Fatigue Charactristics of Sn-3.5Ag-X (X=Bi, Cu, Zn and In) solder Alloy", Journal of Electronic Materials, Vol.27, No.11, (1998), pp.1223-1228.

[11] Taneda, M., Oku, Y. and Kaminishi, K., "A method for solder fatigue life prediction by strain-range partitioning approach", Trans. Jpn. Soc. Mech. Eng., (in Japanese), Vol.58, No.549, A, (1992), pp.669-675.

[12] Sloan, S.W., "An implementation of Watson's algorithm for computing two-dimensional delaunay triangulations", Advances in Engineering Software, Vol.6, (1984), pp.192-193.

[13] Yagawa, G., Kanto, Y. and Ando, Y., "Analysis of dynamic contact problems using penalty function method", Trans. Jpn. Soc. Mech. Eng., (in Japanese), Vol.49, No.448, A, (1983), pp.1581-1589.

[14] Iino, M., Kaminishi, K. and Taneda, M., "Fatigue crack nucleation and growth in microelectronics solder joints", Proc. ASME Int. Intersociety Electronic Packaging Conf., EEP-19-2, (1997), pp.1575-1582.

Multi-phase Flow with Reaction

Hideo Kawarada[1] and Hiroshi Suito[1]

Department of Urban Environment Systems, Chiba University,
Yayoicho 1-33, Inage-ku, Chiba, 263-8522, Japan
Email:{kawarada, suito}@tu.chiba-u.ac.jp

Abstract. In order to study the effects of spilled oil on coastal ecosystem, multiphase flow with reaction is modeled mathematically. In this procedure, Discontinuous Interface Generating Method plays an important role to formulate the decomposition phenomena of oil into water and soluble components. This mathematical model is numerically solved by use of finite difference method and the numerical results are presented.

1 Introduction

1.1 Motivation

In this paper, the effects of spilled oil on coastal ecosystem are discussed from the viewpoint of fluid dynamical studies. A tanker was stranded in the Japan Sea on January 2nd, 1997 and oil flowed out into sea from the tanker. Spilled oil drifted on the shore of Mikuni district, adhered to sand beach and penetrated into sand accompanied with the ebbing tide. The local people and many volunteers cooperated to remove spilled oil from seashore. At that time, serious influences of spilled oil to exert on ecosystem were discussed from a lot of viewpoints. Such an incident mentioned above promoted strongly to organize national projects to investigate and to study environmental pollution.

A part of this study is under cooperation with "Research for the Future Program" of the Japan Society for the Promotion of Science (Research theme: Evaluation and restoration of the effects of oil pollution on coastal ecosystem, Principal investigator: Prof. Mitsumasa Okada, Department of Environmental Science, Hiroshima University). Main themes included in this project are;

1. Process of drifting ashore of spilled oil.
2. Penetrating process of spilled oil into tidal flats and seabeds.
   (a) Effect of wave motion to the penetration.
   (b) Effect of tidal motion to the penetration.
   (c) Transport process of oil in sand.
3. Decomposition process of spilled oil after the penetration.
4. Influence for the penetration to exert on soluble components in seawater.

5. Evaluation of influence for the penetration to exert on ecosystem and benthos.
6. Restorative techniques for the influence due to oil pollution.

This study is focused on theme 3. from fluid dynamical point of view. The phenomena of theme 3. describe the decomposition of penetrated oil into discharged materials by bacteria in the tidal flats, main components of which are water and soluble ones. These phenomena are mathematically modeled into a multi-phase flow formulation with reactions. One of characteristics of this model is an appearance of velocity jump on the reaction surface between phases of oil and water based on the difference of densities among them. In order to treat this difficulty, Discontinuous Interface Generating Method (DIGM) has been proposed by the authors, an original idea of which was invented through repeating try and error of numerical experiments. Flow equations for all phases are unified into a single flow equation to construct a numerical model. The fictitious domain method plays an important role in this procedure. Also it should be noted that adhesion and penetration as characteristic properties of oil are taken into consideration through boundary conditions of friction type prescribed on the boundary between oil and sand. Finally, numerical results are presented by solving the unified model by means of the finite difference method.

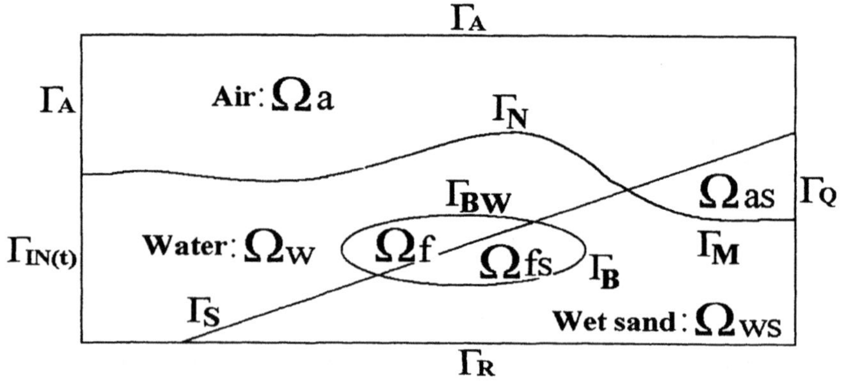

Fig. 1. Geometry of multi-phase flow with reaction

1.2 Notations

x_i : Cartesian coordinates$(i = 1, 2, 3)$
u_i : Velocity vector$(i = 1, 2, 3)$
u_n : Normal component of velocity on boundaries
u_T : Tangential component of velocity on boundaries
p : Pressure
ρ : Density
η : Yield value of Bigham fluid
t : Time
D_{ij} : Rate of strain tensor
D_{Π} : An invariant of rate of strain tensor
Ω_α : Domain for α-phase
$\Omega = \bigcup_\alpha \Omega_\alpha$: Total domain
Γ_β : Boundary between different phases
ν^α : Kinematic viscosity of α-phase
μ^α : Viscosity of α-phase
χ^α : Characteristic function representing a domain Ω_α
σ_{ij}^α : Stress tensor of α-phase
ρ^α : Density of α-phase
c : Coefficient of registance force receiving from sand
g_T : Friction coefficient
K_i : ith component of external force

where $\alpha = \{a, w, f, as, ws, fs\}$ and $\beta = \{N, M, B, BW, IN, Q, R, S, A\}$.

1.3 Geometry

The domain(Ω) is made of six parts.

1. The first part(Ω_a) is occupied with air.
2. The second part(Ω_w) is filled with water.
3. The third part(Ω_{ws}) is occupied with water penetrated into sand from the sloping beach.
4. The fourth part(Ω_{as}) is dry part of sand filled with air.
5. The fifth part(Ω_f) is occupied with spilled oil drifted on the sloping beach.
6. The sixth part(Ω_{fs}) is occupied with spilled oil penetrated into sand.

Remark 1.1

$\Omega_a \cup \Omega_{as}$ is regarded as the fictitious domain and $\Omega_{wf} = \Omega \backslash \left(\overline{\Omega_a} \cup \Omega_{as}\right)$.

2 Mathematical model

2.1 Conservation of mass for total flow system

Total mass of multi-phase flow system is conserved;

$$\frac{\partial \rho}{\partial t} + \frac{\partial}{\partial x_j}(\rho u_j) = 0, \qquad \text{in } \Omega_{wf}. \tag{1}$$

Here, $\rho = \sum_l \rho^{(l)} \chi^{(l)}$, $u_i = \sum_l u_i^{(l)} \chi^{(l)}$ and l means $\{w, f, ws, fs\}$.

Fluids for each phase are assumed to be incompressible, i.e., $\rho^{(l)}$ =const. Then we have

$$\sum_l \rho^{(l)} \left\{ \frac{\partial \chi^{(l)}}{\partial t} + u_j^{(l)} \frac{\partial \chi^{(l)}}{\partial x_j} \right\} + \sum_l \rho^{(l)} \chi^{(l)} \frac{\partial u_j^{(l)}}{\partial x_j}$$
$$= 0 \text{ in } \Omega_{wf},\ t > 0. \tag{2}$$

From (2) follows,

$$\frac{\partial u_j^{(l)}}{\partial x_j} = 0 \text{ in } \Omega_l, \text{ for each } l,\ \ t > 0 \tag{3}$$

which means the incompressibility condition to fluids for each phase and

$$\frac{\partial \chi^{(l)}}{\partial t} + u_j^{(l)} \frac{\partial \chi^{(l)}}{\partial x_j} = 0 \text{ in } \Omega, \text{ for each } l,\ \ t > 0 \tag{4}$$

which means the interface motion equation.

2.2 Conservation of momentum for total flow system

Equations of motion for sea water The motion equations for sea water on the beach are described as Navier-Stokes equations;

$$\frac{\partial u_i}{\partial t} + u_j \frac{\partial u_i}{\partial x_j} = -\frac{1}{\rho} \frac{\partial p}{\partial x_i} + \nu \Delta u_i + K_i. \tag{5}$$

The motion equations for sea water in sand are represented by the ones to include the resistance force $-cu_i$ from the sand. Here c means the inverse of porosity.

Equations of motion for spilled oil On the other hand, spilled oil is treated as Bingham fluid[2], which behaves like a solid body ($D_{ij} = 0$) until the stress $\sigma_{\Pi}^{1/2}$ reaches the yield value η of Bingham fluid and behaves like a viscous fluid when $\sigma_{\Pi}^{1/2}$ exceeds η, i.e.,

$$\sigma_{\Pi}^{1/2} < \eta \Longleftrightarrow D_{ij} = 0, \tag{6}$$
$$\sigma_{\Pi}^{1/2} \geq \eta \Longleftrightarrow D_{ij} = \frac{1}{2\mu}(1 - \eta/\sigma_{\Pi}^{1/2})(\sigma_{ij} + p\delta_{ij}). \tag{7}$$

$\sigma_{\Pi}^{1/2} = \eta + 2\mu D_{\Pi}^{1/2}$ and $D_{\Pi} = \frac{1}{2} D_{ij} D_{ij}$ (strain energy). From (7), we have

$$\sigma_{ij} = -p\delta_{ij} + 2\left(\mu_f + \frac{\eta}{\sqrt{4D_{\Pi}}}\right) D_{ij}, \tag{8}$$

where an effective viscosity becomes a plastic viscosity μ_f when $D_{\Pi} \to +\infty$. Concretely, (8) is regularized as follows;

$$\sigma_{ij} = -p\delta_{ij} + 2\left(\mu_f + \frac{\eta}{\sqrt{4D_{\Pi} + \varepsilon_b^2}}\right) D_{ij}, \tag{9}$$

where ε_b is a small positive parameter.

By use of (9), the equations of motion for Bingham fluid are written as follows;

$$\frac{\partial u_i}{\partial t} + u_j \frac{\partial u_i}{\partial x_j} = -\frac{1}{\rho_f}\frac{\partial p}{\partial x_i} + K_i$$
$$+\frac{\partial}{\partial x_j}\left\{\left(\nu_f + \frac{\tau}{\sqrt{4D_{\Pi} + \varepsilon_b^2}}\right)\left(\frac{\partial u_i}{\partial x_j} + \frac{\partial u_j}{\partial x_i}\right)\right\}, \tag{10}$$

where $\tau = \frac{\eta}{\rho_f}$.

The motion equations for oil in the sand are represented by the ones to include the resistance force $-cu_i$ received from the sand.

Fictitious domain method Regard the domain occupied by air as the fictitious domain because of large discrepancy of density between air and water.

In the fictitious domain, we assume that air satisfies singularly perturbed parabolic equation. As a matter of course, that does not need the incompressibility condition. Then pressure p takes the value of atmospheric pressure P_∞ there.

$$\frac{\partial u_i}{\partial t} = \varepsilon\left(\Delta u_i + \frac{\partial^2 u_j}{\partial x_i \partial x_j}\right), \tag{11}$$

$$p = P, \tag{12}$$

where $\varepsilon (> 0)$ is a sufficiently small parameter.

The unified equations of motion Conservation of momentum for the total system of multi-phase flow is represented by use of motion equations for each flow and the fictitious domain method as follows. Here the domain occupied with quasi-air is regarded as the fictitious domain.

$$\rho\left(\frac{\partial u_i}{\partial t} + (\chi_{wt} + \chi_{fl}) u_j \frac{\partial u_i}{\partial x_j}\right) =$$

$$-\frac{\partial\{(\chi_{wt}+\chi_{fl})(p-P_\infty)\}}{\partial x_i}-c(\chi_{ws}+\chi_{fs})u_i+\rho K_i$$
$$+\frac{\partial}{\partial x_j}\left\{\left(\varepsilon\chi_{air}+\mu_{wt}\chi_{wt}+\left(\mu_f+\frac{\eta}{\sqrt{4D_{\Pi}+\varepsilon_b^2}}\right)\chi_{fl}\right)\right.$$
$$\left.\cdot\left(\frac{\partial u_i}{\partial x_j}+\frac{\partial u_j}{\partial x_i}\right)\right\}\quad \text{in } \Omega,\ t>0, \tag{13}$$

$$(\chi^{(wt)}+\chi^{(fl)})\frac{\partial u_j}{\partial x_j}=0 \qquad \text{in } \Omega,\ t>0. \tag{14}$$

2.3 Interfacial interactions

Adhesion and sliding phenomena of oil in sand As an interfacial interaction, we formulate adhesion and sliding phenomena of oil in sand on the basis of simplified Coulomb law for friction. Let $[\sigma_T]$ be the jump of tangential stress defined on the boundary between oil and water in sand[4].

$$\begin{cases} |[\sigma_T]| \le g_T, \\ g_T\cdot|u_T|+[\sigma_T]\cdot u_T=0, \end{cases} \quad \text{on } \Gamma_B. \tag{15}$$

$$\begin{cases} |[\sigma_T]| < g_T \longmapsto u_T=0 & \text{(Adhesion)} \\ |[\sigma_T]| = g_T \longmapsto u_T=0 \text{ or } u_T\neq 0 & \text{(Sliding)} \end{cases} \tag{16}$$

where g_T is the friction coefficient. For $A=(a_1,a_2,a_3)$, $|A|$ means $\sqrt{a_1^2+a_2^2+a_3^2}$. By use of the definition of the subdifferential, (15) and (16) are formulated by

$$-[\sigma_T]=g_T\cdot\partial(|u_T|) \qquad \text{on } \Gamma_B. \tag{17}$$

In order to avoid the difficulty due to singularity arises in the numerical treatment of (17), (17) is regularized as follows;

$$-[\sigma_{T_j}]=g_T\cdot\frac{u_{T_j}}{\sqrt{|u_T|^2+\varepsilon_g^2}} \quad \text{on } \Gamma_B,\ (j=1,2), \tag{18}$$

where ε_g is a small positive parameter.

Decomposition of oil into water Assume $\Omega_{fs}\cap\Omega_b\neq\{0\}$, that means the occurrence of the decomposition. Interface motion equation for Γ_B between Ω_{fs} and Ω_{ws} is described as follows;

$$\frac{\partial\chi^{(fs)}}{\partial t}+(u_j^{(fs)}+k_f n_j^{(fs)}\chi^{(b)})\frac{\partial\chi^{(fs)}}{\partial x_j}=0 \text{ in } \Omega,\ t>0, \tag{19}$$

or

$$\frac{\partial\chi^{(ws)}}{\partial t}+(u_j^{(ws)}+k_w n_j^{(ws)}\chi^{(b)})\frac{\partial\chi^{(ws)}}{\partial x_j}=0 \text{ in } \Omega,\ t>0. \tag{20}$$

Here, k_f and k_w mean the rate of consumption for oil and the rate of production for water, respectively. Therefore, k_f is a negative constant and k_w is a positive constant. $\chi^{(b)}$ is characteristic function of Ω_b.

Remark 2.1

1. $\rho_w > \rho_f$,
2. $\rho_w \cdot k_w + \rho_f \cdot k_f = 0$ *(Mass transference condition)*,
3. $-k_f > k_w > 0$.

Reactivity condition Assume the nonoccurrence of separation of oil and water phases. Then there holds

$$u_j^{(fs)} + k_f n_j^{(fs)} \chi^{(b)} = u_j^{(ws)} + k_w n_j^{(ws)} \chi^{(b)} \quad \text{on } \Gamma_B, \quad (j = 1, 2, 3), \tag{21}$$

from which follows,

$$\left[u \cdot n^{(ws)}\right] = u^{(ws)} n^{(ws)} - u^{(fs)} n^{(ws)} = -(k_f + k_w)\chi^{(b)} \text{ on } \Gamma_B, \tag{22}$$

$$\left[u \cdot T^{(fs)}\right] = u^{(fs)} T^{(fs)} - u^{(ws)} T^{(fs)} = 0 \qquad \text{on } \Gamma_B. \tag{23}$$

(22) and (23) are called by reactivity condition and $-(k_f + k_w)$ is reactivity constant. The meaning of the reactivity condition is easily understood. (See figure 2.)

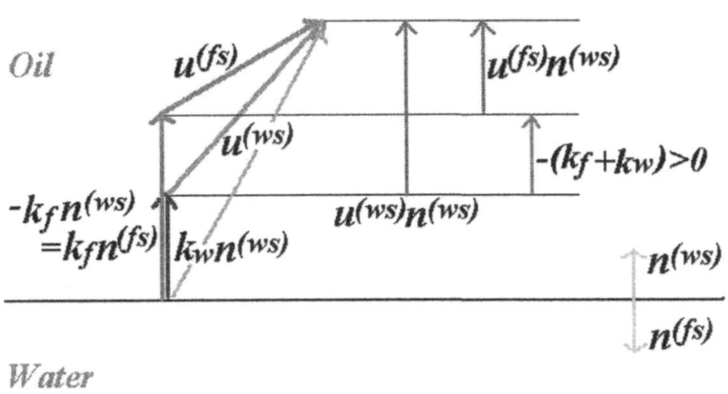

Fig. 2. Reactivity condition

2.4 Biological contribution to satisfy reactivity condition on the reaction surface

We propose the following conjecture as one of possible interpretations of adding dipole moment distribution along the reaction surface.

The torque caused by the difference of the bouyancy between the bacteria pairs, which are closely located along the reaction surface, produces the distribution of dipole moments on the reaction surface. Let γ be the reaction surface which is assumed to be smooth. Then the dipole moment defined on γ is represented by $\frac{\partial}{\partial \nu}\delta(\gamma)$, where $\delta(\gamma)$ is delta measure supported on γ. On the other hand, we have $\frac{\partial}{\partial \nu}\delta(\gamma) = \Delta\chi$. Define the volume of a bacteria by V_B. Then there holds

$$(\rho_w - \rho_f)V_B \cdot g = -(k_f + k_w) \cdot \frac{\rho_f}{k_w} V_B \cdot g. \tag{24}$$

3 Discontinuous Interface Generating Method

3.1 The statement of Theorem

Let us note that the interfacial interactions stated in 2.3 are represented by the jump boundary conditions for the Dirichlet type (the reactivity condition) and the Neumann type (the frictional condition). In order to introduce such jump conditions into an unified model for multi-phase flow system, we have developed Discontinuous Interface Generating Method (DIGM). For simplicity, we discuss the case of steady Stokes problem defined in $\Omega \subset R^3$. Ω is divided into subdomains Ω_1 and Ω_2, whose interface is denoted by Γ. (See figure 3.) The two phase Stokes problem (S_1) with jump boundary conditions on Γ is defined as follows.

$$(S_1)\quad \begin{cases} -\Delta u^{(1)} + \nabla p^{(1)} = f^{(1)} & \text{in } \Omega_1, \\ \text{div } u^{(1)} = 0 & \text{in } \Omega_1, \\ -\Delta u^{(2)} + \nabla p^{(2)} = f^{(2)} & \text{in } \Omega_2, \\ \text{div } u^{(2)} = 0 & \text{in } \Omega_2, \\ [\sigma_n] = \sigma_n^{(1)} - \sigma_n^{(2)} = a & \text{on } \Gamma, \\ [\sigma_{T_j}] = \sigma_{T_j}^{(1)} - \sigma_{T_j}^{(2)} = b_j \ (j=1,2) & \text{on } \Gamma, \\ [u_n] = u_n^{(1)} - u_n^{(2)} = c & \text{on } \Gamma, \\ [u_{T_j}] = u_{T_j}^{(1)} - u_{T_j}^{(2)} = d_j \ (j=1,2) & \text{on } \Gamma, \\ u^{(1)} = 0, & \text{on } \partial\Omega, \end{cases} \tag{25}$$

Let χ be the characteristic function of Ω_2 in Ω. Then (S_1) is settled into a single equation (S_2) in the following way;

$$(S_2)\quad \begin{cases} -\Delta u + \nabla p + a \cdot \nabla\chi + b_1 \cdot \Sigma_1\chi + b_2 \cdot \Sigma_2\chi \\ \qquad + c \cdot \boldsymbol{n} \cdot \Delta\chi + d_1 \cdot T_1 \cdot \Delta\chi + d_2 \cdot T_2 \cdot \Delta\chi = f, & \text{in } \Omega, \\ \text{div}\boldsymbol{u} = 0, & \text{in } \Omega\backslash\Gamma, \\ u = 0, & \text{on } \partial\Omega, \end{cases} \tag{26}$$

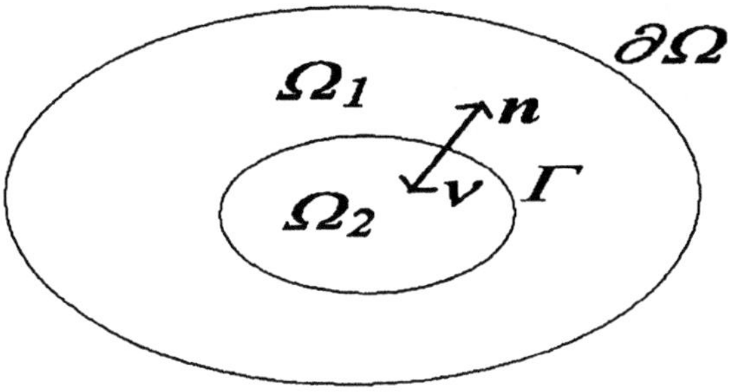

Fig. 3. Geometry for Stokes problem

where

$$f = (1-\chi)f^{(1)} + \chi f^{(2)}, \tag{27}$$

and

$$\nabla\chi = \left(\frac{\partial\chi}{\partial x_1}, \frac{\partial\chi}{\partial x_2}, \frac{\partial\chi}{\partial x_3}\right), \tag{28}$$

$$\Sigma_1\chi = \left(\frac{\partial\chi}{\partial x_2}, -\frac{\partial\chi}{\partial x_1}, 0\right), \tag{29}$$

$$\Sigma_2\chi = \left(\frac{\partial\chi}{\partial x_1}\frac{\partial\chi}{\partial x_3}, \frac{\partial\chi}{\partial x_2}\frac{\partial\chi}{\partial x_3}, \left(\frac{\partial\chi}{\partial x_3}\right)^2 - |\nabla\chi|^2\right), \tag{30}$$

$$\boldsymbol{n} = -\frac{\nabla\chi}{|\nabla\chi|}, \quad T_1 = \frac{\Sigma_1\chi}{|\Sigma_1\chi|}, \quad T_2 = \frac{\Sigma_2\chi}{|\Sigma_2\chi|}. \tag{31}$$

Here, $\dfrac{\partial\chi}{\partial x_j}$ $(j = 1, 2, 3)$ is a distribution supported on Γ, multiplication and division of them are defined by convolution of distribution with support. Summarizing the above mentioned facts, we have

Theorem 3.1

(S_1) is equivalent to (S_2).

Remark 3.1

The surface distribution of the dipole moment with the normal direction to the surface brings about the jump of the normal component of velocity for the fluid flowing across the interface.

3.2 Proof of Theorem 3.1

The case of the jump condition for the Dirichlet type Let us consider the problem $(Pr)_c$ to generate $[u_n] = c$. For simplicity, let us put $c = 1$.

$$(Pr)_c \begin{cases} \Delta u - \nabla p = \tilde{n}\Delta\chi & \text{in } \Omega, \\ \text{div} u = 0 & \text{in } \Omega, \\ u = 0 & \text{on } \partial\Omega. \end{cases} \tag{32}$$

where $\Delta\chi = \frac{\partial}{\partial\nu}\delta(\Gamma)$ is a dipole moment, $\Gamma = \partial\Omega_2$ is smooth and $\tilde{n}$ is smoothly extended into R^3 so as to satisfy $\left.\frac{\partial \tilde{n}_j}{\partial \nu}\right|_{\Gamma} = 0$ $(j = 1, 2, 3)$ and $\Delta\tilde{n} = 0$ in Ω_2.

[1st step] Potential flow is defined to satisfy

$$(Pr)_{c1} \begin{cases} \Delta U - \nabla P = \tilde{n}\cdot\Delta\chi & \text{in } R^3, \\ \text{div}\, U = 0 & \text{in } R^3. \end{cases} \tag{33}$$

We define fundamental solution: $\boldsymbol{E}^k = \left\{E_j^k\right\}_{j=1}^{3}$ $(k = 1, 2, 3)$ s.t.

$$\begin{cases} \Delta \boldsymbol{E}^k - \nabla q^k = \delta(x-y)\cdot \boldsymbol{e}_k & \text{in } R^3, \\ \text{div}\, \boldsymbol{E}^k = 0 & \text{in } R^3, \end{cases} \tag{34}$$

where $\boldsymbol{e}^k$ is an unit vector of kth axis $(k = 1, 2, 3)$ and $x = (x_1, x_2, x_3)$, $y = (y_1, y_2, y_3)$.

$$\begin{cases} E_j^k(x,y) = -\dfrac{\delta_j^k}{4\pi\,|x-y|} + \dfrac{(x_k - y_k)(x_j - y_j)}{|x-y|^3} \\ \qquad = \delta_j^k E(x-y) + F_j^k(x-y) \in L_y^1(R^3), \\ q^k(x,y) = \dfrac{\partial}{\partial x_k} E(x-y) \in L_y^1(R^3). \end{cases} \tag{35}$$

Potential $\{U, P\}$ is represented in the following way;

$$\begin{cases} \boldsymbol{U}(x) = \boldsymbol{E} * \tilde{n}\Delta\chi = \boldsymbol{E} * \Delta(\tilde{n}\chi) & \text{in } \mathcal{D}'(R^3), \\ P(x) = \boldsymbol{q} * \tilde{n}\Delta\chi = \boldsymbol{q} * \Delta(\tilde{n}\chi) & \text{in } \mathcal{D}'(R^3). \end{cases} \tag{36}$$

$$\begin{aligned} U_j(x) &= \int_{\Omega_2} \Delta_y \left(E_j^k(x-y)\cdot \tilde{n}_k(y)\right) d\Omega_y, \\ &= \int_{\Omega_2} \text{div}_y \nabla_y \left(E_j^k(x-y)\cdot \tilde{n}_k(y)\right) d\Omega_y, \end{aligned}$$

$$= \int_{\Gamma} \frac{\partial}{\partial n_y} E_j^k(x-y) \cdot \tilde{n}_k(y) d\sigma_y,$$

$$= \int_{\Gamma} \frac{\partial}{\partial n_y} E(x-y) \cdot \tilde{n}_j(y) d\sigma_y + \int_{\Gamma} \frac{\partial}{\partial n_y} F_j^k(x-y) \cdot \tilde{n}_k(y) d\sigma_y. \quad (37)$$

The second term of the right hand side dose not contribute to make the jump because

$$F_j^k = E * \frac{\partial^2}{\partial x_k \partial x_j} \delta = \frac{\partial^2}{\partial x_k \partial x_j} E. \quad (38)$$

The first term is a double layer potential. Then U satisfies

$$[\boldsymbol{U}] = \boldsymbol{U}|_{\Gamma_+} - \boldsymbol{U}|_{\Gamma_-} = \boldsymbol{n}, \quad (39)$$

$$\boldsymbol{n} \cdot [\boldsymbol{U}] = \boldsymbol{n} \cdot \boldsymbol{n} = 1. \quad (40)$$

Similarly, P satisfies

$$P(x) = \int_{\Omega_2} \triangle \left\{ q^k(x,y) \cdot \tilde{n}_k(y) \right\} d\Omega_y, \quad (41)$$

$$= \int_{\Gamma} \frac{\partial}{\partial n_y} q^k(x,y) \cdot n_k(y) d\sigma_y, \quad (42)$$

$$= \frac{\partial}{\partial x_k} \int_{\Gamma} n_k(y) \frac{\partial}{\partial n_y} E(x-y) d\sigma_y. \quad (43)$$

Then there holds

$$[P] = P|_{\Gamma_+} - P|_{\Gamma_-} = 0. \quad (44)$$

Let us note that $\boldsymbol{U}$ and P are real analytic in $R^3 \backslash \Gamma$.

[2nd step]

Let $u = U + \tilde{u}$ and $p = P + \tilde{p}$. Then $(\tilde{u}, \tilde{p})$ satisfies

$$(Pr)_{c2} \begin{cases} \Delta \tilde{u} - \nabla \tilde{p} = 0 \text{ in } \Omega, \\ \mathrm{div} \tilde{u} = 0 \quad \text{in } \Omega, \\ \tilde{u} = -U \quad \in \left\{ H^{\frac{1}{2}}(\partial \Omega) \right\}^2. \end{cases} \quad (45)$$

where $\int_{\Omega} \mathrm{div} U \, d\Omega = \int_{\Gamma} U \cdot n \, d\sigma = 0$. Note that there exists a unique solution $\{\tilde{u}, \tilde{p}\}$ for $(Pr)_{c2}$ satisfying

1. $\tilde{u} \in \left\{ H^1(\Omega) \right\}^2$,

2. $\tilde{p} \in L^2(\Omega)\backslash R$.

Then $n \cdot [u] = n\,[U] + n\,[\tilde{u}] = n\,[U] = 1$ in $\left\{H^{\frac{1}{2}}(\Gamma)\right\}^2$.

Theorem 3.2

The solution of $(Pr)_c$ *satisfies that* $n \cdot [u] = 1$ *in* $\left\{H^{\frac{1}{2}}(\Gamma)\right\}^2$.

The case of the jump condition for the type of $[\sigma_n] = a$ Let us consider the problem $(Pr)_a$ to generate $[\sigma_n] = a(=1)$.

$$(Pr)_a \begin{cases} \Delta u - \nabla p = \nabla \chi \text{ in } \Omega, \\ \operatorname{div} u = 0 \qquad\quad \text{in } \Omega, \\ u = 0 \qquad\qquad\;\, \text{on } \partial\Omega. \end{cases} \tag{46}$$

[1st step] Potential flow is defined to satisfy,

$$(Pr)_{a1} \begin{cases} \Delta U - \nabla P = \nabla \chi & \text{in } R^3, \\ \operatorname{div} U = 0 & \text{in } R^3. \end{cases} \tag{47}$$

It is obvious that

$$\begin{cases} U(x) = 0 & \text{in } R^3, \\ P(x) = -\chi & \text{in } R^3. \end{cases} \tag{48}$$

is the solution of $(Pr)_{a1}$.

[2nd step]

$u = U$ and $p = P$ satisfies $(Pr)_a$. Then we have

Theorem 3.3

$$\left[\frac{\partial u_n}{\partial n} - p\right] = 1 \text{ in } \left\{H^{-\frac{1}{2}}(\Gamma)\right\}^2.$$

The case of the jump condition for the type of $[\sigma_T] = b(=1)$ Let us consider the problem $(Pr)_b$ to generate $[\sigma_T] = b$.

$$(Pr)_b \begin{cases} \Delta u - \nabla p = \Sigma \chi \text{ in } \Omega, \\ \operatorname{div} u = 0 \qquad\quad \text{in } \Omega, \\ u = 0 \qquad\qquad\;\, \text{on } \partial\Omega. \end{cases} \tag{49}$$

[1st step] Potential flow is defined to satisfy

$$(Pr)_{b1} \begin{cases} \Delta U - \nabla P = \Sigma \chi & \text{in } R^3, \\ \operatorname{div} U = 0 & \text{in } R^3. \end{cases} \tag{50}$$

Then we have

$$U_j(x) = \int_{R^3} E_j^k(x-y) \cdot T_k(y) \cdot \delta(y-\Gamma)\, dy$$
$$= \int_{\Gamma} E_j^k(x-y) \cdot T_k(y)\, d\sigma_y \tag{51}$$

and

$$U_{\tilde{T}}(x) = \int_{\Gamma} E_j^k(x-y) \cdot T_j(x) \cdot T_k(y)\, d\sigma_y. \tag{52}$$

from which follows

$$(\tilde{n} \cdot \nabla)\, U_{\tilde{T}} = -\int_{\Gamma} \frac{\partial}{\partial n_y} E(x-y) \cdot T_j(x) \cdot T_j(y)\, d\sigma_y + \cdots \tag{53}$$

By use of non contribution of F_j^k to make the jump, we have

$$\left[\frac{\partial}{\partial n} U_T\right] = -1. \tag{54}$$

[2nd step]

Repeating similar arguments as before, we have

Theorem 3.4

$$\left[\frac{\partial u_T}{\partial n}\right] = -1 \text{ in } \left\{H^{-\frac{1}{2}}(\Gamma)\right\}^2.$$

3.3 The Stokes equation with a variable viscosity

In this section, we shall deal with Stokes equation with a variable viscosity in place of the one with a constant viscosity discussed in 3.1 and 3.2 as follows;

$$\begin{cases} \dfrac{\partial}{\partial x_j}\left(\nu(x)\dfrac{\partial u_i}{\partial x_j}\right) - \dfrac{\partial p}{\partial x_i} = f_i \text{ in } \Omega \text{ for } i = 1, 2, \\ \dfrac{\partial u_j}{\partial x_j} = 0 \qquad \text{in } \Omega, \end{cases} \tag{55}$$

where $\nu(x) = \nu_1(x)(1-\chi) + \nu_2(x)\chi$ and $\nu_j(x) \in C(\Omega_j)$ $(j = 1, 2)$.

According to the replacement of the Laplacian with the divergence form, the jumped boundary conditions defined on Γ should be modified;

$$[\sigma_n] = \nu_1 \frac{\partial u_{1n}}{\partial n} - p_1 - \left(\nu_2 \frac{\partial u_{2n}}{\partial n} - p_2\right), \tag{56}$$

$$[\sigma_T] = \nu_1 \frac{\partial u_{1T}}{\partial n} - \nu_2 \frac{\partial u_{2T}}{\partial n}, \tag{57}$$

$$[u_n] = \nu_1 u_{1n} - \nu_2 u_{2n}, \tag{58}$$

$$[u_T] = \nu_1 u_{1T} - \nu_2 u_{2T}. \tag{59}$$

However, if the relation $u_{1n} - u_{2n} = c$ on Γ is required in place of $[u_n] = c$, then $n \cdot \Delta\chi$ should be replaced with $n \cdot \frac{\partial}{\partial x_j}\left(H \frac{\partial}{\partial x_j}\chi\right)$. In fact,

$$\nu_1 u_{1n} - \nu_2 u_{2n} = \nu_1(u_{1n} - u_{2n}) + (\nu_1 - \nu_2)u_{2n} \ \text{ on } \Gamma. \tag{60}$$

Then H should be defined on Γ in the following way;

$$H = c\nu_1 + (\nu_1 - \nu_2)u_{2n} \ \text{ on } \Gamma. \tag{61}$$

The additional terms except $n \cdot \Delta\chi$ in (S_2) brings about the same jumps as in the statement of Theorem 3.1.

Remark 3.2 *DIGM for Stokes equation with a variable viscosity is proved by treating the transmission problem defined on Γ under the weak formulation of the equation, that will be shown in the succeeding paper.*
Remark 3.3 In the case of time dependent Stokes equations, we can show the same result as obtained in the steady case.

4 Unified model for multi-phase flow with interfacial interactions

Incompressibility condition for multi-phase flow system is defined as follows;

$$(\chi^{(wt)} + \chi^{(fl)}) \left\{ \frac{\partial u_j}{\partial x_j} + (k_f + k_w)\chi_b \frac{\partial \chi^{(fl)}}{\partial n^{(fl)}} \right\} = 0, \ \text{ in } \Omega, \ t > 0. \tag{62}$$

An existence of the second term in the brace of the left hand side is due to the reactivity condition.

Conservation of momentum for the total flow system is settled up into the single equation in the following way;

$$\rho\left(\frac{\partial u_i}{\partial t} + (\chi_{wt} + \chi_{fl})u_j \frac{\partial u_i}{\partial x_j}\right) + c(\chi_{ws} + \chi_{fs})u_i = -\frac{\partial(\chi_{wt} + \chi_{fl})(p - P_\infty)}{\partial x_i}$$
$$+ \frac{\partial}{\partial x_j}\left\{\left(\varepsilon\chi_{air} + \mu_{wt}\chi_{wt} + \left(\mu_f + \frac{\eta}{\sqrt{4D_\Pi + \varepsilon_b^2}}\right)\chi_{fl}\right)\left(\frac{\partial u_i}{\partial x_j} + \frac{\partial u_j}{\partial x_i}\right)\right\}$$

$$+\rho\chi_s g_T\left(\frac{u_{T_1}}{\sqrt{|u_T|^2+\varepsilon_g^2}}(\Sigma_1\chi_{fl})_i+\frac{u_{T_2}}{\sqrt{|u_T|^2+\varepsilon_g^2}}(\Sigma_2\chi_{fl})_i\right)+\rho g\,\delta_{i,3}$$
$$-\chi^{(b)}\frac{\partial}{\partial x_j}\left(H\frac{\partial}{\partial x_j}\chi_{fs}\right)\cdot n_i^{(fs)}\ \text{in}\,\Omega,\ \ t>0, \tag{63}$$

where $H=-\mu_{wt}(k_f+k_w)+\left\{\mu_{wt}-\left(\mu_f+\frac{\eta}{\sqrt{4D_\Pi+\varepsilon_b^2}}\right)\right\}u_j$. The third term in (63) represents the adhesion and sliding phenomena occured on the interface between oil and sand. The fourth term means the dipole moment distribution along the reaction surface included in the bacteria zone, which plays an important role to satisfy the reactivity condition.

Motion equation of free surface for air is;

$$\frac{\partial\chi^{(air)}}{\partial t}+u_j\frac{\partial\chi^{(air)}}{\partial x_j}=0\ \text{ in}\ \Omega,\ t>0. \tag{64}$$

Motion equation of free surface for oil is;

$$\frac{\partial\chi^{(fl)}}{\partial t}+u_j^{(fl)}\frac{\partial\chi^{(fl)}}{\partial x_j}=k_f\cdot\left|\nabla\chi^{(fl)}\right|\chi^{(b)}\ \text{ in}\ \Omega,\ t>0, \tag{65}$$

Finaly, the outer boundary conditions are prescribed as follows;

$$\begin{cases}u_n=u_0\sin\omega t,\\ u_T=0,\end{cases}\quad\text{on}\ \Gamma_{IN}(t),\qquad\begin{cases}u_n=0,\\ \dfrac{\partial u_T}{\partial n}=0,\end{cases}\quad\text{on}\ \Gamma_R, \tag{66}$$

$$\begin{cases}u_n=0,\\ u_T=0,\end{cases}\quad\text{on}\ \Gamma_Q,\qquad\begin{cases}\sigma_n=0,\\ u_T=0,\end{cases}\quad\text{on}\ \Gamma_A. \tag{67}$$

5 Numerical procedure

5.1 Poisson equation for the pressure

Poisson equation for the pressure p is derived by operating the divergence to momentum equations for total system;

$$\frac{\partial}{\partial x_j}\left\{\frac{1}{\rho}\frac{\partial}{\partial x_j}(\chi_{wt}+\chi_{fl})(p-P_\infty)\right\}-\frac{1}{\varepsilon}\chi_{air}\,(p-P_\infty)$$
$$=-\frac{\partial}{\partial x_j}\frac{\partial u_j}{\partial t}+\cdots\ \text{ in}\,\Omega,\ \ t>0, \tag{68}$$

where

$$\frac{\partial}{\partial x_j}\frac{\partial u_j}{\partial t}=\frac{\partial}{\partial t}\frac{\partial u_j}{\partial x_j}=\frac{\partial}{\partial t}\left\{\chi_{wf}\frac{\partial u_j}{\partial x_j}+(1-\chi_{wf})\frac{\partial u_j}{\partial x_j}\right\}$$

$$= \frac{\chi_{wf}^{n+1}\dfrac{\partial u_j^{n+1}}{\partial x_j} - \chi_{wf}^{n}\dfrac{\partial u_j^{n}}{\partial x_j}}{\Delta t} + \frac{(1-\chi_{wf}^{n})\dfrac{\partial u_j^{n}}{\partial x_j} - (1-\chi_{wf}^{n-1})\dfrac{\partial u_j^{n-1}}{\partial x_j}}{\Delta t}$$

$$= \frac{-\chi_{wf}^{n}(k_f+k_w)\chi_b\dfrac{\partial \chi_{fl}^{n}}{\partial n_{fl}} - \chi_{wf}^{n}\dfrac{\partial u_j^{n}}{\partial x_j}}{\Delta t}$$

$$+\frac{(1-\chi_{wf}^{n})\dfrac{\partial u_j^{n}}{\partial x_j} - (1-\chi_{wf}^{n-1})\dfrac{\partial u_j^{n-1}}{\partial x_j}}{\Delta t}. \qquad (69)$$

Here we used the relation;

$$\frac{\partial u_j}{\partial x_j} = -(k_f+k_w)\chi_b\frac{\partial \chi_{fl}}{\partial n_{fl}} \text{ in } \Omega_{wf}, \qquad (70)$$

where the right hand side of the above equation is due to the reactivity condition. Superscript means time steps.

Boundary conditions for p is derived from the momentum equations and boundary conditions for them. Penalty term, which is the second term in the left hand side in (68), works in the air region, i.e., the fictitious domain, to satisfy $p = P_\infty$. ε is a sufficiently small positive parameter.

5.2 Numerical treatment

- **Discretization**
  Finite difference method
- **Overall scheme**
  Advection term : Third order upwind scheme
  Space discretization except for advection term : Second order central difference scheme
  Time integration : First order semi-implicit scheme
- **Mesh structure**
  Time-independent equi-spaced orthogonal mesh
- **Iteration procedure**
  GP-BiCG : Poisson equation for the pressure
  Gauss-Seidel : Equations for momentum and free surfaces

6 Numerical results

Figure 4 shows the time-sequence of decomposition of oil into water. Top figure shows the initial condition, in which air, oil and water in layers are at rest. A rectangle located near the center of the figure shows the habitat of bacteria. We can see that the jump of velocity component normal to the boundary between oil and water occurrs while the decomposition proceeds. Figure 5 shows the 3D case with same situation.

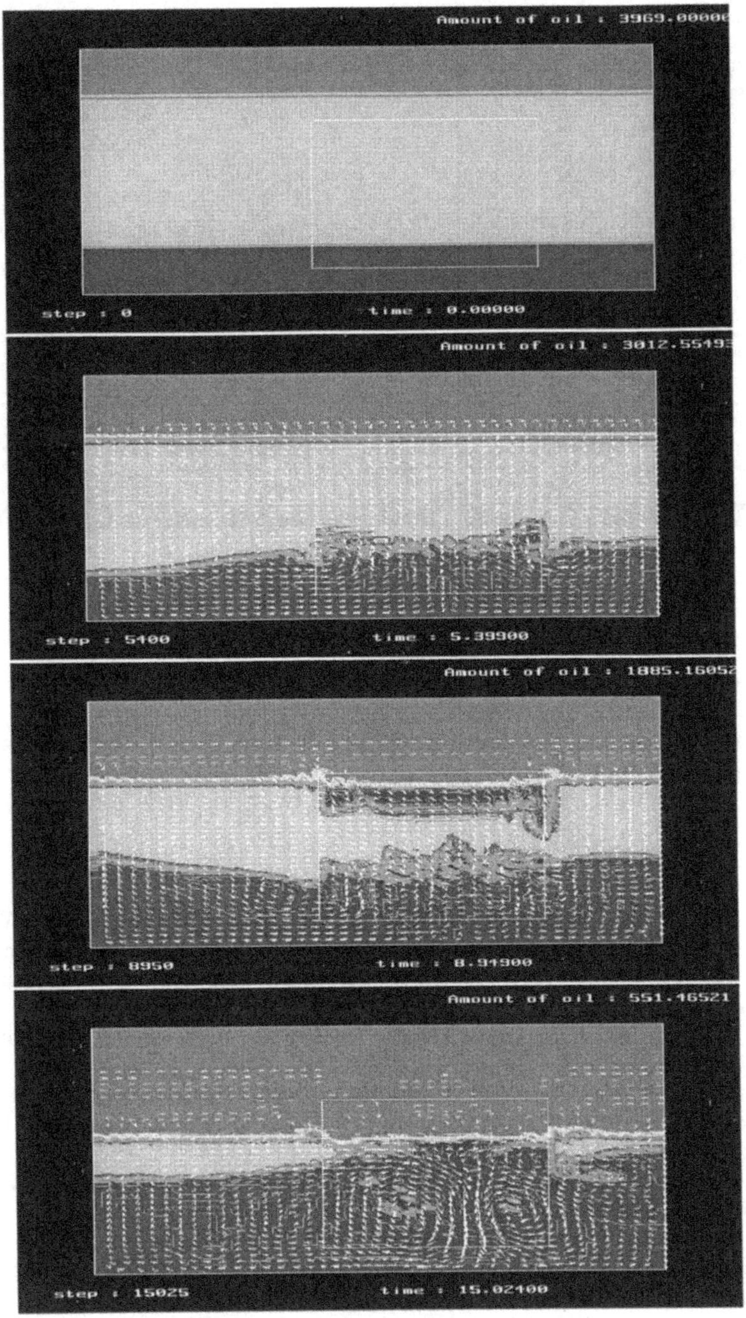

Fig. 4. Decomposition of oil in 2D case

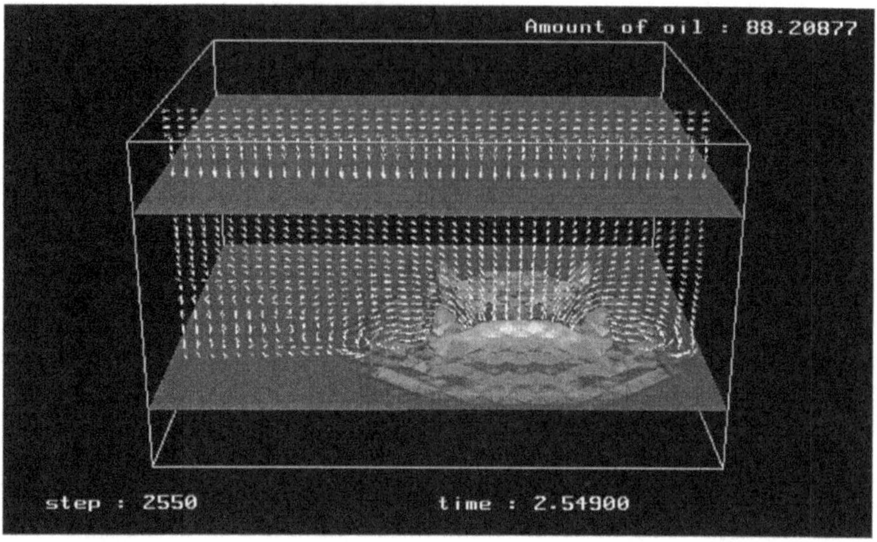

Fig. 5. Decomposition of oil in 3D case

Acknowledgements

The authors express their sincere gratitude to Professors H. Fujita, M. Okada and E. Baba for their guidance.

References

1. E. Baba and C. Cheong (1999): Effect of Spilled Oil on Exchange of Sea Water over the Sand Beach and Tidal-flat, Proceedings of Workshop on Environmental Fluid Mechanics for Coastal Ecosystems, p. 11, Massachusetts Institute of Technology, USA
2. G. Duvaut and J. L. Lions, (1976): Inequalities in Mechanics and Physics, Springer-Verlag.
3. H. Fujita, H. Kawahara and H. Kawarada (1995): Distribution Theoretic Approach to Fictitious Domain Method for Neumann Problems, East-West J. Numer. Math., Vol. 3, No.2, pp.111-126.
4. R. Glowinski, J.L. Lions and R. Tremolieres (1981): Numerical Analysis of Variational Inequalities, North-Holland publishing Company.
5. H. Kawarada, H. Fujita and H. Suito (1998): Wave motion Breaking upon the Shore, GAKUTO International Series, Mathematical Sciences and Applications, Vol. 11, pp.145-159.
6. H. Kawarada and H. Suito (1997): Numerical Method for a Free Surface Flow on the bases of the Fictitious Domain Method, East-West J. Numer. Math., Vol. 5, No. 1, pp.57-66.

Universal and Simultaneous Solution of Solid, Liquid and Gas in Cartesian-Grid-Based CIP Method

Takashi Yabe

Tokyo Institute of Technology, Tokyo 152-8552, JAPAN

Abstract. We present a review of the CIP method that is known as a general numerical solver for solid, liquid and gas. This method is a kind of semi-Lagrangean scheme and has been extended to treat incompressible flow in the framework of compressible fluid. Since it uses primitive Euler representation, it suits for multi-phase analysis. The recent version of this method guarantees the exact mass conservation even in the framework of semi-Lagrangean scheme. Comprehensive review is given for the strategy of the CIP method that has a compact support and subcell resolution including front capturing algorithm with functional transformation. Some practical applications are also reviewed such as milk crown or coronet.

1 Introduction

Solving all phases of matter together by one universal scheme is a grand challenge to the field of computational mathematics. For these types of problems such as melting and deformation, and evaporation, we need to treat topology and phase changes of the materials simultaneously, where the grid system aligned to the solid or liquid surface has no meaning and sometimes the mesh is distorted and even broken up. A universal treatment of all phases by one simple algorithm is thus essential and we are at the turning point of attacking this goal. Even without phase change, problems of surface capturing and structure-fluid interaction are not easy task. In most of cases, the grid can not always be adapted to those surfaces. Therefore, the description of moving surfaces of complicated shape in the Cartesian grid system will be a challenging subject.

In order to attack the problems mentioned above, we must first find a method to treat a sharp interface and to solve the interaction of compressible gas with incompressible liquid or solid. Toward this goal, we take Eulerian-approach based on the CIP(cubic-interpolated propagation) method [1–7] which does not need adaptive grid system and therefore removes the problems of grid distortion caused by structural break up and topology change. The material surface can be captured by almost one grid throughout the computation[8]. Furthermore, the scheme can treat all the phases of matter from solid state through liquid and two phase state to gas without restriction on the time step from high sound speed [9].

Pressure-based algorithm coupled with semi-Lagrangian approach like the CIP proved to be stable and robust in analyzing these subjects. The only disadvantage of this method was the lack of conservative property. Recent version of the CIP-CSL4[10] can overcome this difficulty and povide exactly conservative semi-Lagrangian scheme. Since these scheme do not use the cubic polynomial but use different orders of polynomial, we re-define the name of these CIP families as "Constrained Interpolation Profile" and still keep the abbreviation, CIP. This means that various constraints such as the time evolution of spatial gradient, that is used in the original CIP method, or spatially integrated conservative quantities can be used to construct the profile. In this paper, we shall give a brief review of the CIP family and give some examples applied to various phases of matter.

2 CIP Family

2.1 CIP method

Although the nature is in a continuous world, digitization process is unavoidable in order to be implemented in numerical simulations. Primary goal of numerical algorithm will be to retrieve the lost information inside the grid cell between these digitized points. Most of numerical schemes proposed before, however, did not take care of real solution inside the grid cell and resolution has been limited to the grid size. The CIP method proposed by one of the authors tries to construct a solution inside the grid cell close enough to this real solution of the given equation with some constraints. We here explain its strategy by using an advection equation,

$$\frac{\partial f}{\partial t} + u\frac{\partial f}{\partial x} = 0. \tag{1}$$

When the velocity is constant, the solution of Eq.(1) gives a simple translational motion of wave with a velocity u. The initial profile (solid line of Fig.1(a)) moves like a dashed line in a continuous representation. At this time, the solution at grid points is denoted by circles and is the same as the exact solution. However, if we eliminate the dashed line as in Fig.1(b), then the information of the profile inside the grid cell has been lost and it is hard to imagine the original profile and it is natural to imagine a profile like that shown by solid line in (c). Thus, numerical diffusion arises when we construct the profile by the linear interpolation even with the exact solution as shown in Fig.1(c). This process is called the first-order upwind scheme. On the other hand, if we use quadratic polynomial for interpolation, it suffers from overshooting. This process is the Lax-Wendroff scheme or Leith scheme.

What made this solution worse ? It is because we neglect the behavior of the solution inside a grid cell and merely follow after the smoothness of the solution. From this experience, we understand that a method incorporating the real solution into the profile within a grid cell is quite an important subject.

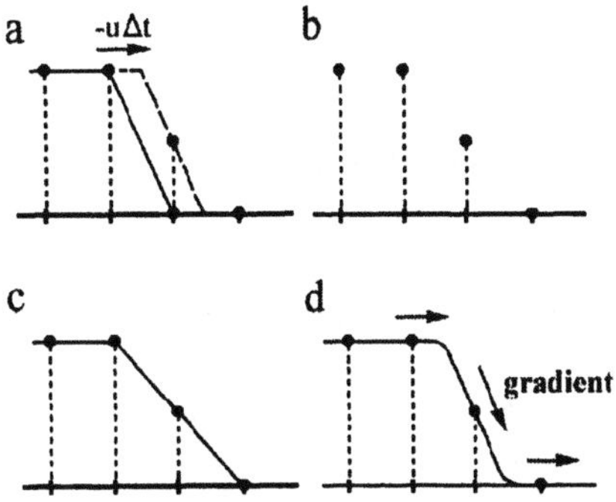

Fig. 1. The principle of the CIP method. (a) solid line is initial profile and dashed line is an exact solution after advection, whose solution (b) at discretized points. (c) When (b) is linearly interpolated, numerical diffusion appears. (d) In the CIP, spatial derivative also propagates and the profile inside a grid cell is retrieved.

We propose to approximate the profile as shown below. Let us differentiate Eq.(1) with spatial variable x, then we get

$$\frac{\partial g}{\partial t} + u\frac{\partial g}{\partial x} = -\frac{\partial u}{\partial x}g, \tag{2}$$

where $g \equiv \partial f/\partial x$ stands for the spatial derivative of f. In the simplest case where the velocity u is constant, Eq.(2) coincides with Eq.(1) and represents the propagation of spatial derivative with a velocity u. By this equation, we can trace the time evolution of f and g on the basis of Eq.(1). If g could be predicted to propagate like that shown by the arrows in Fig.1(d), the profile after one step would be limited to a specific profile. It is easy to imagine that by this constraint, the solution becomes much closer to the initial profile that is the real solution. Most importantly, the solution thus created gives a profile consistent with Eq.(1) even inside the grid cell.

If both the values of f and g are given at two grid points, the profile between these points can be interpolated by cubic polynomial $F(x) = ax^3 + bx^2 + cx + d$. Thus, the profile at n+1 step is readily obtained by shifting the profile by $u\Delta t$ like $f^{n+1} = F(x - u\Delta t), g^{n+1} = dF(x - u\Delta t)/dx$.

$$\begin{aligned} a_i &= \frac{g_i + g_{iup}}{\Delta x_i^2} + \frac{2(f_i - f_{iup})}{\Delta x_i^3}, \\ b_i &= \frac{3(f_{iup} - f_i)}{\Delta x_i^2} - \frac{2g_i + g_{iup}}{\Delta x_i}, \\ \Delta x_i &= x_{iup} - x_i \\ iup &= i - \mathrm{sgn}(u_i) \end{aligned} \tag{3}$$

$$\begin{aligned} f_i^{n+1} &= a_i\xi_i^3 + b_i\xi_i^2 + g_i^n\xi_i + f_i^n, \\ g_i^{n+1} &= 3a_i\xi_i^2 + 2b_i\xi_i + g_i^n, \end{aligned} \tag{4}$$

where we define $\xi_i = -u_i \Delta t$ and $\mathrm{sgn}(u)$ stands for the sign of u.

2.2 Interface tracking: a sharpness preserving method

Treatment of interface that lies between materials of different properties remains a formidable challenge to the computation of multi-phase fluid dynamics. Eulerian methods have proven robust in simulating flows with interfaces of complex topology. Generally, Eulerian methods use color function to distinguish the regions where different materials fall in. To accurately reproduce the physical processes across the interface transition region, keeping the compact thickness of the interface is of great importance. The finite difference schemes constructed on an Eulerian grid, however, intrinsically produce numerical diffusions to the solution of advection equation by which the interface is predicted temporally. Thus, the direct implementation of finite difference schemes (even of high order) can not maintain the compactness of the interface.

Various kinds of methods have been developed so far to achieve a compact and correctly defined interface by introducing extra programming. Among those mostly used algorithms are the level set methods and the VOF(volume of fluid) methods for front capturing, and others for front tracking [11]. Level set method that was firstly proposed by Osher and Sethian[12] gets around the computation of interfacial discontinuity by evaluating the field in higher dimensions. The interface of interest is then recovered by taking a subset of the field. Practically, the interface is defined as the zero level set of a distance function from the interface.

In a VOF kind method on the other hand, the interface needs to be reconstructed based on the volume fraction of fluid. VOF methods are mainly classified as SLIC(simple line interface calculation)algorithm[13] and PLIC (piecewise linear interface calculation) algorithm[14] according to the interpolation function used to represent the interface. The SLIC makes use of piecewise constant reconstruction and the interfaces are approximated by lines aligned with mesh coordinates. The PLIC estimates the interface with a truly piecewise linear approximation that improves largely the geometrical faithfulness of the method.

In [8] and [15], we devised an interface tracking technique which appears efficient, geometrically faithful and diffusionless. The method is a combination of the CIP advection solver and a tangent function transformation.

Consider K kinds of impermeable materials occupying closed areas $\{\Omega_k(t), k = 1, 2, \cdots, K\}$ in computational domain $D \in \mathbf{R}^3(x, y, z)$, we identify them with color functions or density functions $\{\phi_k(x, y, z, t), k = 1, 2, \cdots, K\}$ by the following definition

$$\phi_k(x, y, z, t) = \begin{cases} 1, & (x, y, z) \in \Omega_k(t), \\ 0, & \text{otherwise.} \end{cases}$$

Suppose these materials move at the local speed, the color functions evolve then according to the following advection equation

$$\frac{\partial \phi_k}{\partial t} + \mathbf{u} \cdot \nabla \phi_k = 0, \quad k = 1, 2, \cdots, K \tag{5}$$

where $\mathbf{u}$ is the local velocity.

It is known that solving the above equation by finite difference schemes in an Eulerian representation will produce numerical diffusion and tend to smear the initial sharpness of the interfaces. In our method, rather than the original variable ϕ_k itself, its transformation, say $F(\phi_k)$, is calculated by the CIP method. We specify $F(\phi_k)$ to be a function of ϕ_k only, which means that the new function $F(\phi_k)$ is also governed by the same equation as (5). Hence, we have

$$\frac{\partial F(\phi_k)}{\partial t} + \mathbf{u} \cdot \nabla F(\phi_k) = 0, \tag{6}$$

and all the algorithms proposed for ϕ_k (schemes for advection equation) can be used to $F(\phi_k)$. Hopefully, by the considerable simplicity, this kind of techniques would be very attractive for practical implementation. We here use a transformation of a tangent function for $F(\phi_k)$, that is,

$$F(\phi_k) = \tan[(1-\epsilon)\pi(\phi_k - 1/2)], \tag{7}$$

$$\phi_k = \tan^{-1} F(\phi_k)/[(1-\epsilon)\pi] + 1/2, \tag{8}$$

where ϵ is a small positive constant. The factor $(1-\epsilon)$ makes us get around $-\infty$ for $\phi_k = 0$ and ∞ for $\phi_k = 1$ and enables us to tune for a desired steepness of the transition layer.

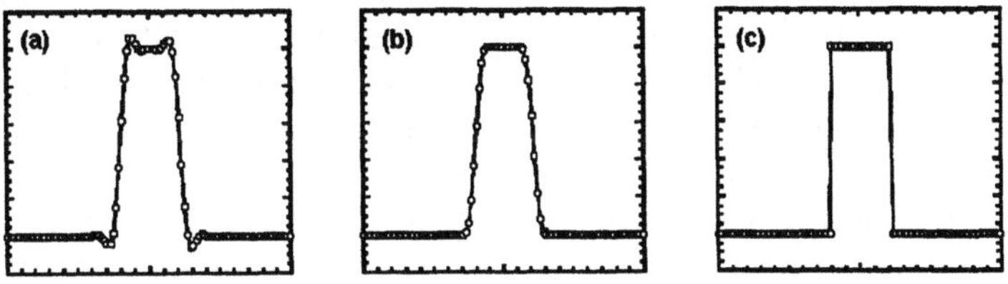

Fig. 2. Square wave propagation by (a) CIP method, (b) Rational CIP method, and (c) tangent-transformed CIP

Although ϕ_k experiences a rapid change from 0 to 1 at the interface, $F(\phi_k)$ shows a quite regular behavior. Because most of the values of $F(\phi_k)$ are concentrated near $\phi_k = 0$ and 1, the function transformation improves locally the spatial resolution near the large gradients. Thus, the sharp discontinuity can be described quite easily. The transformation of this kind is effective only for the case where the value of ϕ_k is limited to a definite range throughout the calculation, like the color function defined before. This method does not involve any interface construction procedure and is quite economical in computational complexity. It should be also notified that the presented method is more attractive in 3-D computation since the extension of the scheme to 3-D is straightforward.

Figure 2(c) shows a 1D square wave propagation computed by the CIP method together with the tangent transformation. The initial sharpness is well preserved and the discontinuities are advected with a correct speed.

2.3 Conservative Semi-Lagrangian Scheme

It is well known that the CIP method shows good conservation of mass, although the method is written in a non-conservative form. However, in some special cases, there still exist problems which require exact conservation of mass. For example, when we treat the black-hole formation and plasma dynamics, small fraction of mass and charge generates a gravity wave and a large electric field, respectively, and therefore, the exact conservation of mass is necessary to success the numerical analysis. For the solution of Vlasov equation, it is possible to constitute and improve the CIP method so as to exactly conserve the mass [16]. However, it is impossible to apply this numerical technique to the solution of general hyperbolic equations. Therefore, the development of the conservative CIP method is desired earnestly.

Under such situation, recently, authors have succeeded in the development of new conservative schemes called as CIP-CSL4 [10] and CIP-CSL2 [17] which are based on the concept of the CIP scheme and succeeded the excellent numerical features of the CIP scheme. In order to include these various families of the schemes, we here extend the name CIP to mean Constrained Interpolation Profile and CSL means Conservative Semi-Lagrangian. CSL4 and CSL2 use the 4-th order and quadratic polynomial, respectively. The scheme has been applied to many problems of the linear and nonlinear one-dimensional hyperbolic equations in the previous papers [10,17].

Since semi-Lagrangian schemes [18–20] can be used for high CFL (Courant-Friedrichs-Lewy) condition in explicit form and are stable for multi-phase flow calculations [21] but only shortcoming is the lack of exact mass conservation, exactly conservative semi-Lagrangian schemes like the CIP-CSL2 and CSL4 have many promising future applications.

As already seen, the CIP adopted additional constraint, that is spatial gradient, to represent the profile inside the grid cell. For being endowed with the conservative property, we here add another constraint as

$$\rho_{i-1/2}^{n} = \int_{x_{i-1}}^{x_i} f^n dx. \tag{9}$$

Therefore the spatial profile must be constructed to satisfy this additional constraint. If this could be realized, f would be advanced in the non-conservative form with exact conservation in a form of ρ which could be advanced maintaining mass conservation.

Keeping this point in mind, then the ith function piece $F_i(x)$ must be determined so as to satisfy the following constraints:

$$\begin{aligned} &F_i(x_{i-1}) = f(x_{i-1}), \quad F_i(x_i) = f(x_i) \\ &\partial F_i(x_{i-1})/\partial x = g(x_{i-1}) \\ &\partial F_i(x_i)/\partial x = g(x_i) \\ &\int_{x_{i-1}}^{x_i} F_i(x)dx = \rho_{i-1/2}. \end{aligned} \tag{10}$$

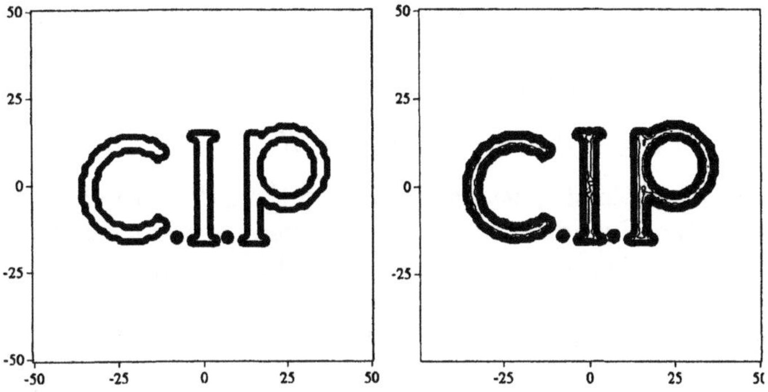

Fig. 3. Contour plots after one complete revolution of a solid-body which consists of three characters of "C.I.P" and all the lines composing the charecters are thiner than 3 grid points. (Left) Initial profile (Right) profile after one complete revolution.

In order to meet the above constraints, the 4th-order polynomial can be chosen as the interpolation function $F_i(x)$. Thus the time development of f and g is calculated simply by shifting the interpolation function $F_i(x)$ by $u\Delta t$ in the same way as Eq.(4) of the CIP method. This method is called CIP-CSL4 because it uses the 4-th order polynomial. Figure 3 demonstrates the advantage of improved accuary by 4-th order and exact conservation. These characters in Fig.3 is rotated within fixed grid system and dots after the character "C" and "I" are one-grid size. It is surprising, material of one grid size has been preserved even after revolution.

In the CIP, the time evolution of f and $g = \partial f/\partial x$ is used as constraints to define a cubic polynomial, while in the CIP-CSL4, constraints are now $f, \partial f/\partial x$ and $\int f dx$ giving 4-th order polynomial. It would be intresting to find a way to apply the CIP to the integrated value of f instead of f itself.

The motivation to employ this analogy stems from the following advection equation.

$$\frac{\partial D}{\partial t} + u\frac{\partial D}{\partial x} = 0. \tag{11}$$

Interestingly, if we take a spatial derivative of Eq.(11) and define $D' \equiv \partial D/\partial x$, we obtain a conservative-type equation

$$\frac{\partial D'}{\partial t} + \frac{\partial (uD')}{\partial x} = 0. \tag{12}$$

Then we come to an idea to use $D' = f$ in Eq.(12) and $D = \int f\ dx$ in Eq.(11). This procedure is exactly the same as Eq.(1) by simply replacing f by $\int f\ dx$, together with Eq.(2) in which g is replaced by f. Thus all the CIP procedure can be used for a pair of $\int f dx$ and f instead of f and $\partial f/\partial x$

By this analogy, we shall introduce a function :

$$D_i(x) = \int_{x_i}^{x} f(x')dx'. \tag{13}$$

We shall use a cubic polynomial to approximate this profile.

$$D_i(x) = A1_i X^3 + A2_i X^2 + f_i^n X \tag{14}$$

where $X = x - x_i$. The role of spatial gradient g in the CIP method is now played by f that is spatial gradient of $D(x)$ in the present scheme. By using the above relation, a profile of $f(x)$ between x_i and x_{iup} is then given by taking the derivative of Eq.(14).

Then we apply the splitting algorithm of the CIP to

$$\frac{\partial f}{\partial t} + u\frac{\partial f}{\partial x} = -f\frac{\partial u}{\partial x}, \tag{15}$$

in which advection part is calculated by

$$f_i^{n+1} = 3A1_i\xi^2 + 2A2_i\xi + f_i^n, \tag{16}$$

where $\xi = -u\Delta t$. Although we separately treat the conservative equation Eq.(12), mass conservation is recovered by Eq.(9) in constructing the spatial profile inside a grid cell.

Although this scheme is quite promising, we will not use it for the calculations given in the following sections, since the original CIP is sufficient for these applications.

2.4 Hydrodynamic Equations

In order to solve all the materials in a universal form, we must find an appropriate equation for solid, liquid and gas. We use full hydrodynamic equations for these materials, which can be written in a form :

$$\frac{\partial \mathbf{f}}{\partial t} + (\mathbf{u} \cdot \nabla)\mathbf{f} = \mathbf{S} \tag{17}$$

where $\mathbf{f} = (\rho, \mathbf{u}, T)$, $\mathbf{S} = (-\rho\nabla\cdot\mathbf{u}+Q_m, -\nabla p/\rho+Q_u, -P_{TH}\nabla\cdot\mathbf{u}/\rho C_v+Q_E)$, and ρ is the density, $\mathbf{u}$ the velocity, p the pressure, T the temperature, Q_m represents the mass source term, Q_u represents viscosity, elastic stress tensor, surface tension etc., and Q_E represents viscous heating, thermal conduction and heat source.

Here, C_v is the specific heat for constant volume and we define $P_{TH} = T(\partial p/\partial T)_\rho$ which is derived from the first principle of thermodynamics as

$$\begin{aligned} TdS = dU + pdV &= \left(\frac{\partial U}{\partial T}\right)_V dT + \left(\frac{\partial U}{\partial V}\right)_T dV + pdV \\ &= C_v dT + T\left(\frac{\partial p}{\partial T}\right)_V dV \end{aligned} \tag{18}$$

where S is the entropy, U the internal energy, $V = 1/\rho$ the specific volume. The last relation in Eq.(18) is derived from the Helmholtz free energy $F = U - TS$ and thermodynamic consistency : $(\partial p/\partial T)_V = -(\partial^2 F/\partial V \partial T) = (\partial S/\partial V)_T = 1/T\,[p + (\partial U/\partial V)_T]$. Here, P_{TH} is not merely the pressure. In the special case of ideal fluid, however, P_{TH} is exactly the pressure p because the pressure linearly depends on temperature. Next simpler example is the two phase flow described by the Clausius-Clapeyron relation :

$$p = p_0 \exp\left(-\frac{L}{RT}\right), \quad P_{TH} = T\left(\frac{\partial p}{\partial T}\right) \propto L \tag{19}$$

where R is the gas constant. In this case, P_{TH} becomes proportional to the latent heat L. Therefore, P_{TH} describes the heat loss due to latent heat when the ratio of gas increases in two-phase flow. More general form of C_v and P_{TH} will be given by semi-analytical formula or tabulated data.

The CIP method solves the equations like Eq.(17) by dividing those into non-advection and advection phases as given in previous papers. A cubic-interpolated profile propagates in space in the advection phase and then nonadvection phase is calculated by finite difference methods.

As shown in the previous papers, we can trace shock waves correctly with the CIP method although it uses fluid equations written in a non-conservative form or in primitive Euler representation.

2.5 Pressure-based algorithm

The CIP method uses the primitive Euler method to solve Eq.(17), thus the formulation into a simultaneous solution of incompressible and compressible fluid is readily obtained. In order to get an idea of this strategy, we shall start at first examining how it has been difficult to solve them together. In ordinary compressible fluid, the density ρ is solved by the mass conservation equation and then the temperature T is obtained by energy equation. After that, from the equation of state (EOS), the pressure $p = p(\rho, T)$ is calculated.

In the low density side, $p \propto \rho T$ like ideal fluid and dependence is relatively weak, but at solid or liquid density p steeply rises as the density. This means that extremely high pressure is need to compress solid or liquid even slightly. In other words, for solid or liquid, the sound speed $C_s = (\partial p/\partial \rho)^{1/2}$ is quite large. Therefore, if we choose the process in which density is calculated at first, only a small amount of error on density of 10%, for example, causes a large pressure pulse by 3-4 orders of magnitude.

In such a situation, incompressible approximation is normaly adopted, that is, pressure equation to ensure $\nabla \cdot \mathbf{u} = 0$ is derived from equation of motion and mass conservation. This scheme is called pressure-based scheme and MAC[22], SMAC[23], SIMPLE[24], SIMPLER[25] etc. are some of the typical examples.

In order to extend this idea to compressible fluid, we had better modify the EOS. Let us rotate the EOS by 90 degree, then the steep pressure curve becomes now flat density curve. This means that if we could solve the pressure at first and then estimate the density later on by this EOS in terms of $\rho(p, T)$, the problem at the liquid density will be removed. Adding to this, since the EOS in lower density gas depends linearly on other quantities, no problem occur there by this reverse procedure.

Then how we realize this reverse procedure? For this purpose, we should predict how the pressure reacts to the change of density and temperature. Such a unified procedure to incorporate compressible fluid with incompressible fluid has been initiated by Harlow as the ICE(Implicit Continuous Eulerian) [26]. The ICE has been improved by the PISO [27](Pressure Implicit with Splitting of Operators). In both cases, however, conservative equations are used as a starting point. Main difference between ICE and PISO comes from the treatment of convection term.

On the other hand, the CCUP [9] uses primitive Euler equations and splits the advection term from the other terms related sound waves. By this simplification, pressure equation becomes quite simple and the ability to attack the multi-phase flow has been greatly improved. One year after this proposal, Zienkiewicz et al.[28] proposed similar method but applied to Finite Element Method. Unfortunately, however, their scheme is not so simple to remove the difficulty stemming from large density ratio at the boundary between liquid and gas as will be discussed later on.

Let us now start with the description of the ICE in a most compact and generalized way. In both the ICE and the PISO, conservation equations of mass and momentum are used in a finite difference form

$$\frac{\rho^{n+1} - \rho^n}{\Delta t} = -\frac{\partial (\rho u)'}{\partial x} \tag{20}$$

$$\frac{(\rho u)' - (\rho u)^n}{\Delta t} = -\frac{\partial p^{n+1}}{\partial x} + H(u) \tag{21}$$

$$H(u) \equiv -\frac{\partial (\rho u^2)}{\partial x} \tag{22}$$

Substituting Eq.(21) into Eq.(20), we get

$$\frac{\partial^2 p^{n+1}}{\partial x^2} = \frac{\rho^{n+1} - \rho^n}{\Delta t^2} + \frac{1}{\Delta t}\left(\frac{\partial \rho u}{\partial x}\right)^n + \frac{\partial H}{\partial x} \tag{23}$$

Nextly if we assume that density changes in proportion to pressure change,

$$\Delta p = \left(\frac{\partial p}{\partial \rho}\right)_T \Delta\rho = C_s^2 \Delta\rho \tag{24}$$

then density change on the right hand side of Eq.(23) is replaced by pressure change,

$$\frac{\partial^2 p^{n+1}}{\partial x^2} = \frac{p^{n+1} - p^n}{C_s^2 \Delta t^2} + \frac{1}{\Delta t}\left(\frac{\partial \rho u}{\partial x}\right)^n + \frac{\partial H}{\partial x} \tag{25}$$

In the ICE, the term H is estimated at the step n, while in the PISO H is predicted by an equation of motion

$$\frac{\rho^n u^p - (\rho u)^n}{\Delta t} = -\frac{\partial p^n}{\partial x} + H(u^p) \tag{26}$$

and finaly get an equation :

$$\frac{\partial^2 (p^{n+1} - p^n)}{\partial x^2} = \frac{p^{n+1} - p^n}{C_s^2 \Delta t^2} + \frac{1}{\Delta t}\left(\frac{\partial \rho^n u^p}{\partial x}\right)^n \tag{27}$$

Original PISO is more complicated because it repeats this predictor-corrector algorithm by a few times and some complication appears to diagonalize H term to solve Eq.(26) in terms of u^p.

2.6 CCUP method

Yabe et al.[9] darely used primitive Euler form to construct pressure equation instead of conservative form. Furthermore, advection part is separated from the other terms, since the advection term can be processed being free from CFL condition in semi-Lagrangian procedure. Fortunately, this splitting led to an unexpected advantage to the solution in multi-phase flow as shown below.

Original CCUP method [9] was proposed only for a special equation of state like Eq.(24), but here we rebuild it with more general EOS [30]. That is, for small change of density and temperature, the pressure change can be linearly proportional to them as

$$\Delta p = \left(\frac{\partial p}{\partial \rho}\right)_T \Delta\rho + \left(\frac{\partial p}{\partial T}\right)_\rho \Delta T \tag{28}$$

where Δp means the pressure change $p^{n+1} - p^*$ during one time step and * is the profile after advection. This applies also to ρ, T. From this relation, once $\Delta\rho, \Delta T$ are predicted, Δp will be predicted based on Eq.(28). Needless to say, $\partial p/\partial\rho, \partial p/\partial T$ are given by EOS.

Since the CIP separates the non-advection terms from the advection, we can concentrate on the non-advection terms related to sound waves which are the primary cause of the difficulty in liquid having large sound speed and hence ρ, T are simply given by

$$\begin{aligned} \Delta\rho &= -\rho^*\nabla\cdot\mathbf{u}^{n+1}\Delta t \\ \rho^* C_v \Delta T &= -P_{TH}\nabla\cdot\mathbf{u}^{n+1}\Delta t \end{aligned} \tag{29}$$

where C_v is the specific heat ratio at constant volume. $\mathbf{u}^{n+1}$ in this equation is given by equation of motion as

$$\Delta\mathbf{u} = -\frac{\nabla p^{n+1}}{\rho^*}\Delta t \tag{30}$$

Since $\Delta\mathbf{u} = \mathbf{u}^{n+1} - \mathbf{u}^*$, Eqs.(28)-(30) leads to a pressure equation [9,30]

$$\nabla\left(\frac{1}{\rho^*}\nabla p^{n+1}\right) = \frac{p^{n+1} - p^*}{\Delta t^2(\rho C_s^2 + \frac{P_{TH}^2}{\rho C_v T})} + \frac{\nabla\cdot\mathbf{u}^*}{\Delta t} \tag{31}$$

Then substituting the given p^{n+1} into Eq.(30), we obtain the velocity $\mathbf{u}^{n+1}$ and then density ρ^{n+1} from Eq.(29). From this procedure, density can be solved in terms of pressure which is analogous to rotate Fig.3 by 90 degree. Equation (31) has many important features in the following points. This equation shows that, at the sharp discontinuity, $\mathbf{n}\cdot(\nabla p/\rho)$ is continuous. Since $\nabla p/\rho$ is the acceleration, it is essential that this term is continuous since the density changes by several orders of magnitude at the boundary between liquid and gas. In this case, the denominator of $\nabla p/\rho$ changes by several orders and pressure gradient must be caulcuated accurately enough to ensure the continuous change of acceleration. Equations (25) and (27) derived by the ICE and the PISO seems to be quite similar to Eq.(31) but the continuity of $\nabla p/\rho$ in the formers is not guaranteed. Thus, the method works robustly even with a density ratio larger that 1000.

It is interesting to examine the meaning of this pressure equation. If $\nabla\cdot\mathbf{u}$ term is absent, this equation is merely the diffusion equation. The origin of this term is as follows. During time step Δt, the sound wave propagate for a distance $C_s\Delta t$. In the next step, the signal also propagates backwardly and forwardly since sound wave should isotropically propagate . Then statistically, 50% propagates backwardly and another 50% forwardly. This process is similar to random walk. The diffusion coefficient of the random walk is given by the quivering distance $\Delta x = C_s\Delta t$ as $D = \Delta x^2/\Delta t$. This leads to the diffusion equation for pressure. From this consideration, we understand how the effect of sound waves is implemented.

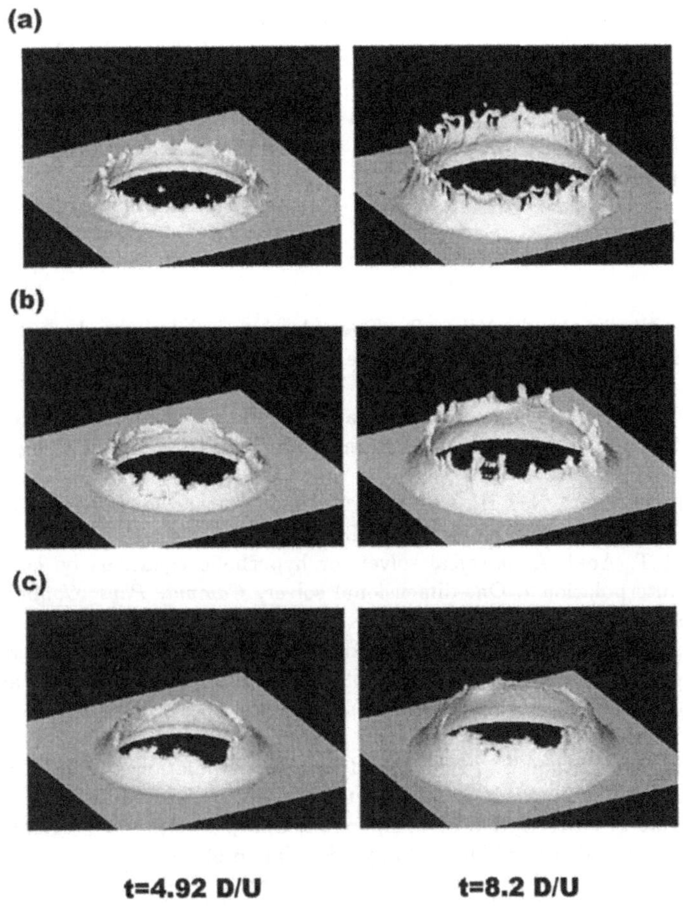

Fig. 4. Water surface plot in the coronet formation process. 100×100×34 Cartesian grid is used. Left and right figures show the plots at $t = 4.92D/U$ and $8.2D/U$, respectively.Here D is the diameter of drop and U is its velocity. Ambient gas density used in the simulation is $\rho_{gas}/\rho_{Liq} = 0.002$(top), 0.02(middle) and 0.03(bottom). The irregular structure dissappears as the gas density increases.

3 Summary

We have proposed a new tool to attack the simultaneous solution of all the materials. The success of the code is due to a high ability of tracing sharp interface even with fixed grid and flexibility of extension to various materials and physics. Before closing this paper, we should remind the reader that the code has been applied to various problems which have never been attacked by conventional schemes. Figure 4 shows a snap shot of Milk-crown formation that has been published before [31,32] and the crown formation is quite sensitive to ambient gas pressure.

References

1. Takewaki,H., Nishiguchi, A., Yabe, T. : The cubic-interpolated pseudo-particle (CIP) method for solving hyperbolic-type equations, J. Comput. Phys. **61** (1985)261.
2. Takewaki,H., Yabe,T. : The cubic-interpolated pseudo particle (CIP) method: application to nonlinear and multi-dimensional hyperbolic equations, J. Comput. Phys. **70** (1987)355.
3. Yabe, T. , Takei, E. : A New Higher-Order Godunov Method for General Hyperbolic Equations. J.Phys.Soc.Japan **57** (1988)2598.
4. T. Yabe and T. Aoki, A universal solver for hyperbolic equations by cubic-polynomial interpolation I. One-dimensional solver, *Comput. Phys. Commun.* **66**, 219 (1991).
5. T. Yabe, T. Ishikawa, P. Y. Wang, T. Aoki, Y. Kadota, F. Ikeda, A universal solver for hyperbolic equations by cubic-polynomial interpolation II. Two- and three-dimensional solver, *Comput. Phys. Commun.* **66**, 233 (1991).
6. The special issue of the CIP method, *CFD Journal* **8** (1999).
7. Yabe,T., Xiao,F., Utsumi,T. : The Constrained Interpolation Profile (CIP) Method for Multi-Phase Analysis. J.Comput.Phys. (2001) in press.
8. Yabe, T. , Xiao,F. : Description of Complex and Sharp Interface during Shock Wave Interaction with Liquid Drop. J.Phys. Soc.Japan **62** (1993)2537.
9. Yabe, T. , Wang,P.Y. : Unified Numerical Procedure for Compressible and Incompressible Fluid. J.Phys.Soc.Japan **60** (1991)2105.
10. Tanaka,R., Nakamura,T., Yabe,T. : Constructing exactly conservative scheme in non-conservative form. Comput. Phys. Commun. **126** (2000)232.
11. Unverdi, S.O. , Tryggvasson,G.A. : A front-tracking method for viscous, incompressible, multi-fluid flows. J. Comp. Phys. **100** (1992)25.
12. Osher, S. , Sethian,J.A. : Front propagating with curvature-dependent speed: algorithms based on Hamilton-Jacobi formulations. J. Comp. Phys. **79** (1988)12.
13. Hirt, C.W. , Nichols,B.D. : Volume of fluid (VOF) method for the dynamics of free boundaries. J. Comp. Phys. **39** (1981)201.
14. Youngs,D.L.: Time-dependent multi-material flow with large fluid distortion. Numer. Methods for Fluids Dynamics edited by K.W.Morton and M.J.Baines, (1982)273.
15. Yabe, T. , Xiao,F. : Description of complex and sharp interface with fixed grids in incompressible and compressible fluid. Computer Math. Applic. **29** (1995)15.

16. Nakamura,T. , Yabe,T. : Cubic interpolation scheme for solving the hyper-dimensional Vlasov-Poisson Equation in phase space. Comput. Phys. Commun. **120** (1999)122.
17. Yabe,T., Tanaka,R. , Nakamura,T., Xiao,F. : Exactly Conservative Semi-Lagrangian Scheme (CIP-CSL) in One-Dimension. Mon.Wea. Rev. **129** (2001) 332.
18. Williamson,D.L., Rasch,P.J.: Two-dimensional semi-Lagrangian transport with shape-preserving interpolation. Mon. Wea. Rev. **117** (1989)102.
19. Staniforth, A. , Côté,J. : Semi-Lagrangian integration scheme for atmospheric model-A review. Mon. Wea. Rev. **119** (1991)2206.
20. Bermejo,R. , Staniforth,A. : The conversion of semi-Lagrangian advection schemes to quasi-monotone schemes. Mon. Wea. Rev. **120** (1992)2622.
21. Karni,S. : Multicomponent flow calculations by a consistent primitive algorithm. J.Comput.Phys. **112** (1994) 31.
22. Harlow, F.H. , Welch,J.E. : Numerical calculation of time dependent viscous incompressible flow with fee surface. Phys. Fluids **8** (1965) 2182.
23. Amsden, A.A., Harlow, F.H.: A Simplified MAC Technique for Incompressible Fluid Flow Calculations. J.Comp.Phys. **6** (1970)322 .
24. Patanker,S.V., Spalding, D.B.: A calculation procedure for heat, mass and momentum transfer in three-dimensional parabolic flows. Int. J. Heat Mass Transfer **15** (1972)1787.
25. Patanker,S.V. : *Numerical heat transfer and fluid flow.* , Mcgraw-Hill, New York (1980).
26. Harlow,F.H., Amsden,A.A. : Numerical simulation of almost incompressible flow. J.Comput.Phys. **3** (1968)80.
27. Issa,R.I. : Solution of the implicitly discretised fluid flow equations by operator-splitting. J. Comput. Phys. **62** (1985)40.
28. Zienkiewicz,O.C., Wu,J. : A general explicit or semi-explicit algorithm for compressible and incompressibe flows. Int. J.Num. Meth. Eng. **35** (1992)457.
29. Karki,K.C., Patankar,S.V. : Pressure based calculation procedure for viscous flows at all speeds in arbitrary configurations. AIAA Journal **27** (1989)1167.
30. Xiao, F. *et al* : An Efficient Model for Driven Flow and Application to GCB. Comput. Model. & Sim. Eng. **1** (1996)235.
31. Yabe,T. , Zhang,Y. , Xiao,F. : A Numerical Procedure -CIP- to Solve All Phases of Matter Together (Invited Lecture) *Lecture Note in Physics* pp.439-457 (Springer, 1998)
32. Zhang,Y., Yabe,T. : Effect of ambient gas on three-dimensional breakup in coronet formation. CFD Journal **8** (1999)378.
33. Yabe,T. *et al.*: Anomalous Crater Formation in Pulsed-Laser-Illuminated Aluminum Slab and Debris Distribution, Research Report NIFS (National Institute for Fusion Science) Series, NIFS-417, May (1996).

Subgrid Phenomena and Numerical Schemes

Franco Brezzi and Donatella Marini

Dipartimento di Matematica, Università di Pavia, and I.A.N.-C.N.R., via Ferrata 1, 27100 Pavia (Italy)

Abstract. In recent times, several attempts have been made to recover some information from the subgrid scales and transfer them to the computational scales. Many stabilizing techniques can also be considered as part of this effort. We discuss here a framework in which some of these attempts can be set and analyzed.

1 Introduction

In the numerical simulation of a certain number of problems, there are physical effects that take place on a scale which is much smaller than the smallest one representable on the computational grid, but have a strong impact on the larger scales, and, therefore, cannot be neglected without jeopardizing the overall quality of the final solution.

In other cases, the discrete scheme lacks the necessary stability properties because it does not treat in a proper way the smallest scales allowed by the computational grid. As a consequence, some "smallest scale mode" appears as abnormally amplified in the final numerical results. Most types of numerical instabilities are produced in this way, as the checkerboard pressure mode for nearly incompressible materials, or the fine-grid spurious oscillations in convection-dominated flows. See for instance [19] and the references therein for a classical overview of several types of these and other instabilities of this nature.

In the last decade it has become clear that several attempts to recover stability, in these cases, could be interpreted as a way of improving the simulation of the effects of the smallest scales on the larger ones. By doing that, the small scales can be *seen* by the numerical scheme and therefore be kept under control.

These two situations are quite different, in nature and scale. Nevertheless it is not unreasonable to hope that some techniques that have been developed for dealing with the latter class of phenomena might be adapted to deal with the former one. In this sense, one of the most promising technique seems to be the use of Residual-Free Bubbles (see e.g. [10], [18].) In the following sections, we are going to summarize the general idea behind it, trying to underline its potential and its limitations. In Section 2 we present the continuous problems in an abstract setting, and provide examples of applications, related to advection dominated flows, composite materials, and viscous incompressible flows. For application of these concepts to other problems we refer, for instance, to [13], [14], [16], [18], [24]. In Section 3 we introduce the basic features of the RFB method. Starting from a given discretization (that might possibly be unstable), we discuss the suitable *bubble space* that can be added to the original finite element space. Increasing the space with bubbles leads to the *augmented problem*, usually infinite dimensional, which, in the end, will have to be solved in some suitable approximate way. In Section 4 we give an idea of how error estimates can be deduced for the augmented problem. In Section 5 we discuss the related computational aspects, and we present several strategies that can be used to deal with the augmented problem, in order to minimize the computational cost. We shall see in particular that several other methods that are known in the literature can actually be seen as variants of the RFB procedure, in which one or another of the above strategies is employed. This includes, for advection dominated problems, the classical SUPG methods (as it was already well known, see, e.g., [4]) as well as the older Petrov-Galerkin methods based on suitable operator dependent choices of test and trial functions [25]. For composite materials, this includes both the multiscale methods of [22], [23], and the upscaling methods of [1], [2]. Finally, in Section 6 we draw some conclusions.

2 The continuous problem

We consider the following continuous problem

$$\begin{cases} \text{find } u \in V \text{ such that} \\ \mathcal{L}(u,v) = < f, v > \qquad \forall v \in V, \end{cases} \tag{2.1}$$

where V is a Hilbert space, and V' its dual space, $\mathcal{L}(u,v)$ is a continuous bilinear form on $V \times V$, and $f \in V'$ is the forcing term. We assume that, for all $f \in V'$, problem (2.1) has a unique solution. Various problems of interest for the applications can be written in the variational form (2.1), according to different choices of the space V and the bilinear form $\mathcal{L}$. Typical choices for V, when V is a space of scalar functions, are the following: if $\mathcal{O} \subset \boldsymbol{R}^d$, $(d = 1,2,3)$ denotes a generic domain, V could be, for instance, $L^2(\mathcal{O})$, $H^1(\mathcal{O})$, $H^1_0(\mathcal{O})$, $H^2(\mathcal{O})$ or $L^2_0(\mathcal{O})$, the last one being the space of L^2−functions having zero mean value. In the case where V is a space of vector valued functions, a first choice could be to take the cartesian product of the previous scalar spaces. Other typical choices for V can be:

$$\begin{aligned} H(\text{div};\mathcal{O}) &:= \{\boldsymbol{\tau} \in (L^2(\mathcal{O}))^d \text{ such that } \nabla \cdot \boldsymbol{\tau} \in L^2(\mathcal{O})\}, \\ H_0(\text{div};\mathcal{O}) &:= \{\boldsymbol{\tau} \in H(\text{div};\mathcal{O}) \text{ such that } \boldsymbol{\tau} \cdot \mathbf{n} = 0 \text{ on } \partial\mathcal{O}\}, \end{aligned}$$

or also, for a generic domain $\mathcal{O} \subset \boldsymbol{R}^3$,

$$H(\mathbf{curl};\mathcal{O}) := \{\boldsymbol{\tau} \in (L^2(\mathcal{O}))^3 \text{ such that } \nabla \wedge \boldsymbol{\tau} \in (L^2(\mathcal{O}))^3\}$$
$$H_0(\mathbf{curl};\mathcal{O}) := \{\boldsymbol{\tau} \in H(\mathbf{curl};\mathcal{O}) \text{ such that } \boldsymbol{\tau} \wedge \mathbf{n} = 0 \text{ on } \partial\mathcal{O}\}.$$

Product spaces are also used quite often: for instance, $H(\mathrm{div};\mathcal{O}) \times L^2(\mathcal{O})$, or $(H_0^1(\mathcal{O}))^d \times L_0^2(\mathcal{O})$, etc. Next, we provide some classical examples of problems and we indicate the corresponding space V, the bilinear form $\mathcal{L}$, and the variational formulation.

Ex 2.1: Advection-dominated scalar equations:

$$-\varepsilon \Delta u + \mathbf{c} \cdot \nabla u = f \quad \text{in } \Omega; \quad u = 0 \quad \text{on } \partial\Omega$$

$$V := H_0^1(\Omega);\ \mathcal{L}(u,v) := \int_\Omega \varepsilon \nabla u \cdot \nabla v\, dx + \int_\Omega \mathbf{c} \cdot \nabla u\, v\, dx;\ \langle f, v\rangle := \int_\Omega f v\, dx$$

$$\mathcal{L}(u,v) = < f, v > \qquad \forall v \in V$$

Ex 2.2: Linear elliptic problems with composite materials:

$$-\nabla \cdot (\alpha(x) \nabla u) = f \quad \text{in } \Omega; \quad u = 0 \quad \text{on } \partial\Omega$$

$$V := H_0^1(\Omega); \quad \mathcal{L}(u,v) := \int_\Omega \alpha(x) \nabla u \cdot \nabla v\, dx; \qquad < f, v > := \int_\Omega f v\, dx$$

$$\mathcal{L}(u,v) = < f, v > \qquad \forall v \in V$$

(where $\alpha(x) \geq \alpha_0 > 0$ might have a very fine structure).

Ex 2.3: Composite materials in mixed form, i.e., the same problem of the previous example, but now with:

$$\boldsymbol{\sigma} = -\alpha \nabla \psi \quad \text{in } \Omega; \qquad \nabla \cdot \boldsymbol{\sigma} = f \quad \text{in } \Omega; \qquad \psi = 0 \quad \text{on } \partial\Omega$$

$$V := \boldsymbol{\Sigma} \times \boldsymbol{\Phi}; \qquad \boldsymbol{\Sigma} := H(\mathrm{div};\Omega); \qquad \boldsymbol{\Phi} := L^2(\Omega)$$

$$a_0(\boldsymbol{\sigma}, \boldsymbol{\tau}) := \int_\Omega \alpha^{-1} \boldsymbol{\sigma} \cdot \boldsymbol{\tau}\, dx, \qquad b(\boldsymbol{\tau}, \varphi) := \int_\Omega \nabla \cdot \boldsymbol{\tau}\, \varphi\, dx$$

$$\mathcal{L}((\boldsymbol{\sigma}, \psi), (\boldsymbol{\tau}, \varphi)) := a_0(\boldsymbol{\sigma}, \boldsymbol{\tau}) - b(\boldsymbol{\tau}, \psi) + b(\boldsymbol{\sigma}, \varphi);\ < f, (\boldsymbol{\tau}, \varphi) > := \int_\Omega f \varphi\, dx$$

$$\mathcal{L}((\boldsymbol{\sigma}, \psi), (\boldsymbol{\tau}, \varphi)) = < f, (\boldsymbol{\tau}, \varphi) > \qquad \forall (\boldsymbol{\tau}, \varphi) \in V$$

Ex 2.4: Stokes problem for viscous incompressible fluids:

$$-\Delta \mathbf{u} + \nabla p = \mathbf{f} \quad \text{in } \Omega; \qquad \nabla \cdot \mathbf{u} = 0 \quad \text{in } \Omega; \qquad \mathbf{u} = 0 \quad \text{on } \partial\Omega$$

$$V := \mathbf{U} \times Q; \quad \mathbf{U} := (H_0^1(\Omega))^d; \quad Q := L_0^2(\Omega)$$

$$a_1(\mathbf{u}, \mathbf{v}) := \int_\Omega \nabla \mathbf{u} : \nabla \mathbf{v}\, dx \quad b(\mathbf{v}, q) := \int_\Omega \nabla \cdot \mathbf{v}\, q\, dx$$

$$\mathcal{L}((\mathbf{u}, p), (\mathbf{v}, q)) := a_1(\mathbf{u}, \mathbf{v}) - b(\mathbf{v}, p) + b(\mathbf{u}, q);\ < f, (\mathbf{v}, q) > := \int_\Omega \mathbf{f} \cdot \mathbf{v}\, dx$$

$$\mathcal{L}((\mathbf{u}, p), (\mathbf{v}, q)) = < f, (\mathbf{v}, q) > \qquad \forall (\mathbf{v}, q) \in V$$

3 From the discrete problem to the augmented problem

Let $\mathcal{T}_h$ be a decomposition of the computational domain Ω, with the usual nondegeneracy conditions [12], and let $V_h \subset V$ be a finite element space. The original discrete problem is then:

$$\begin{cases} \text{find } u_h \in V_h \text{ such that} \\ \mathcal{L}(u_h, v_h) =< f, v_h > \qquad \forall v_h \in V_h. \end{cases} \tag{3.1}$$

Note that we do not assume that (3.1) has a unique solution. Indeed, the stabilization that we are going to introduce can, in some cases, take care of problems originally ill-posed. Our aim is, essentially, to solve in the end a final linear system having as many equations as the number of degrees of freedom of V_h. Apart from that, we are ready to pay some extra work, in order to have a better method. In some cases, the total amount of additional work will be small. In other cases, it can be huge. However, we want to be able to perform the extra work independently in each element so that we can do it, as a pre-processor, *in parallel.* This implies that we are ready to add as many degrees of freedom as we want at the interior of each element. For that, to V and $\mathcal{T}_h$ we associate the **maximal space of bubbles**

$$B(V;\mathcal{T}_h) = \prod_K B_V(K), \qquad \text{with } B_V(K) = \{v \in V : \; supp(v) \subseteq \overline{K}\}.$$

Let us give some examples of the dependence of $B_V(K)$ on V.

- if $V = H^1_0(\Omega)$ then $B_V(K) = H^1_0(K)$
- if $V = H^1(\Omega)$ then $B_V(K) = \{v \in H^1(K),\, v = 0 \; on \; \partial K \cap \Omega\}$
- if $V = L^2(\Omega)$ then $B_V(K) = L^2(K)$
- if $V = L^2_0(\Omega)$ then $B_V(K) = L^2_0(K)$
- if $V = H^2_0(\Omega)$ then $B_V(K) = H^2_0(K)$
- if $V = H_0(\text{div};\Omega)$ then $B_V(K) = H_0(\text{div};K)$
- if $V = H(\text{div};\Omega)$ then $B_V(K) = \{\boldsymbol{\tau} \in H(\text{div};K),\, \boldsymbol{\tau}\cdot\mathbf{n} = 0 \; on \; \partial K \cap \Omega\}$

Similar definitions and properties hold for the spaces $H(\mathbf{curl};\mathcal{O})$, but we are not going to use them here.

Let us now turn to the choice of the local bubble space $B_h(K)$. If possible, we would like to augment the space V_h by adding, in each element K, the whole $B_V(K)$. This would change V_h into $V_h + B(V;\mathcal{T}_h)$. However, some conditions are needed, as we shall see below. This might forbid, in some cases, to take the whole $B_V(K)$ in the augmentation process: some components of $B_V(K)$ have to be discarded. This will become more clear in the examples below. At this very abstract and general level, we assume that, in each $K \in \mathcal{T}_h$, we choose a subspace $B_h(K) \subseteq B_V(K)$ and, for the moment, "the bigger the better". A first condi tion that we require is that, for every $g \in V'$, the auxiliary problem

$$\begin{cases} \text{find } w_{B,K} \in B_h(K) \text{ such that} \\ \mathcal{L}(w_{B,K}, v) = < g, v > \qquad \forall v \in B_h(K) \end{cases} \tag{3.2}$$

has a unique solution. We point out that the choice "the bigger the better" for $B_h(K)$ is made (so far) in order to understand the full potential of the method. As we shall see, in practice we will need to solve (3.2) a few times in each K. This implies that a finite dimensional choice for $B_h(K)$ will be, in the end, necessary.

Having chosen $B_h(K)$, we can now write the **augmented problem**. For that, let

$$V_A := V_h + \Pi_K B_h(K). \tag{3.3}$$

Two requirements have to be fulfilled: first of all, in (3.3) we must have a direct sum, and, second, for every $f \in V'$, the augmented problem

$$\begin{cases} \text{find } u_A \in V_A \text{ such that} \\ \mathcal{L}(u_A, v_A) = < f, v_A > \qquad \forall v_A \in V_A \end{cases} \tag{3.4}$$

must have a unique solution. To summarize, in the augmentation process three conditions have to be fulfilled:

1) the local problems (3.2) must have a unique solution;
2) in (3.3) we must have a direct sum;
3) the augmented problem (3.4) must have a unique solution.

These are then the requirements that can guide us in choosing $B_h(K)$ in the various cases.

Examples of choices of $B_h(K)$.

Ex 3.1 - Referring to Examples 2.1 and 2.2 of the previous section, suppose that V_h is made of continuous piecewise linear functions. In this case it is easy to check that the choice $B_h(K) = B_V(K) \equiv H_0^1(K)$ verifies all of the three conditions.

Ex 3.2 - Suppose now that, always referring to Examples 2.1 and 2. 2, V_h is made of continuous piecewise cubic functions. The choice $B_h(K) = B_V(K)$ is not viable anymore, as clearly condition 2) is violated: V_h contains functions of $B_V(K)$. In situations like this we should then choose a different $B_h(K)$, but we could also *reduce* the original space V_h. This is actually the simplest strategy, and we are going to follow it. Here, for instance, we can just remove the cubic bubble from $V_{h|K}$ and take a reduced space, still denoted by V_h with an abuse of notation, as a space of any serendipity cubic element (see, for instance, the element described in [12], page 50). Or we might take V_h as the space of functions v_h that are polynomials of degree ≤ 3 at the interelement boundaries and verify $Lv_h = 0$ separately in each K. Notice that these two choices produce the same augmented space V_A, and hence the same solution u_A to (3.4).

Ex 3.3 - Let us consider the problem of Example 2.3, and assume that $V_h = \Sigma_h \times U_h$ is made by lowest order Raviart-Thomas elements (see for instance [3]). For this problem we have

$$B_V(K) = \{\boldsymbol{\tau} \in H(\text{div}; K),\, \boldsymbol{\tau} \cdot \mathbf{n} = 0 \text{ } on \text{ } \partial K \cap \Omega\} \times L^2(K).$$

we notice now that taking $B_h(K) = B_V(K)$ would not guarantee that problems (3.2) have a unique solution. Indeed, for internal elements K, the Inf-sup condition is not verified, since $\int_K div\tau\, v\, dx = 0$ forall v constant on K. Condition 2) would also be violated by the choice $B_h(K) = B_V(K)$: in fact, U_h being the space of piecewise constants, $U_{h|K}$ contains bubbles of $L^2(K)$. A possible remedy in this case is to take

$$B_h(K) = H_0(\text{div}; K) \times L_0^2(K) \subset B_V(K).$$

With this choice V_h remains the same, and B_h is the space of all pairs $(\tau, v) \in V$ such that τ has zero normal component at the boundary of each element, and v has zero mean value in each element. The same choice for B_h would be suitable also in the case of higher order Raviart-Thomas spaces (or, say, for BDM spaces; see always [3]), but then V_h should lose all internal degrees of freedom, apart from the piecewise constant scalars.

Ex 3.4 - Let us now examine the Stokes problem of Example 2.4, and assume that V_h is made of piecewise quadratic velocities in $(H_0^1(\Omega))^d$, and discontinuous piecewise linear pressures in $L_0^2(\Omega)$, a choice which is known not to be stable, but can be stabilized with the present technique. Actually, in this case one can see that $B_V(K) = (H_0^1(K))^d \times L_0^2(K)$. Taking $B_h(K) = B_V(K)$ would violate condition 2), but we can reduce the space V_h, taking it to be the space of quadratic velocities and *constant* pressures. It is easy to check that with this last choice we have a direct sum in (3.3). Moreover, problem (3.4) has a unique solution, because the Inf-sup condition is now verified in V_A.

Ex 3.5 - Let us consider again the Stokes problem of Example 2.4, but now with $V_h = U_h \times Q_h$ made of piecewise linear continuous velocities in $(H_0^1(\Omega))^d$, and piecewise constant pressures in $L_0^2(\Omega)$. It is well known that for this choice the Inf-sup condition does not hold. Moreover, if we augment V_h with bubble functions, no matter how, the augmented problem (3.4) will **never** verify the Inf-sup condition. To see that, augment as much as you can the velocity space: $U_A = U_h + \Pi_K(H_0^1(K))^d$, and augment as little as you can the pressure space: $Q_A = Q_h + \{0\}$. For every $v \in (H_0^1(K))^d$ and for every constant q in K, we clearly have $(\operatorname{div} v, q) = 0$. Hence, for $q \in Q_h$:

$$\sup_{v \in V_A} \frac{(\operatorname{div} v, q)}{||v||_1} = \sup_{v \in U_h} \frac{(\operatorname{div} v, q)}{||v||_1},$$

and we know that the last quantity cannot bound $||q||_0$ for all $q \in Q_h$. We clearly see that, in cases like this, our strategy is totally useless, and should not be applied.

4 An example of error estimates

To give an idea of how to proceed to obtain error estimates, let us consider, as an example, a general singular perturbation problem where

$$\mathcal{L}(u, v) := \varepsilon a_1(u, v) + a_0(u, v)$$

with

$$a_1(v, v) \geq \alpha ||v||_V^2 \quad \forall v \in V, \qquad a_1(u, v) \leq ||u||_V \, ||v||_V \quad \forall u, v \in V \tag{4.1}$$

$$a_0(v, v) \geq 0 \quad \forall v \in V, \qquad a_0(u, v) \leq M \, ||u||_V \, ||v||_H \quad \forall u, v \in V \tag{4.2}$$

where H is a space such that $V \subset H$ with continuous embedding. We set $e := u - u_A$ and $\eta := u - u_I$, u_I being some interpolant of u in V_h. Proceeding as usual we have

$$\varepsilon\alpha ||e||_V^2 \leq \mathcal{L}(e, e) = \mathcal{L}(e, \eta) = \varepsilon a_1(e, \eta) + a_0(e, \eta), \tag{4.3}$$

and the term $a_0(e, \eta)$ is the source of all difficulties, since it does not contain ε as an explicit factor. In order to estimate it, let $\eta = \eta_B + \eta_H$ be any decomposition of η with $\eta_B \in B_h$ and $\eta_H \in H$. Notice that $\eta_B \in B_h \subset V_A$, so that, by Galerkin orthogonality,

$$\varepsilon a_1(e, \eta_B) = -a_0(e, \eta_B). \tag{4.4}$$

Using this and the bounds (4.1)-(4.2) we can proceed as in [9] and deduce:

$$\begin{aligned} a_0(e, \eta) &= a_0(e, \eta_B) + a_0(e, \eta_H) = -\varepsilon a_1(e, \eta_B) + a_0(e, \eta_H) \\ &\leq \varepsilon ||e||_V ||\eta_B||_V + M ||e||_V ||\eta_H||_H \\ &\leq \varepsilon^{1/2} \Big(\varepsilon^{1/2} ||e||_V ||\eta_B||_V + M \varepsilon^{-1/2} ||e||_V ||\eta_H||_H \Big) \\ &\leq \varepsilon^{1/2} (1 + M) ||e||_V \Big(\varepsilon^{1/2} ||\eta_B||_V + \varepsilon^{-1/2} ||\eta_H||_H \Big). \end{aligned} \tag{4.5}$$

Taking now the supremum over all possible decompositions $\eta = \eta_B + \eta_H$, and then over $\varepsilon > 0$ we obtain

$$a_0(e, \eta) \leq \varepsilon^{1/2} (1 + M) ||e||_V \sup_{\varepsilon > 0} \Big[\sup_{\eta_B + \eta_H = \eta} \Big(\varepsilon^{1/2} ||\eta_B||_V + \varepsilon^{-1/2} ||\eta_H||_H \Big) \Big]. \tag{4.6}$$

By definition (see [7]) the double supremum is the norm of η in a suitable interpolation space, usually denoted by $[B_h, H]_{\frac{1}{2},\infty}$, that for brevity we shall denote by F. Hence, (4.6) becomes

$$a_0(e,\eta) \le \varepsilon^{1/2}(1+M)||e||_V||\eta||_F. \tag{4.7}$$

Inserting (4.7) in (4.3) gives

$$\varepsilon\alpha||e||_V^2 \le \varepsilon a_1(e,\eta) + a_0(e,\eta) \le \varepsilon^{1/2}||e||_V(\varepsilon^{1/2}||\eta||_V + (1+M)||\eta||_F),$$

and finally

$$\varepsilon^{1/2}\alpha||u-u_A||_V \le \varepsilon^{1/2}||u-u_I||_V + (1+M)||u-u_I||_F. \tag{4.8}$$

Notice that an estimate for $\varepsilon^{1/2}||u-u_A||_V$ is not as bad as we are used to. For instance, with an argument similar to the one used before, using (4.4)-(4.5), from (4.8) we can see that

$$\begin{aligned}
||A_0(u-u_A)||_{F'} &:= \sup_{\varphi} \frac{a_0(u-u_A,\varphi)}{||\varphi||_F} \\
&= \sup_{\varphi} \frac{a_0(u-u_A,\varphi_B) + a_0(u-u_A,\varphi_H)}{||\varphi||_F} \\
&= \sup_{\varphi} \frac{-\varepsilon a_1(u-u_A,\varphi_B) + a_0(u-u_A,\varphi_H)}{||\varphi||_F} \\
&\le (1+M)\varepsilon^{1/2}||u-u_A||_V \sup_{\varphi} \frac{\varepsilon^{1/2}||\varphi_B||_V + \varepsilon^{-1/2}||\varphi_H||_H}{||\varphi||_F} \\
&\le (1+M)\varepsilon^{1/2}||u-u_A||_V \le C\,(\varepsilon^{1/2}||u-u_I||_V + ||u-u_I||_F),
\end{aligned}$$

which is a typical estimate that can be obtained with stabilized methods (see, e.g., [22], [27]). We refer to [6], [9], [28] for the error analysis for residual-free bubbles methods for advection dominated problems.

5 Computational aspects

Let us now examine the structure of the abstract augmented problem (3.4). Since we constructed the space V_A as a direct sum:

$$V_A := \Pi_K B_h(K) \oplus V_h$$

we have then the unique splittings: $u_A = u_B + u_h$, $v_A = v_B + v_h$. The augmented problem can then be written as

$$\begin{cases} \text{find } u_A = u_B + u_h \in V_A \text{ such that} \\ \mathcal{L}(u_B + u_h, v_B + v_h) =< f, v_B + v_h > \qquad \forall v_B \in B_h,\ \forall v_h \in V_h. \end{cases} \tag{5.1}$$

The associated system will therefore have the form:

$$\begin{pmatrix} L_{B,B} & L_{B,h} \\ L_{h,B} & L_{h,h} \end{pmatrix} \begin{pmatrix} u_B \\ u_h \end{pmatrix} = \begin{pmatrix} f_B \\ f_h \end{pmatrix} \qquad \text{with } L_{B,B} \text{ block diagonal.}$$

There are different strategies for solving the (still infinite dimensional) problem (5.1). All of them are based on the (approximate) solution of the problems

$$\begin{cases} \text{find } w_B^i \in B_h \text{ such that} \\ \mathcal{L}(w_B^i, v_B) = \mathcal{L}(v_i, v_B) \equiv < Lv_i, v_B > \qquad \forall v_B \in B_h, \end{cases} \tag{5.2}$$

where the $\{v_i\}$'s are a basis for V_h, plus, if necessary, the solution of the problem

$$\begin{cases} \text{find } w_B^f \in B_h \text{ such that} \\ \mathcal{L}(w_B^f, v_B) = < f, v_B > \qquad \forall v_B \in B_h. \end{cases} \tag{5.3}$$

As we shall see, what is actually needed, for all strategies, is the computation (for $i, j = 1, ..., dim(V_h)$) of the quantities

$$S_{j,i} := \mathcal{L}(w_B^i, v_j) \equiv < w_B^i, L^* v_j >, \quad \text{and} \quad T_j := \mathcal{L}(w_B^f, v_j) \equiv < w_B^f, L^* v_j >, \tag{5.4}$$

where L^* is the adjoint operator of L. In turn, the computation of the solution of the problems (5.2) amounts to solve, in each K, the local bubble problem

$$\begin{cases} \text{find } w_{B,K}^i \in B_h(K) \text{ such that} \\ \mathcal{L}(w_{B,K}^i, b) = < Lv_i, b > \qquad \forall b \in B_h(K). \end{cases} \tag{5.5}$$

The same is obviously true for (5.3). Moreover, f can often be approximated, in each K, by elements of $LV_{h|K}$, so that the solution of (5.3) can be easily obtained from the solutions of the problems (5.2).

A careful inspection of the local problems (5.5) suggests several observations that are computationally relevant.

- For each v_i, the computation of w_B^i can be done in parallel.
- In each element K, the dimension of $span\{Lv_{i|K}\}$ will be small. In general, it will be less than or equal to the number of degrees of freedom of V_h in K.
- Finally, as we already pointed out, only the quantities $S_{j,i} = < w_B^i, L^* v_j >$ are actually needed. Hence, only some averages of w_B^i will be used, and therefore a rough approximation might often be sufficient.
- The same considerations clearly hold for the contributions T_j to the right-hand side.

5.1 First strategy

Let us see in more detail how the whole procedure can be applied in practice. For this, consider problem (5.1) and note that u_B is the solution of

$$\mathcal{L}(u_B, v_B) = -\mathcal{L}(u_h, v_B) + < f, v_B > \qquad \forall v_B \in B_h,$$

and can be seen as an (affine) function of u_h and f:

$$u_B = L_{B,B}^{-1}(f - Lu_h).$$

Substituting into (5.1), and taking now v_h as a test function, gives

$$\mathcal{L}(u_h, v_h) + \mathcal{L}(L_{B,B}^{-1}(f - Lu_h), v_h)) =< f, v_h > \qquad \forall v_h \in V_h, \tag{5.6}$$

which is an equation in terms of u_h alone, where the additional term

$$\mathcal{L}(L_{B,B}^{-1}(f - Lu_h), v_h) \equiv \mathcal{L}(u_B, v_h) \tag{5.7}$$

represents the effect of the small scales onto the coarse ones. To see how to compute the additional term (5.7) let us write $u_h := \sum_i U_i v_i$ and take v_j as a test function. We have

$$\begin{aligned}\mathcal{L}(u_B, v_j) &= \mathcal{L}(L_{B,B}^{-1}(f - Lu_h), v_j) = \mathcal{L}(L_{B,B}^{-1}f, v_j) - \sum_i \mathcal{L}(L_{B,B}^{-1}Lv_i, v_j)U_i \\ &= \mathcal{L}(w_B^f, v_j) - \sum_i \mathcal{L}(w_B^i, v_j)U_i = T_j - \sum_i S_{j,i}U_i,\end{aligned}$$

that clearly shows the use of the auxiliary terms T_j and $S_{j,i}$. Indeed, setting

$$K_{j,i} = \mathcal{L}(v_i, v_j), \qquad \text{and} \qquad F_j =< f, v_j >, \tag{5.8}$$

we have from (5.6) that the U_i's can be obtained as the solution of the following linear system of equations:

$$\sum_i (K_{j,i} - S_{j,i})\, U_i = F_j - T_j \qquad j = 1, ..., dim(V_h). \tag{5.9}$$

Example - To see how this strategy can be applied, let us go back to the advection-dominated equation, that we recall here:

$$-\varepsilon\Delta u + \mathbf{c}\cdot\nabla u = f \quad \text{in } \Omega; \quad u = 0 \quad \text{on } \partial\Omega,$$

$$V := H_0^1(\Omega); \quad \mathcal{L}(u,v) := \int_\Omega \varepsilon\nabla u\cdot\nabla v\, dx + \int_\Omega \mathbf{c}\cdot\nabla u\, v\, dx.$$

Assume that the original finite element space V_h is made of piecewise linear continuous functions. Assume moreover that both the source term f and the convective term $\mathbf{c}$ are piecewise constant. Then, it is easy to see that for all

v_i the terms Lv_i and L^*v_i are constant in each K. Consequently, all the w_B^i can be computed by solving a **single** problem in each K, that is

$$\begin{cases} \text{find } b_K \in H_0^1(K) \text{ such that} \\ \mathcal{L}(b_K, b) = <1, b> \qquad \forall b \in H_0^1(K). \end{cases} \tag{5.10}$$

With some computations, the problem becomes now (see, e.g., [4]):

$$\begin{cases} \text{find } u_h \in V_h \text{ such that, for all } v_h \in V_h : \\ \mathcal{L}(u_h, v_h) - \sum_K \frac{\int_K b_K\, dx}{|K|} \int_K (f - \mathbf{c} \cdot \nabla u_h)\mathbf{c} \cdot \nabla v_h\, dx = < f, v_h > . \end{cases} \tag{5.11}$$

This coincides with the *SUPG* method with $\tau_K = \dfrac{\int_K b_K\, dx}{|K|}$ (see [11], [16]).

5.2 Alternative computational strategies

Another possibility is to change the space V_h: for every basis function $v_i \in V_h$, define

$$\widetilde{v}_i := v_i - w_B^i, \tag{5.12}$$

and remember that w_B^i was defined by

$$\mathcal{L}(w_B^i, v_B) = \mathcal{L}(v_i, v_B) \qquad \forall v_B \in B_h. \tag{5.13}$$

Therefore,

$$\mathcal{L}(\widetilde{v}_i, v_B) = 0 \qquad \forall v_B \in B_h. \tag{5.14}$$

Set now $\widetilde{V}_h =$ span $\{\widetilde{v}_i\}$, and notice that, again, $V_A = \widetilde{V}_h \oplus B_h$. Split then u_A as $u_A = \widetilde{u}_h + \widetilde{u}_B$, with $\widetilde{u}_h$ in $\widetilde{V}_h$, and $\widetilde{u}_B$ in B_h. Then, thanks to (5.14), $\widetilde{u}_B$ is the solution of

$$\mathcal{L}(\widetilde{u}_B, v_B) \equiv \mathcal{L}(u_A, v_B) = < f, v_B > \quad \forall v_B \in B_h. \tag{5.15}$$

Hence $\widetilde{u}_B$ equals w_B^f, solution of (5.3), and can be computed **before** knowing $\widetilde{u}_h$. Finally, $\widetilde{u}_h$ can be computed as the solution of

$$\mathcal{L}(\widetilde{u}_h, v_h) + \mathcal{L}(\widetilde{u}_B, v_h) = < f, v_h > \quad \forall v_h \in V_h, \tag{5.16}$$

with the same number of unknowns and equations as the dimension of V_h. It is interesting to observe that the difference between this and the first strategy is mainly psycological. Indeed, setting $\widetilde{u}_h := \sum_i \widetilde{U}_i \widetilde{v}_i$, we have from (5.12), (5.8), and (5.4)

$$\begin{aligned} \mathcal{L}(\widetilde{u}_h, v_j) &= \sum_i \mathcal{L}(\widetilde{v}_i, v_j)\widetilde{U}_i = \sum_i \mathcal{L}(v_i - w_B^i, v_j)\widetilde{U}_i = \sum_i (K_{j,i} - S_{j,i})\, \widetilde{U}_i, \\ \mathcal{L}(\widetilde{u}_B, v_j) &= \mathcal{L}(w_B^f, v_j) = T_j, \end{aligned} \tag{5.17}$$

so that, inserting (5.17) into (5.16) we obtain

$$\sum_i (K_{j,i} - S_{j,i})\,\widetilde{U}_i = F_j - T_j \qquad j = 1, ..., dim(V_h), \tag{5.18}$$

which is exactly (5.9).

A third possibility would be, assuming that the adjoint problem of (5.13) is uniquely solvable, to define $\hat{w}_B^i$ solution of

$$\mathcal{L}(v_B, \hat{w}_B^i) = \mathcal{L}(v_B, v_i) \qquad \forall v_B \in B_h, \tag{5.19}$$

and to associate to any v_i, basis function in V_h, the function

$$\hat{v}_i = v_i - \hat{w}_B^i. \tag{5.20}$$

Therefore, $\hat{v}_i$ is the solution of

$$\mathcal{L}(v_B, \hat{v}_i) \equiv < v_B, L^* \hat{v}_i >= 0 \qquad \forall v_B \in B_h. \tag{5.21}$$

Set then $V_h^* =$ span $\{\hat{v}_i\}$, and notice that, in general, V_h^* will be different from $\widetilde{V}_h$, unless the bilinear form $\mathcal{L}$ is symmetric. We have again $V_A = V_h^* + B_h$, always with a direct sum. Take now in (5.1) for u_A the same splitting as before, that is, $u_A = \widetilde{u}_h + \widetilde{u}_B$, with $\widetilde{u}_h \in \widetilde{V}_h$, $\widetilde{u}_B \in B_h$, and for v_A take instead the splittig $v_A = \hat{v}_h + v_B$, with $\hat{v}_h \in V_h^*$, $v_B \in B_h$, always without changing the final solution u_A. Substituting in (5.1) shows that $\widetilde{u}_B$ is again the solution of (5.15). Hence, as before, $\widetilde{u}_B$ equals w_B^f, and can be computed before knowing $\widetilde{u}_h$. Finally, $\widetilde{u}_h$ can be computed as the solution of

$$\mathcal{L}(\widetilde{u}_h, \hat{v}_h) =< f, \hat{v}_h > \quad \forall \hat{v}_h \in V_h^*. \tag{5.22}$$

The matrix associated with (5.22) is however given by

$$\mathcal{L}(\widetilde{v}_i, \hat{v}_j) = \mathcal{L}(\widetilde{v}_i, v_j - \hat{w}_B^j) = \mathcal{L}(\widetilde{v}_i, v_j) = K_{j,i} - S_{j,i} \tag{5.23}$$

(having used (5.20), (5.14), and (5.17)). On the other hand,

$$< f, \hat{v}_j >=< f, v_j - \hat{w}_B^j >= F_j - < f, \hat{w}_B^j >, \tag{5.24}$$

and, using (5.3), (5.19), and (5.4),

$$< f, \hat{w}_B^j >= \mathcal{L}(w_B^f, \hat{w}_B^j) = \mathcal{L}(w_B^f, v_j) = T_j. \tag{5.25}$$

We are therefore back to the system (5.18). It is somehow remarkable that the solution of (5.22) can be computed without actually computing the functions $\hat{v}_j$.

Remark Although the above strategies, as we have seen, do coincide in practice, this is not often recognized in the literature. For instance, formulations (5.16) and (5.22), when applied to advection dominated problems coincide with the classical so-called Petrov-Galerkin methods in which suitable trial and test functions, depending on the operator, were used (see [25], and see, in Fig. 1, the typical shape of the basis functions in $\widetilde{V}_h$ and V_h^*). The above computation shows that these methods coincide with SUPG when the choice of the stabilization parameter τ_K is made as in (5.11). On the other hand, when applied to problems related to composite materials, as in Example 2.2 (respectively, Example 2.3), the formulation (5.22) reproduces the multiscale methods of [22], [23] and the upscaling method of [1], [2], respectively.

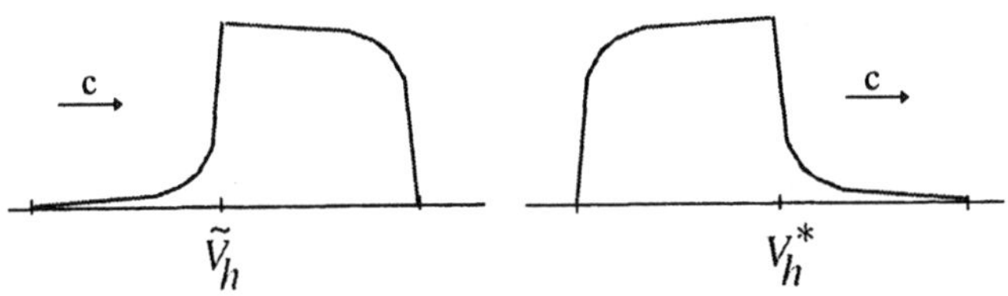

Fig. 1. Typical shape of the basis functions in $\widetilde{V}_h$ and V_h^*

So far, we assumed that we were able to compute the solutions of the local bubble problems (5.2). As anticipated, these solutions cannot be computed exactly, but require some suitable approximation. Let us see, in the particular case of advection dominated problems, how this approximate solutions can be carried out in practice.

We recall that, in this case, solving (5.15) amounts in practice to compute, in each K, the "unitary bubble" b_K, solution of

$$-\varepsilon \Delta b_K + \mathbf{c} \cdot \nabla b_K = 1 \quad \text{in each } K. \tag{5.26}$$

Actually, what we really need is its mean value in each K (see (5.11)).

Several tricks can be used to compute $\int_K b_K \, dx$.

• A possibility is to solve by hand the pure convective problem, as advocated in [10]:

$$\begin{cases} \text{find } \widetilde{b}_K \in H^1(K) \text{ such that} \\ \mathbf{c} \cdot \nabla \widetilde{b}_K = 1 \ \text{ in } K, \\ \widetilde{b}_K = 0 \ \text{ on } \partial K^- \ (= \text{inflow}) \end{cases}$$

Notice that the integral of $\widetilde{b}_K$ on K is just the volume of a pyramid, as shown in Fig. 2.

• Another possibility is to solve (5.26) on a subgrid with very few degrees of freedom, but well chosen (e.g., Pseudo RFB [8], Shishkin [17], etc, see Fig. 3). Typically few nodes in the element boundary layer are needed.

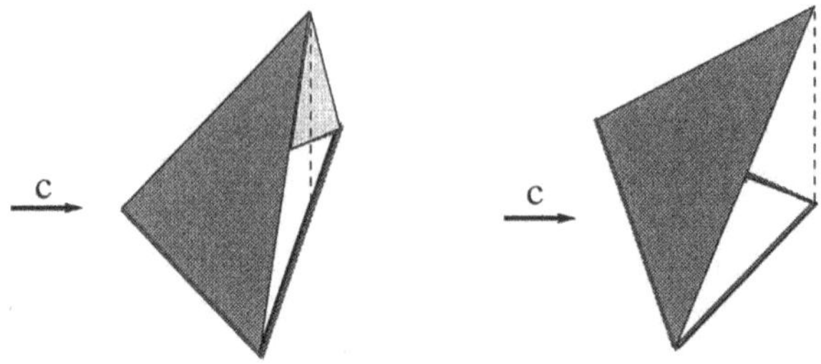

Fig. 2. Possible shapes of $\widetilde{b}_K$; here $\mathbf{c} = (1, 0)$

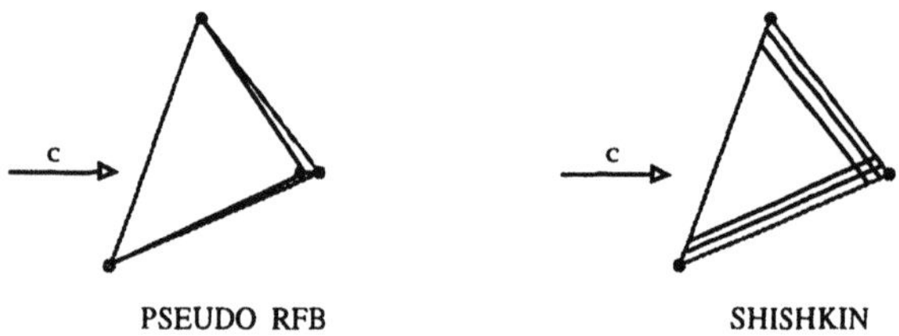

Fig. 3. Example of meshes

• As an alternative, one could use subgrid artificial viscosity; that means solving, instead of (5.26), the problem

$$-(\varepsilon + \varepsilon_A)\Delta b_K + \mathbf{c} \cdot \nabla b_K = 1 \quad \text{in each } K$$

on a very rough grid (typically, one node), where ε_A is a suitably chosen artificial viscosity, in general $\simeq h_K$ (see [20]). Unfortunately, the problem of the optimal choice for ε_A is rather delicate. Indeed, using a one-dimensional space $B_h(K) =$ span $\{\beta_K(x)\}$ results in an SUPG method with

$$\tau_K = \frac{(\int_K \beta \, dx)^2}{|K|(\varepsilon + \varepsilon_A) \int_K |\nabla \beta|^2 \, dx},$$

as shown in [5]. This implies that the bigger is ε_A the smaller is τ_K, that is, we add artificial viscosity for stabilizing and we decrease the stabilization parameter.

6 Conclusions

The Residual Free Bubble approach offers a unified framework for setting and analyzing several two-level and/or stabilized methods. It consists, essentially, in augmenting a given finite element space with spaces of functions having support in a single element. The necessary requirements for this augmentation process have been introduced and discussed for several examples. The split nature of the bubble space allows to eliminate the additional unknowns with an element by element procedure, that can be carried out in parallel. The elimination process involves in general the approximate solution of a partial differential equation in each element. We have seen however that in many cases a rough approximation can be sufficient.

The use of this type of approach for stabilizing unstable finite element formulations were already well known. Here we presented the method in a very general setting, and this allowed us to show that several other methods for stabilizing and, mostly, for dealing with subgrid phenomena, can actually be seen as a particular case of the RFB approach. This includes, on one side, old methods like the Petrov Galerkin methods with special, operator dependent, trial and test functions for advection dominated problems, as well as more recent approaches like the multiscale method or the upscaling method for problems with composite materials.

Other developments and applications to different problems are surely worth further investigations, as well as some recent variants like the use of non-conforming bubbles, the possibility of adding edge-bubbles, or the connections with domain decomposition methods.

References

1. T. Arbogast, " Numerical subgrid upscaling of two-phase flow in porous media,"in "Multiphase flows and transport in porous media: State of the art", (Z. Chen, R.E. Ewing, and Z.-C. Shi eds.), Lecture Notes in Physics, Springer, Berlin, 2000.
2. T. Arbogast, S.E. Minkoff, and P.T. Keenan, "An operator-based approach to upscaling the pressure equation," in: Computational Methods in Water Resources XII, v.1, V.N. Burganos et als., eds., Computational Mechanics Publications, Southampton, U.K., 1998.
3. F. Brezzi, M. Fortin, "Mixed and Hybrid Finite Element Methods," Springer Verlag, New York, Springer Series in Computational Mathematics **15**, 1991.
4. F. Brezzi, L.P. Franca, T.J.R. Hughes, and A. Russo, " $b = \int g$," Comput. Methods Appl. Mech. Engrg. **145**, 329-339 (1997). Methods Appl. Mech. Engrg. **166**, 25-33 (1998).
5. F. Brezzi, P. Houston, L.D. Marini, and E. Süli, "Modeling subgrid viscosity for advection-diffusion problems," Comput. Methods Appl. Mech. Engrg. **190**, 1601-1610 (2000).
6. F. Brezzi, T.J.R. Hughes, L.D. Marini, A. Russo, and E. Süli, "A priori error analysis of a finite element method with residual-free bubbles for advection-dominated equations," SIAM J. Num. Anal. **36**, 1933-1948 (1999)
7. J. Bergh, J. Löfström "Interpolation Spaces" Springer Verlag, Berlin, 1976.
8. F. Brezzi, D. Marini, and A. Russo, "Applications of pseudo residual-free bubbles to the stabilization of convection-diffusion problems," Comput. Methods Appl. Mech. Engrg. **166**, 51-63 (1998).

9. F. Brezzi, D. Marini, and E. Süli, "Residual-free bubbles for advection-diffusion problems: the general error analysis," Numer. Math. **85**, 31-47 (2000).
10. F. Brezzi, A. Russo, "Choosing bubbles for advection-diffusion problems," Math. Mod. and Meth. in Appl. Sci. **4**, 571-587 (1994).
11. A.N. Brooks, T.J.R. Hughes, "Streamline Upwind/Petrov-Galerkin formulations for convection dominated flows with particular emphasis on the incompressible Navier-Stokes equations, " Comput. Methods Appl. Mech. Engrg. **32**, 199-259 (1982).
12. Ph.G. Ciarlet, "The finite element method for elliptic problems," North-Holland, 1978.
13. C. Farhat, A. Macedo, and M. Lesoinne, "A two-level domain decomposition method for the iterative solution of high frequency exterior Helmholtz problems," Numer. Math. **85**, 283-308 (2000)
14. L.P. Franca, C. Farhat, A.P. Macedo and M. Lesoinne, "Residual-Free Bubbles for the Helmholtz Equation," Int. J. Num. Meth. in Eng. **40**, 4003-4009 (1997).
15. L.P. Franca, S.L. Frey and T.J.R. Hughes, "Stabilized finite element methods: I. Applications to advective-diffusive model," Comput. Methods Appl. Mech. Engrg. **95**, 253-276 (1992).
16. L.P. Franca, A.P. Macedo, "A Two-Level Finite Element Method and its Application to the Helmholtz Equation," Int. J. Num. Meth. in Eng. **43**, 23-32 (1998).
17. L.P. Franca, A. Nesliturk and M. Stynes, "On the Stability of Residual-Free Bubbles for Convection-Diffusion Problems and Their Approximation by a Two-Level Finite Element Method," Comput. Methods Appl. Mech. Engrg. **166**, 35-49 (1998).
18. L.P. Franca, A. Russo, "Deriving upwinding, mass lumping and selective reduced integration by residual-free bubbles." Appl. Math. Lett. **9**, 83-88 (1996).
19. D.F. Griffiths, A.R. Mitchell, " Spurious behavior and nonlinear instability in discretised partial differential equations," In: The dynamics of numerics and the numerics of dynamics. Inst. Math. Appl. Conf. Ser., New Ser. **34**, 215-242 (1992).
20. J.L. Guermond, "Stabilization of Galerkin approximations of transport equations by subgrid modeling," Math. Mod. Num. Anal. **33**(6), 1293-1316 (1999).
21. P. Hansbo, C. Johnson, "Streamline diffusion finite element methods for fluid flow," von Karman Institute Lectures, 1995.
22. T.Y. Hou, X.H. Wu, "A multiscale finite element method for elliptic problems in composite materials and porous media," J. Comput. Phys. **134**, 169-189 (1997).
23. T.Y. Hou, X.H. Wu, and Z. Cai, "Convergence of a multiscale finite element method for elliptic problems with rapidly oscillating coefficients," Math. of Comp. **68**, 913-943 (1999).
24. T.J.R. Hughes, "Multiscale phenomena: Green's functions, the Dirichlet to Neumann formulation, subgrid scale models, bubbles and the origins of stabilized methods," Comput. Methods Appl. Mech. Engrg. **127**, 387-401 (1995).
25. A.R. Mitchell, D.F. Griffiths, "Generalised Galerkin methods for second order equations with significant first derivative terms," In: Proc. bienn. Conf., Dundee 1977, Lect. Notes Math **630**, 90-104 (1978).
26. H.-G. Roos, M. Stynes, and L. Tobiska, "Numerical methods for singularly perturbed differential equations: convection diffusion and flow problems," Springer-Verlag, 1996.

27. A. Russo, "A posteriori error estimators via bubble functions," Math. Models Methods Appl. Sci. **6**, 353-360 (1997).
28. G. Sangalli, "Global and local error analysis for the Residual Free Bubble method applied to advection-dominated problems," Submitted to Numer. Math.

Two Scale FEM for Homogenization Problems*

Christoph Schwab

Seminar for Applied Mathematics, ETH-Zentrum
CH-8092 Zürich, Switzerland

Abstract. We analyze generalized Finite Element Methods for the numerical solution of elliptic problems with coefficients or geometries which are oscillating at a small length scale ε. Two-scale elliptic regularity results which are uniform in ε are presented. Two-scale FE spaces are introduced with error estimates that are uniform in ε. They resolve the ε scale of the solution with work independent of ε and without analytical homogenizations. Numerical experiments confirming the theory are presented.

1 Introduction

The accurate and efficient numerical solution of partial differential equations involving length scales that differ by many orders of magnitude has received increasing attention recently, due in part to the increasing miniaturization and manufacturing capabilities in engineering. For example, lattice materials or electronic circuitboards are assembled out of many basic building blocks of small size into larger macroscopic structures. *Scale resolution*, i.e. the direct numerical solution of boundary value problems in multiple scale structures by standard methods such as the Finite Element Method (FEM) is infeasible if the difference in scales is sufficiently large and the FE mesh is refined to the smallest scale.

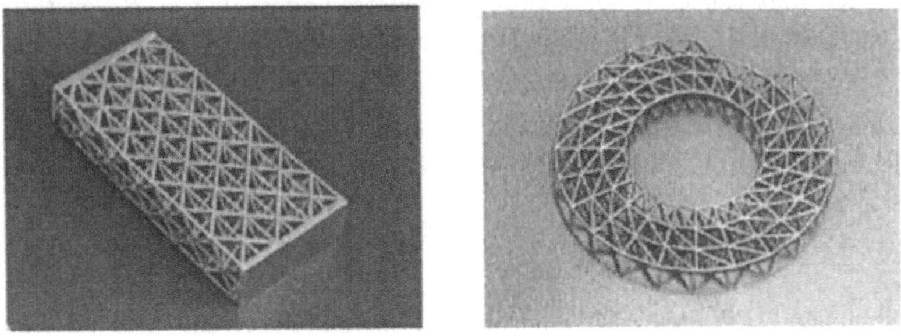

Fig. 1. Lattice block materials

* Research supported by the Swiss National Foundation under Project Number BBW 21-58754.99

Modelling approaches try to account for the effect of the small scales in the problem on the macroscopic solution behaviour analytically and to circumvent the requirement of scale resolution in the numerical solution of the problem. In this process, information on the fine scale behaviour of the solution is lost and cannot be recovered numerically since the modelling error is not a discretization error.

Here we consider a different approach if the solution contains several length scales differing by orders of magnitude and these scales are *separated*, i.e. the spatial variation of the solution is concentrated at length scales which are a-priori known or can be estimated. In between these scales, only a small part of the solution energy is concentrated. If, in addition, the fine-scale data contains regular patterns, the resolution of this scale is possible with substantially fewer degrees of freedom than (uniform or adaptive) mesh refinement which assumes a rather uniform distribution of solution scales. In the present paper, we illustrate this idea for the numerical solution of elliptic homogenization problems in divergence form. The present paper is a short version of [6] where full proofs shall be given.

1.1 Homogenization Problem

Exemplarily, we consider the scalar model problem

$$L^\varepsilon\left(\frac{x}{\varepsilon},\partial_x\right)u^\varepsilon := -\nabla\cdot A\left(\frac{x}{\varepsilon}\right)\nabla u^\varepsilon + a_0\left(\frac{x}{\varepsilon}\right)u^\varepsilon = f(x) \tag{1}$$

(everything works also for strongly elliptic systems in divergence form, see e.g. [3]). We assume that $A(y)$, $a_0(y)$ are 1-periodic in each variable and

$$A(\cdot)\in L^\infty_{\mathrm{per}}(\widehat{Q})^{n\times n}_{\mathrm{symm}},\quad a_0(\cdot)\in L^\infty_{\mathrm{per}}(\widehat{Q}) \tag{2}$$

satisfy, for some $\gamma > 0$,

$$\xi^\top A(y)\xi \ge \gamma|\xi|^2,\quad a_0(y)\ge\gamma\quad \forall\xi\in\mathbb{R}^n,\ \text{a.e. } y\in\widehat{Q}, \tag{3}$$

where the unit cell $\widehat{Q}\subset[0,1]^n$ has Lipschitz boundary $\partial\widehat{Q} = \widehat{\Gamma}_{\mathrm{per}}\cup\widehat{\Gamma}_N$ with $\widehat{\Gamma}_{\mathrm{per}} = \partial\widehat{Q}\cap\partial[0,1]^n$, and $\widehat{\Gamma}_N = \partial\widehat{Q}\backslash\widehat{\Gamma}_{\mathrm{per}}$ is the (possibly empty) Neumann boundary (see Figure 2).

We consider (1) in a bounded Lipschitz domain Ω covered by a pavement of cells of the form $\varepsilon(k+\widehat{Q})$, with $k\in\mathbb{Z}^n$. Set thus $\Omega_\varepsilon = \Omega_\varepsilon^\infty\cap\Omega$, where

$$\Omega_\varepsilon^\infty = \bigcup_{\mathbf{Z}^n}\varepsilon(k+\widehat{Q}),\quad \Gamma^\infty_{N,\varepsilon} := \bigcup_{\mathbf{Z}^n}\varepsilon(k+\widehat{\Gamma}_N). \tag{4}$$

We complete (1) in Ω_ε by Dirichlet boundary conditions on $\partial\Omega$, i.e.,

$$u^\varepsilon = 0\quad \text{on } \partial\Omega_\varepsilon\cap\partial\Omega, \tag{5}$$

and, if $\widehat{\Gamma}_N\neq\emptyset$, by Neumann boundary conditions on the hole boundaries

$$\gamma_1 u^\varepsilon := n\cdot A\left(\frac{x}{\varepsilon}\right)\nabla u^\varepsilon = 0\quad \text{on } \partial\Omega_\varepsilon\backslash\partial\Omega = \partial\Omega_\varepsilon\cap\Gamma^\infty_{N,\varepsilon}. \tag{6}$$

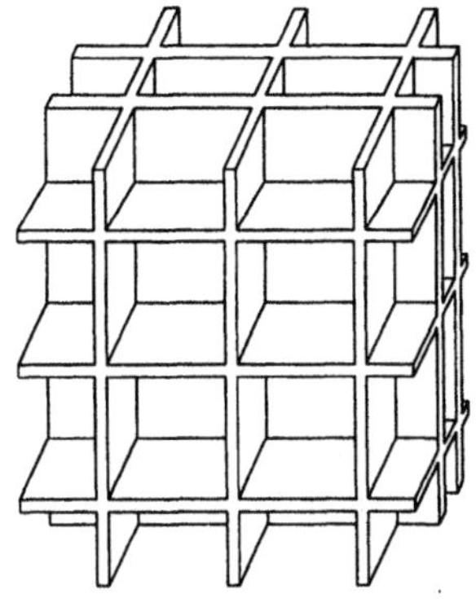

Fig. 2. Lattice material with rectangular, periodic pattern.

1.2 Finite Element Discretization

The variational form of (1), (5), (6) reads

$$\text{Find } u^\varepsilon \in H^1_D(\Omega_\varepsilon) \ : \ a(u^\varepsilon, v) = (f, v) \quad \forall v \in H^1_D(\Omega_\varepsilon), \tag{7}$$

where $H^1_D(\Omega_\varepsilon) := \{u \in H^1(\Omega_\varepsilon) \ : \ (5) \text{ holds for } u\}$. By (3), (7) admits a unique solution $u^\varepsilon \in H^1_D(\Omega_\varepsilon)$ for every $\varepsilon > 0$ and every $f \in L^2(\Omega)$.

Let $V^\varepsilon_N \subset H^1_D(\Omega_\varepsilon)$ be any subspace of dimension $N = \dim(V^\varepsilon_N) < \infty$. The Finite Element Method for (7)

$$u^\varepsilon_N \in V^\varepsilon_N \ : \ a(u^\varepsilon_N, v) = (f, v) \quad \forall v \in V^\varepsilon_N \tag{8}$$

defines a unique FE solution u^ε_N and

$$\|u^\varepsilon - u^\varepsilon_N\|_{H^1(\Omega_\varepsilon)} \le C \min_{v \in V^\varepsilon_N} \|u^\varepsilon - v\|_{H^1(\Omega_\varepsilon)}, \tag{9}$$

where $C > 0$ is independent of ε, i.e. the FE-error is bounded by the best approximation of u^ε from V^ε_N. Finite Element convergence of u^ε_N is therefore related to regularity of u^ε in dependence on the scale parameter ε.

Even if the right hand side f, the domain Ω_ε and the coefficients A and a_0 are smooth, for $\varepsilon/\text{diam}(\Omega) \ll 1$ the solution u^ε exhibits oscillations on the ε-scale. These in turn stall the convergence of standard FEM: consider for illustration $\widehat{Q} = [0,1]^n$. Then $\Omega_\varepsilon = \Omega$ and

$$\|u\|_{L^2(\Omega)} \le C, \quad \|D^\alpha u\|_{L^2(\Omega)} \le C(\alpha)\varepsilon^{1-|\alpha|}, \quad \forall \alpha \in \mathbb{N}^n, \ |\alpha| > 0$$

where $C, C(\alpha)$ are independent of ε. Denoting by $V^\varepsilon_N = V_N = S^{p,1}(\Omega, \mathcal{T}_H) \subset H^1(\Omega)$, the standard FE-space of continuous piecewise polynomials of degree $p \ge 1$ on a quasiuniform mesh $\mathcal{T}_H$ of meshwidth H, it holds

$$\min_{v \in S^{p,1}(\Omega, \mathcal{T}_H)} \|u^\varepsilon - v\|_{H^1(\Omega_\varepsilon)} \le C H^p \|D^{p+1} u\|_{L^2(\Omega)} \le C(H/\varepsilon)^p.$$

Trivially, we have also that

$$\min_{v \in S_H^{p,1}(\Omega,\mathcal{T})} \|u^\varepsilon - v\|_{H^1(\Omega^\varepsilon)} \le C\|u^\varepsilon\|_{H^1(\Omega_\varepsilon)} \le C\|f\|_{L^2(\Omega)}.$$

It follows therefore that the FE error with respect to the usual FE space $V_N = S^{p,1}(\Omega, \mathcal{T}_H)$ satisfies the following *a-priori* bounds

$$\|u^\varepsilon - u_N^\varepsilon\|_{H^1(\Omega_\varepsilon)} \le C \min(1, (H/\varepsilon)^p),$$

with $C = C(p, \Omega, f, A, a_0) > 0$ a constant independent of ε and H. Standard FEM, as e.g., piecewise linears on a quasiuniform mesh $\mathcal{T}_H$ of size H, thus converge only if $H < \varepsilon$, i.e., if $N = \dim V_N^\varepsilon = O(\varepsilon^{-n})$. This *scale resolution requirement* is often prohibitive, especially if $n \ge 3$.

1.3 Scale separation and outline of the paper

In view of (9), the key to robust approximations is the design of V_N^ε. Rather than incorporating the asymptotics of u^ε (which is not always defined, see [1,8] and the references there) into the FE-space V_N^ε, we design V_N^ε based on a refined regularity theory of u^ε. To this end, ignoring boundary conditions (5) for the moment, we consider (1) on the unbounded domain $\Omega_\varepsilon^\infty$ in (4). For any $f \in L^2(\mathbb{R}^n)$, (1), (6) admits a unique solution $u^\varepsilon \in H^1_{-\nu}(\Omega_\varepsilon^\infty)$, the weighted H^1-space with weight $exp(\nu|x|)$, $0 < \nu \le \nu_0(\gamma)$. This solution u^ε can be written in the form [7,4,3]

$$u^\varepsilon(x) = \int_{t \in \mathbb{R}^n} \hat{f}(t)\psi(x, \varepsilon, t)\, dt, \quad x \in \Omega_\varepsilon^\infty, \tag{10}$$

i.e. as superposition of the kernel $\psi(x, \varepsilon, t)$ which is solution of

$$L^\varepsilon\left(\frac{x}{\varepsilon}, \partial_x\right)\psi = e^{it\cdot x} \text{ on } \Omega_\varepsilon^\infty, \quad n \cdot A(x/\varepsilon)\nabla\psi = 0 \text{ on } \Gamma_{N,\varepsilon}^\infty. \tag{11}$$

Problem (1) has separated scales, a slow variable x and a fast variable $y = x/\varepsilon$, in the following sense: the kernel ψ in (11) (which is, in a sense, the fine scale response to the coarse scale excitation $e^{it\cdot x}$) can be written in the form $\psi(x, \varepsilon, t) = e^{it\cdot x}\phi(\frac{x}{\varepsilon}, \varepsilon, t)$ where $\phi(y, \varepsilon, t)$ is the solution of the so-called *unit-cell problem*: find $\phi \in H^1_{\text{per}}(\widehat{Q})$ such that

$$\begin{aligned} \mathcal{L}(\varepsilon, t, y; \partial_y)\phi &:= e^{-i\varepsilon t\cdot y} L^\varepsilon(y, \varepsilon^{-1}\partial_y) e^{i\varepsilon t\cdot y}\phi = 1 \text{ in } \widehat{Q}, \\ \mathcal{B}(\varepsilon, t, y; \partial_y)\phi &:= e^{-i\varepsilon t\cdot y} n \cdot A(y)\nabla_y(e^{i\varepsilon t\cdot y}\phi) = 0 \text{ on } \widehat{\Gamma}_N. \end{aligned} \tag{12}$$

Unlike ψ in (11), the kernel ϕ is computable by solving the unit-cell problem (12) numerically, for example (but not necessarily) with finite elements. In the the remainder of this note, we present approaches for the design of FE-spaces V_N^ε which give ε-independent convergence. We proceed as follows: First, based on the representation (10), we see that on $\Omega_\varepsilon^\infty$ (i.e., in the absence of boundary layers) the solution $u^\varepsilon(x)$ can be viewed as a map from the 'slow' variable $x \in \Omega$ (*not* in Ω_ε) into the 'fast' variable $x/\varepsilon \in \widehat{Q}$. Two-scale regularity results on $u^\varepsilon(x)$ which are uniform in ε are obtained by analysing this map and we present these in Section

2. The two-scale point of view of regularity gives rise to a 'natural' FE discretization of (1) by means of a non-standard two-scale FE-space V_N^ε in Ω_ε constructed as follows: Let $\mathcal{T}_H$ be a quasiuniform 'macro' mesh in Ω (**not** in Ω_ε, i.e., the fine structure of the coefficients is ignored) of meshwidth H and denote by $S^p(\Omega, \mathcal{T}_H)$ the usual FE-space of continuous, piecewise polynomials of degree p on $\mathcal{T}_H$ (we assume also for convenience that $\mathcal{T}_H$ is aligned with the periodic pattern in Ω_ε). We discretize the unit-cell problem (12) by a FEM in $\widehat{Q}$, based on the mesh $\widehat{\mathcal{T}}_h$ (for simplicity also quasiuniform of width h), and the space $S^\mu_{\text{per}}(\widehat{Q}, \widehat{\mathcal{T}}_h)$.The 2-scale FE space V_N^ε in (9) is then the Bochner space

$$V_N^\varepsilon = S^p(\Omega, \mathcal{T}_H;\ S^\mu_{\text{per}}(\widehat{Q}, \widehat{\mathcal{T}}_h)). \tag{13}$$

Since $\{1\} \subseteq S^\mu_{\text{per}}(\widehat{Q}, \widehat{\mathcal{T}}_h)$, $S^p(\Omega, \mathcal{T}_H) \subseteq V_N^\varepsilon$ and V_N^ε is a generalized FE-space. With V_ε^N in (8) robust convergence rates as $h, H \to 0$ can be achieved for u_N^ε as we shall show in Section 3. These 2-scale approximation results are quite general and applicable whenever the solution has the 2-scale regularity; in particular, the representation (10) which is valid only in a linear setting is not necessary. In contrast, in [3–5] a different (in general smaller) space V_N^ε than (13) was proposed. In that approach the kernel $\phi(y, \varepsilon, t)$ in (12) is incorporated directly in the FE-space via shape functions $\phi(y, \varepsilon, t)$ sampled at suitable points t_j in the frequency domain.

2 Two scale regularity

As in the two-scale asymptotics in e.g. [1,8], we separate the slow from the fast scales. We do not expand $u^\varepsilon(x)$ asymptotically, however, but rather interpret it as map from the "slow variable" x into the "fast variable" $y = x/\varepsilon$:

$$u^\varepsilon(x) = U^\varepsilon(x, x/\varepsilon), \text{ where } U^\varepsilon(x, y) \in H^r(\Omega, H^s_{\text{per}}(\widehat{Q})) \tag{14}$$

for $r, s \geq 0$ depending on the regularity of the coefficients and of the data f and where the ε-dependence of $U^\varepsilon(x, y)$ is smooth. We consider here only the case when the unit cell problem admits maximal elliptic regularity and therefore take in (14) as target space $H^s_{\text{per}}(\widehat{Q})$. Then the 2-scale shift theorem holds in standard Sobolev spaces (if the unit cell has corners as e.g. in Figure 2, the target space $H^s_{\text{per}}(\widehat{Q})$ has to be replaced by a suitable weighted space with weights associated to the corners in the unit cell).

Theorem 1. *Assume that $A(\cdot)$, $a_0(\cdot)$ are smooth and 1-periodic in $y = x/\varepsilon \in \widehat{Q}$ with $\widehat{Q}$ denoting $(0,1)^n$ if $\widehat{\Gamma}_N = \emptyset$ or that $\widehat{\Gamma}_N$ is smooth otherwise. If in addition $f \in H^k_{\text{comp}}(\mathbb{R}^n)$ $(k \geq 0)$, then the solution $u^\varepsilon(x)$ of (1) on the unbounded domain $\Omega^\varepsilon_\infty$ can be written as*

$$u^\varepsilon(x) = U^\varepsilon(x, y)|_{y=x/\varepsilon}, \quad x \in \Omega^\varepsilon_\infty,$$

where $U^\varepsilon(x, y)$ satisfies in $\Omega = \mathbb{R}^n$ the two-scale regularity estimate

$$\|U^\varepsilon\|_{H^r(\Omega, H^s_{\text{per}}(\widehat{Q}))} \leq C(k)\ \|f\|_{H^{r+s-1}(\Omega)} \tag{15}$$

provided $r+s \le k+1$, $r,s \ge 0$, *and*

$$\|\varepsilon^{-1}\nabla_y U^\varepsilon\|_{H^r(\Omega,H^{s-1}_{\mathrm{per}}(\widehat{Q}))} \le C(k)\,\|f\|_{H^{r+s-1}(\Omega)} \tag{16}$$

provided $r+s \le k+1$, $r, s-1 \ge 0$. *Here,* $C(k)$ *is independent of* ε, *but depends on* $r+s$ *(see Remark 12 ahead for this dependence).*

Proof. The proof is based on the Fourier-Bochner integral representation (10) of the solution $u^\varepsilon(x) = U^\varepsilon(x, x/\varepsilon)$ and on two-scale regularity estimates on the Fourier-Bochner integral kernel which are uniform in ε and t. For multiindices α, β with $|\alpha| \le r$, $|\beta| \le s$, the mixed derivative (in the sense of distributions) $D_x^\alpha D_y^\beta U^\varepsilon(x,y)$ can be interpreted as mapping $L^2_{\mathrm{per}}(\widehat{Q})$ into $L^2(\mathbb{R}^n)$. More precisely, for arbitrary $\varphi \in L^2_{\mathrm{per}}(\widehat{Q})$, $\langle D_x^\alpha D_y^\beta U^\varepsilon(x,\cdot), \varphi\rangle_{L^2_{\mathrm{per}}(\widehat{Q})\times L^2_{\mathrm{per}}(\widehat{Q})}$ is the inverse Fourier transform of a $L^2(\mathbb{R}^n)$ functional

$$\langle D_x^\alpha D_y^\beta U^\varepsilon(x,\cdot), \varphi\rangle_{L^2_{\mathrm{per}}(\widehat{Q})\times L^2_{\mathrm{per}}(\widehat{Q})} =$$
$$\frac{1}{(2\pi)^{n/2}} \int_{\mathbb{R}^n} e^{it\cdot x} \hat{f}(t)(it)^\alpha \langle D_y^\beta \phi(y,\varepsilon,t), \varphi(y)\rangle_{L^2_{\mathrm{per}}(\widehat{Q})\times L^2_{\mathrm{per}}(\widehat{Q})}\, dt.$$

By Parseval equation the $L^2(\mathbb{R}^n)$-norm of $\langle D_x^\alpha D_y^\beta U^\varepsilon(x,y), \varphi\rangle_{L^2_{\mathrm{per}}(\widehat{Q})\times L^2_{\mathrm{per}}(\widehat{Q})}$ is equal to

$$\left\|\langle D_x^\alpha D_y^\beta U^\varepsilon(x,y), \varphi\rangle_{L^2_{\mathrm{per}}(\widehat{Q})\times L^2_{\mathrm{per}}(\widehat{Q})}\right\|_{L^2(\mathbb{R}^n)}$$
$$= \left\|(it)^\alpha \hat{f}(t) \langle D_y^\beta \phi(y,\varepsilon,t), \varphi(y)\rangle_{L^2_{\mathrm{per}}(\widehat{Q})\times L^2_{\mathrm{per}}(\widehat{Q})}\right\|_{L^2(\mathbb{R}^n)}.$$

It can be shown ([6]) that there exists a positive constant $C > 0$ independent of ε, t and of the test function φ, such that for all $t \in \mathbb{R}^n$

$$\left|\langle D_y^\beta \phi(y,\varepsilon,t), \varphi(y)\rangle_{L^2_{\mathrm{per}}(\widehat{Q})\times L^2_{\mathrm{per}}(\widehat{Q})}\right| \le C(1+|t|)^{|\beta|-1} \|\varphi\|_{L^2(\widehat{Q})}. \tag{17}$$

Hence, by Parseval's identity again,

$$\left\|\langle D_x^\alpha D_y^\beta U^\varepsilon(x,y), \varphi\rangle_{L^2_{\mathrm{per}}(\widehat{Q})\times L^2_{\mathrm{per}}(\widehat{Q})}\right\|_{L^2(\mathbb{R}^n)} \le C\|f\|_{H^{r+s-1}(\mathbb{R}^n)}\, \|\varphi\|_{L^2(\widehat{Q})},$$

which proves (15). Proceeding in a similar fashion one can prove the two scale regularity estimate on the gradient of the solution in (16), see [6] for full details.

Remark 2. The two-scale regularity result is based on (10), i.e. the solution of (1) on the infinite domain. Such solutions, when restricted to a finite domain Ω_ε, correspond to the case when boundary correctors are absent in the two-scale asymptotics. In an $O(\varepsilon)$ neighborhood of $\partial\Omega$, the scale-separation assumption does not hold and $\mathcal{T}_H$ has to be refined to resolve the $O(\varepsilon)$ scale directly, in general. In the corresponding elements $K \in \mathcal{T}_H$, the subspace $S^\mu_{\mathrm{per}}(\widehat{Q}, \widehat{\mathcal{T}}_h)$ has to be coarsened to the point when $S^\mu_{\mathrm{per}}(\widehat{Q}, \widehat{\mathcal{T}}_h) = \{1\}$, i.e. when no micro shapefunctions are used.

Estimates (15), (16) appear to be suboptimal, in terms of the regularity of the right hand side. They can nevertheless not be improved, if one insists on ε-independence of the constant $C(k)$, as the following example from [6] shows. We consider $n = 1$ and assume that $f \in L^2_{\text{per}}(0,1)$ has the Fourier expansion $f(x) = \sum_{k\in\mathbb{Z}} f_k e^{2\pi i k x}$. Assume further that $a(\cdot)$ is a 1-periodic, L^∞ function and $\varepsilon = 1/M$, with $M \in \mathbb{N}^*$. Let $u^\varepsilon(x) \in H^1_0(0,1)$ be the solution of the following boundary value problem

$$-\frac{d}{dx}\left(a\left(\frac{x}{\varepsilon}\right)\frac{du^\varepsilon}{dx}\right) = f(x) \quad \text{in } \Omega = (0,1), \quad u^\varepsilon\Big|_{\partial\Omega} = 0.$$

Proposition 3. *Assume that $a(\cdot)$ is smooth and 1-periodic in $y = x/\varepsilon \in \widehat{Q}$. Then, for $f \in H^r_{\text{per}}(0,1)$ $(r \geq 0)$, the solution $u^\varepsilon(x)$ of (2) on $(0,1)$ satisfies the two-scale regularity estimates (15), (16)*

$$\begin{aligned} \|U^\varepsilon\|_{H^r(\Omega,\, L^2_{\text{per}}(\widehat{Q}))} &\leq C(r)\|f\|_{H^{r-1}(\Omega)}, \\ \|\varepsilon^{-1}\nabla_y U^\varepsilon\|_{H^r(\Omega,\, L^2_{\text{per}}(\widehat{Q}))} &\leq C(r)\|f\|_{H^r(\Omega)}. \end{aligned} \tag{18}$$

Moreover, the first estimate in (18) is sharp, in the sense that for ε sufficiently small, there exists a constant $c = c(r) > 0$, which does not depend on ε, such that

$$c(r)\|f\|_{H^{r-2}(\Omega)} \leq \|U^\varepsilon\|_{H^r(\Omega,\, L^2_{\text{per}}(\widehat{Q}))}.$$

Remark 4. The proof in [6] reveals that the upper bound in (18) has the form $C(r)\|f\|_{H^{r-2}(\Omega)} + C(\varepsilon,r)\|f\|_{H^{r-1}(\Omega)}$ with $C(\varepsilon,r) > 0$ vanishing as $\varepsilon \to 0$. In the limit $\varepsilon = 0$, we recover the regularity in a smooth domain Ω

$$-\Delta u = f \quad \text{in } \Omega, \quad f \in H^r(\Omega), \quad u|_{\partial\Omega} \text{ smooth}$$

where we have the shift theorem: there exists $C(r,\Omega) > 0$

$$\|u\|_{H^{r+2}(\Omega)} \leq C(r,\Omega)\|f\|_{H^r(\Omega)}, \quad r \geq -1,$$

in the sense that for generic data $\|u\|_{H^{r+2}(\Omega)}$ has a lower bound of the same type $(c(r,\Omega) > 0)$

$$\|u\|_{H^{r+2}(\Omega)} \geq c(r,\Omega)\|f\|_{H^r(\Omega)}.$$

In our case, however, the gap $C(\varepsilon,r)\|f\|_{H^{r-1}(\Omega)}$ can not be removed.

3 Two scale FE convergence

In the previous section we saw that $u^\varepsilon(x)$ admits elliptic regularity independent of the scale parameter in the framework of the two-scale Sobolev spaces $H^r(\mathbb{R}^n, H^s_{\text{per}}(\widehat{Q}))$. The two-scale Finite-Element spaces in the Introduction are, in a sense, natural for the direct discretization of homogenization problems. In the present section we prove robust approximation properties for two-scale FE-spaces under two-scale regularity hypothesis on $u^\varepsilon(x)$. In particular, we will generalize h, p and hp convergence results which are well known for standard FEM to two-scale FEM.

3.1 Tools

Sobolev spaces of mixed order Let $\Omega \subset \mathbb{R}^n$, $\Omega' \subset \mathbb{R}^n$ be two Lipschitz domains. For $\alpha, \beta \in \mathbb{N}^n$ two multiindices we define the Sobolev spaces $\mathcal{H}^{\alpha,\beta}(\Omega \times \Omega')$ of mixed order on the product domain $\Omega \times \Omega'$ as

$$\mathcal{H}^{\alpha,\beta}(\Omega \times \Omega') := \{u \in L^2(\Omega \times \Omega') \ : \ D_x^\gamma D_z^\delta u \in L^2(\Omega \times \Omega'),\ \forall \gamma \leq \alpha, \delta \leq \beta\},$$

in which $\gamma \leq \alpha$ is understood componentwise. These are Hilbert spaces with respect to the norm

$$\|u\|^2_{\mathcal{H}^{\alpha,\beta}(\Omega\times\Omega')} := \sum_{\gamma\leq\alpha,\delta\leq\beta} \|D_x^\gamma D_z^\delta u\|^2_{L^2(\Omega\times\Omega')}.$$

Traces in Sobolev spaces of mixed order For a function $f(x,y) : \Omega \times \Omega \to \mathbb{C}$, we denote by $\mathcal{R}f(x) = f(x,x) : \Omega \to \mathbb{C}$ its restriction to the diagonal $\{(x,y) \in \Omega \times \Omega \mid x = y\}$.

Lemma 5. *Let $\Omega = \Omega' := [0,1]^n$ and denote by $\mathbf{1} \in \mathbb{N}^n$, $\mathbf{0} \in \mathbb{N}^n$, the multiindices $(1,\ldots,1)$, $(0,\ldots,0)$ respectively. Then, the operator*

$$\mathcal{R} : \mathcal{H}^{\mathbf{1},\mathbf{0}}(\Omega \times \Omega) \to L^2(\Omega)$$

is continuous, i.e., there exists a positive constant $C = C(n)$ such that

$$\|\mathcal{R}f\|_{0,\Omega} \leq C \left(\sum_{0\leq\alpha_i\leq 1} \|D_x^\alpha f(x,z)\|_{0,\Omega\times\Omega} \right).$$

Moreover, for any fixed pair of multiindices $\alpha, \beta \in \mathbb{N}^n$ with $\alpha + \beta = \mathbf{1}$ the restriction operator $\mathcal{R} : \mathcal{H}^{\alpha,\beta}(\Omega \times \Omega) \to L^2(\Omega)$ is continuous, i.e., there exists a constant $C = C(n) > 0$ such that

$$\|\mathcal{R}f\|_{L^2(\Omega)} \leq C(n) \|f\|_{\mathcal{H}^{\alpha,\beta}(\Omega\times\Omega)}, \quad \forall f \in \mathcal{H}^{\alpha,\beta}(\Omega \times \Omega).$$

Polynomial approximation results We present some approximation results which are needed for our analysis. We start with the one-dimensional case (see also [9]).

Let $|\cdot|_{H^k(\widehat{\Omega})}$ denote the Sobolev seminorm of order k on $\widehat{\Omega} = (-1,1)$ given by

$$|\hat{u}|_{H^k(\widehat{\Omega})} := \|\hat{u}^{(k)}\|_{L^2(\widehat{\Omega})}, \quad \forall \hat{u} \in H^k(\widehat{\Omega}).$$

Let $\hat{u} \in H^{k+1}(\widehat{\Omega})$ for some $k \geq 1$. Then, for each $p \geq 1$, there exists a polynomial interpolant $\hat{s} = \pi_p \hat{u} \in S^p(\widehat{\Omega})$, with $S^p(\widehat{\Omega})$ denoting the space of polynomials of degree at most p on $\widehat{\Omega}$, such that

$$\|\hat{u}' - \hat{s}'\|^2_{L^2(\widehat{\Omega})} \leq \frac{(p-k)!}{(p+k)!} |\hat{u}|^2_{H^{k+1}(\widehat{\Omega})}$$

$$\|\hat{u} - \hat{s}\|^2_{L^2(\widehat{\Omega})} \leq \frac{1}{p(p+1)} \frac{(p-k)!}{(p+k)!} |\hat{u}|^2_{H^{k+1}(\widehat{\Omega})}.$$

To introduce the polynomial interpolant of degree p ($p \geq 1$) in the multi-dimensional case, we denote by $\widehat{\Pi}_p := \pi_p^{(x_1)} \otimes \cdots \otimes \pi_p^{(x_n)}$ ($n = 2, 3$) the tensor product polynomial interpolant of degree p in the reference element $\widehat{K} := (-1, 1)^n$. The polynomial interpolant Π_p in quadrilateral element $K := F(\widehat{K})$ with curved boundaries obtained via a C^∞-diffeomorphism $F : \widehat{K} \to K$ is given by $\Pi_p u := (\widehat{\Pi}_p(u \circ F)) \circ F^{-1}$.

Lemma 6. *Let $n = 2$ and let $\widehat{\Pi}_p = \pi_p^{(x_1)} \otimes \pi_p^{(x_2)}$ be the tensor product polynomial interpolant of degree p ($p \geq 1$) in the unit square $\widehat{K} = (-1, 1)^2$ in each variable [9]. Then, for all $u \in H^{k+1}(\widehat{K})$, $1 \leq k \leq p$, it holds*

$$\sum_{0 \leq \alpha_j \leq 1} \|D^\alpha(u - \widehat{\Pi}_p u)\|_{L^2(\widehat{K})} \leq C \Phi_2(p, k) \|D^{k+1} u\|_{L^2(\widehat{K})}.$$

where

$$\Phi_n(p, k) = \sqrt{\frac{(p - k + (n-1))!}{(p + k - (n-1))!}}$$

and $C > 0$ is a constant independent of p, k and u.

Affine transformation of the elements, addition of these local estimates gives

Lemma 7. *Assume that $\mathcal{T}_h$ is a quasiuniform, axiparallel quadrilateral mesh in $\Omega := (0, 1)^2$ and let $\Pi_{p,\mathcal{T}_h}$ denote the piecewise polynomial interpolant of degree $p \geq 1$ given by $\Pi_{p,\mathcal{T}_h} u|_K = \widehat{\Pi}_p(u|_K \circ F_K) \circ F_K^{-1}$ in each element $K \in \mathcal{T}_h$ with F_K being the associated affine element mapping. Then for any $u \in H^2(\Omega)$*

$$\|u - \Pi_{p,\mathcal{T}_h} u\|_{H^1(\Omega)} \leq C \sum_{K \in \mathcal{T}_h} h^{s_K} \Phi_2(p, s_K) |u|_{H^{s_K+1}(K)} \tag{19}$$

for $1 \leq s_K \leq p$ such that the right hand side in (19) is finite. The constant $C > 0$ is independent of p, s_K and h.

If $n = 3$ we distinguish between $p = 1$ and $p \geq 2$ as follows. For all $u \in H^3(\Omega)$ and for all $2 \leq s_K \leq p$ such that $u|_K \in H^{s_K+1}(K)$ for all $K \in \mathcal{T}_h$ there exists a positive constant $C > 0$ independent of p, s_K and h such that

$$\|u - \Pi_{p,\mathcal{T}_h} u\|_{H^1(\Omega)} \leq C \sum_{K \in \mathcal{T}_h} h^{s_K} \Phi_3(p, s_K) |u|_{H^{s_K+1}(K)}. \tag{20}$$

If $n = 3$ and $p = 1$, then for all $u \in \mathcal{H}^1(\Omega)$

$$\|u - \Pi_{1,\mathcal{T}_h} u\|_{H^1(\Omega)} \leq Ch|u|_{\mathcal{H}^1(\Omega)}, \tag{21}$$

where $C > 0$ is independent of h and we denoted by $|u|^2_{\mathcal{H}^1(\Omega)} := \sum_{\substack{0 \leq \alpha_j \leq 1 \\ |\alpha| > 0}} \|D^\alpha u\|^2_{L^2(\Omega)}$.

3.2 Two-scale approximation results

Recall that the two scale Bochner Finite Element space is given by

$$S^p(\Omega, \mathcal{T}_H; S^\mu_{\text{per}}(\widehat{Q}, \widehat{\mathcal{T}}_h)). \tag{22}$$

We assume that the domain Ω is axiparallel and we take $\mathcal{T}_H$ to be a quasiuniform triangulation of Ω of affine quadrilateral elements of size H. If $\widehat{Q} = (0,1)^n$ then we take $\widehat{\mathcal{T}}_h$ as well as a quasiuniform mesh in $\widehat{Q}$ of axiparallel quadrilaterals. For the case when the unit-cell domain $\widehat{Q}$ has e.g. interior holes the 'micro' triangulation $\widehat{\mathcal{T}}_h$ is obtained as follows. First one assumes the existence of a partition $\widehat{Q} = \cup_{i=1}^I \widehat{Q}_i$ ($I < \infty$ fixed) of $\widehat{Q}$ in a finite number of patches $\widehat{Q}_i$. Each patch $\widehat{Q}_i$ is obtained by mapping the reference domain $(0,1)^n$ via the C^∞ diffeomorphism $F_i : (0,1)^n \to \widehat{Q}_i$. These mappings satisfy also a compatibility condition in the sense that $F_i \circ F_{i'}^{-1} = Id$ on $Q_i \cap Q_{i'}$ for all $i, i' = 1, \ldots, I$ (F_i can be constructed by blending, see also [11]). The mesh $\widehat{\mathcal{T}}_h$ is obtained as follows:

$$\widehat{\mathcal{T}}_h = \cup_{i=1}^I \widehat{\mathcal{T}}_{h,i}, \quad \widehat{\mathcal{T}}_{h,i} = F_i(\widehat{\widehat{\mathcal{T}}}_h)$$

with $\widehat{\widehat{\mathcal{T}}}_h$ being a uniform, affine quadrilateral mesh in the reference domain $(0,1)^n$. In this case we will mean by $S^\mu_{\text{per}}(\widehat{Q}, \widehat{\mathcal{T}}_h)$ the finite element space given by all piecewise mapped polynomials of degree μ of the form

$$S^\mu_{\text{per}}(\widehat{Q}, \widehat{\mathcal{T}}_h) = \{u \in H^1_{\text{per}}(\widehat{Q}) \mid (u|_{\widehat{Q}_i} \circ F_i)|_{\widehat{\widehat{K}}} \in S^\mu(\widehat{\widehat{K}}) \,\forall \widehat{\widehat{K}} \in \widehat{\widehat{\mathcal{T}}}_h\}.$$

The piecewise polynomial interpolant $\mathcal{I}_{\mu,\widehat{\mathcal{T}}_h} \in S^\mu_{\text{per}}(\widehat{Q}, \widehat{\mathcal{T}})$ is given by $\mathcal{I}_{\mu,\widehat{\mathcal{T}}_h} u|_{\widehat{Q}_i} = (\Pi_{\mu,\widehat{\widehat{\mathcal{T}}}_h}(u|_{\widehat{Q}_i} \circ F_i)) \circ F_i^{-1}$. A similar estimate as in Lemma 7 for the interpolation error $u - \mathcal{I}_{\mu,\widehat{\mathcal{T}}_h} u$ holds.

Lemma 8.

$$\|u - \mathcal{I}_{\mu,\widehat{\mathcal{T}}_h} u\|_{H^1(\widehat{Q})} \leq C h^{\min(\mu,s)} \Phi_n(\mu, s) \sum_{i=1}^I \|u\|_{H^{s+1}(\widehat{Q}_i)} \tag{23}$$

Proof. The result is a direct consequence of the definition of the interpolation operator $\mathcal{I}_{\mu,\widehat{\mathcal{T}}_h}$ with respect to $S^\mu_{\text{per}}(\widehat{Q}, \widehat{\mathcal{T}}_h)$ and Lemma 7

$$\begin{aligned}
\|u - \mathcal{I}_{\mu,\widehat{\mathcal{T}}_h} u\|^2_{H^1(\widehat{Q})} &= \sum_{i=1}^I \|u - \mathcal{I}_{\mu,\widehat{\mathcal{T}}_h} u\|^2_{H^1(\widehat{Q}_i)} \\
&\leq C \textstyle\sum_i \|u \circ F_i - \mathcal{I}_{\mu,\widehat{\mathcal{T}}_h} u \circ F_i\|^2_{H^1([0,1]^n)} \\
&= C \sum_i \|u \circ F_i - \Pi_{\mu,\widehat{\widehat{\mathcal{T}}}_h}(u \circ F_i)\|^2_{H^1([0,1]^n)} \\
&\leq C h^{2\min(s,\mu)} \Phi_n^2(\mu, s) \sum_i \|u|_{\widehat{Q}_i}\|^2_{H^{s+1}(\widehat{Q}_i)}.
\end{aligned}$$

For each element $K \in \mathcal{T}_H$ of the 'macro' triangulation, define $U^{\varepsilon,K}(\hat{x}, y) := U^\varepsilon(F_K(\hat{x}), y)$, with $F_K : \widehat{K} \to K$ being an affine element map on the reference element $\widehat{K} = [0,1]^n$. Then, the interpolation error $E_{\mathcal{I}}^{\varepsilon}(x,y) = U^{\varepsilon}(x,y) - U_{\mathcal{I}}^{\varepsilon}(x,y)$, $x \in \Omega$, $y \in \mathbb{R}^n$, is given by

$$E_{\mathcal{I}}^{\varepsilon}(F_K(\hat{x}), y) := U^{\varepsilon,\,K}(\hat{x}, y) - (\Pi_{p,\hat{x}} \otimes \mathcal{I}_{h,y}) U^{\varepsilon,\,K}(\hat{x}, y),$$

with $\Pi_{p,\hat{x}}$ being the p interpolant in each reference element $\widehat{K}$ and $\mathcal{I}_{h,y}$ the $H^1_{\mathrm{per}}(\widehat{Q})$ projection into $S^{\mu}_{\mathrm{per}}(\widehat{Q}, \widehat{\mathcal{T}}_h)$ in the unit cell $\widehat{Q}$. Then, if H denotes the mesh size of the quasiuniform 'macroscopic' triangulation on Ω and h is the mesh size of the quasiuniform 'micro' triangulation on the unit cell $\widehat{Q}$, we obtain that

Proposition 9. *Assume that $n = 2$. For $p, \mu, k, s \geq 1$ and $H/\varepsilon \in \mathbb{N}$ in (22) it holds*

$$\begin{aligned}\|e_{\mathcal{I}}^{\varepsilon}\|_{L^2(\Omega_\varepsilon)} \leq C\big(&H^{\min(p,k)+1}\Phi_2(p,k)\|U^{\varepsilon}\|_{H^{k+1}(\Omega;\, L^2_{\mathrm{per}}(\widehat{Q}))} \\ &+ h^{\min(\mu,s)}\Phi_2(\mu,s)\|U^{\varepsilon}\|_{H^n(\Omega;\, H^{s+1}_{\mathrm{per}}(\widehat{Q}))}\big),\end{aligned}$$

where $C > 0$ is a positive constant independent of p, μ, k, s and ε.

Proof. We sketch the proof here – for full details, see [6].

Let $K = F_K(\widehat{K}) \in \mathcal{T}_H$ be an element of the 'macro' triangulation, affine image of the reference element $\widehat{K}$ under the element mapping F_K. We split the interpolation error into a 'macro' and a 'micro' error as follows:

$$\begin{aligned} E_{\mathcal{I}}^{\varepsilon}(F_K(\hat{x}), y) := &\, U^{\varepsilon,\,K}(\hat{x}, y) - \Pi_{p,\hat{x}} U^{\varepsilon,\,K}(\hat{x}, y) \\ &+ \Pi_{p,\hat{x}} U^{\varepsilon,\,K}(\hat{x}, y) - (\Pi_{p,\hat{x}} \otimes \mathcal{I}_{h,y}) U^{\varepsilon,\,K}(\hat{x}, y). \end{aligned} \tag{24}$$

To estimate the L^2 norm of the error on K we apply the trace result in Lemma 5. This gives

$$\int_K |e_{\mathcal{I}}^{\varepsilon}(x)|^2 \, dx \leq C H^n \left(\mathrm{I}_K + \mathrm{II}_K\right),$$

where

$$\mathrm{I}_K = \int_{\widehat{K}\times\widehat{Q}} \sum_{0\leq\alpha_j\leq 1} \left| D_{\hat{x}}^{\alpha} \left(U^{\varepsilon,\,K}(\hat{x}, y) - \Pi_{p,\hat{x}} U^{\varepsilon,\,K}(\hat{x}, y) \right) \right|^2 d\hat{x} dy$$

$$\mathrm{II}_K = \int_{\widehat{K}\times\widehat{Q}} \sum_{0\leq\alpha_j\leq 1} \left| D_{\hat{x}}^{\alpha} \left(\Pi_{p,\hat{x}} U^{\varepsilon,\,K}(\hat{x}, y) - (\Pi_{p,\hat{x}} \otimes \mathcal{I}_{h,y}) U^{\varepsilon,\,K}(\hat{x}, y) \right) \right|^2 d\hat{x} dy.$$

By Lemma 6, the 'macro' error I_K can be estimated as follows

$$\begin{aligned}\mathrm{I}_K &= \int_{\widehat{K}\times\widehat{Q}} \sum_{0\leq\alpha_j\leq 1} \left| D_{\hat{x}}^{\alpha} \left(U^{\varepsilon,\,K}(\hat{x}, y) - \Pi_{p,\hat{x}} U^{\varepsilon,\,K}(\hat{x}, y) \right) \right|^2 d\hat{x} dy \\ &\leq C H^{2(k+1)-n} \Phi_2^2(p,k) \int_{K\times\widehat{Q}} \left| \left(D_x^{k+1} U^{\varepsilon} \right)(x, y) \right|^2 dx dy.\end{aligned}$$

Applying now the error estimates in Lemma 8 for the interpolation error in the 'micro' FE space $S^{\mu}_{\rm per}(\widehat{Q}, \widehat{\mathcal{T}}_h)$, the 'micro' error II_K can be estimated as follows

$$\mathrm{II}_K = \int_{\widehat{K}\times\widehat{Q}} \sum_{0\le\alpha_j\le 1} \left| D^{\alpha}_{\hat{x}} \left(\Pi_{p,\hat{x}} U^{\varepsilon,\,K}(\hat{x},y) - (\Pi_{p,\hat{x}} \otimes \mathcal{I}_{h,y}) U^{\varepsilon,\,K}(\hat{x},y) \right) \right|^2 d\hat{x}dy$$
$$\le CH^{-n} h^{2\min(\mu,s)} \Phi_2^2(\mu,s) \|U^\varepsilon\|^2_{H^n(K;\, H^{s+1}_{\rm per}(\widehat{Q}))}.$$

Summing up over all elements $K \in \mathcal{T}_H$ we obtain that

$$\|e^\varepsilon_{\mathcal{I}}\|_{L^2(\Omega_\varepsilon)} \le C\big(H^{\min(p,k)+1} \Phi_2(p,k) \|U^\varepsilon\|_{H^{k+1}(\Omega;\, L^2_{\rm per}(\widehat{Q}))}$$
$$+ h^{\min(\mu,s)} \Phi_2(\mu,s) \|U^\varepsilon\|_{H^n(\Omega;\, H^{s+1}_{\rm per}(\widehat{Q}))}\big).$$

□

A similar result can be derived also for the energy norm of the two scale interpolation error. To this end, we estimate the $L^2(\Omega)$-norm of $\nabla_x e^\varepsilon_{\mathcal{I}}$ in terms of the regularity of the data and of the 'macro', resp. 'micro' triangulations .

Proposition 10. *Assume that $n = 2$, $k, s \ge 1$ and $H/\varepsilon \in N$. Then it holds for any $p, \mu \ge 1$*

$$\|\nabla_x e^\varepsilon_{\mathcal{I}}(x)\|_{L^2(\Omega_\varepsilon)} \le C\, H^{\min(p,k)} \Phi_2(p,k) \Big(\|\varepsilon^{-1} \nabla_y U^\varepsilon\|_{H^k(\Omega;\, L^2_{\rm per}(\widehat{Q}))} + \|U^\varepsilon\|_{H^{k+1}(\Omega;\, L^2_{\rm per}(\widehat{Q}))} \Big)$$
$$+ C\, h^{\min(\mu,s)} \Phi_2(\mu,s) \|\varepsilon^{-1} \nabla_y U^\varepsilon\|_{H^n(\Omega;\, H^s_{\rm per}(\widehat{Q}))}. \tag{25}$$

Similar error estimates for the interpolation error as in Propositions 9, 10 can be obtained in the case $n = 3$.

Theorem 11. *Assume for the solution u^ε of (7) the two-scale regularity (15)–(16) in Ω_ε. Then, the error in the two-scale FEM based on the space (22) can be estimated as follows:*

$$\|u^\varepsilon - u_{FE}\|_{H^1(\Omega_\varepsilon)} \le C_1(k) H^{\min(p,k)} \Phi_2(p,k) \|f\|_{H^k(\Omega)}$$
$$+ C_2(s) h^{\min(\mu,s)} \Phi_2(\mu,s) \|f\|_{H^{n+s}(\Omega)}.$$

Proof. The proof is a direct consequence of Theorem 1 and Propositions 9, 10. □

The previous bounds allow to deduce the convergence rates h and p. Under analyticity assumptions, even exponential convergence results are possible.

Remark 12. Suppose that the solution $U^\varepsilon(x,y)$ is patch-wise analytic on the 'macro' level and analytic on the 'micro' scale; more precisely,

$$\|D^k U^\varepsilon(x,y)\|_{L^2(K;\, L^2(\widehat{Q}))} \le C(d_K)^k k!\, |K|^{1/2}$$
$$\|\varepsilon^{-1} \nabla_y D^k U^\varepsilon(x,y)\|_{L^2(K;\, L^2(\widehat{Q}))} \le C(d_K)^k k!\, |K|^{1/2}$$

hold with constants independent of ε. In this case the estimates in Propositions 9, 10 lead to exponential convergence.

Remark 13. The convergence estimates in Theorem 11 are robust in ε. However, this robustness comes at a price: for $\Omega \subset \mathbb{R}^n$ the number N of degrees of freedom in the two scale FE space (22) grows, as $h = H \to 0$ at fixed p, μ for example, asymptotically as $O(h^{-2n})$. With the two-scale FEM based on (22), scale resolution and ε independent convergence is achieved by inflating the dimension of the approximation: we resolve the fine scales by simultaneously approximating in $(x, y) \in \Omega \times \widehat{Q} \subset \mathbb{R}^{2n}$. Tensor product approximations represent full interactions between scales. The product structure of $\Omega \times \widehat{Q}$ and the anisotropic regularity in Theorem 1 allow, however, to get the convergence in Theorem 11 with substantially fewer degrees of freedom: the scale interaction is 'thinned out' by means of *sparse tensor products.*

4 Implementation of Two-Scale FEM

In order to obtain an efficient algorithm it is essential that the element stiffness and mass matrices can be computed in a complexity independent of ε and to an accuracy which will not compromise the asymptotic convergence rates in Theorem 11. Due to the rapid oscillations of the coefficients and of the micro-shapefunctions, the elemental stiffness matrices on the macro mesh can *not* be evaluated robustly by standard quadratures, or if the macro mesh $\mathcal{T}_H$ is not aligned with the periodic pattern. If $\mathcal{T}_H$ *is* aligned, however, the macro stiffness and mass matrices can be developed from *moments*, i.e., from integrals in the fast variable corresponding to discretization of the unit-cell problem with monomial weighted coefficients, combined with certain lattice summation formulas. We will explain this in Section 4.1 and present in Section 4.2 numerical experiments confirming our error analysis.

Proposition 14. *For any $\varepsilon > 0$ and for any finite dimensional subspace $\mathcal{M}^{\mu}_{\text{per}}(\widehat{Q})$ of $H^1_{\text{per}}(\widehat{Q})$, with $\mathcal{M}^{\mu}_{\text{per}}(\widehat{Q}) = Span\{\Phi_i(y)\}_{i=1}^{\mu}$ of dimension μ independent of ε, the FEM with respect to the two-scale space $S^p(\Omega, \mathcal{T}_H;\ \mathcal{M}^{\mu}_{\varepsilon,\text{per}}(\varepsilon\widehat{Q}))$ ($\mathcal{M}^{\mu}_{\varepsilon,\text{per}}(\varepsilon\widehat{Q}) = Span\{\Phi_i(x/\varepsilon)\}_{i=1}^{\mu}$) can be implemented with a computational work independent of ε.*

The ε-independence is achieved by judiciously exploiting the periodicity in the fast variable.

4.1 Development of Macroelement Stiffness Matrix

We start from the discrete variational formulation:
Find $u \in S^p(\Omega, \mathcal{T}_H;\ \mathcal{M}^{\mu}_{\varepsilon,\text{per}}(\varepsilon\widehat{Q}))$ such that

$$B(u, v) = \int_{\Omega} f v\, dx \quad \forall v \in S^p(\Omega, \mathcal{T}_H; \mathcal{M}^{\mu}_{\varepsilon}(\varepsilon\widehat{Q})),$$

where $\mathcal{M}^{\mu}_{\mathrm{per}}(\widehat{Q}) = \mathrm{Span}\,\{\Phi_i\}$ is any conforming FE discretization of $H^1_{\mathrm{per}}(\widehat{Q})$. For $u, v \in S^p(\Omega, \mathcal{T}_H; \mathcal{M}^{\mu}_{\varepsilon}(\varepsilon\widehat{Q}))$ the bilinear form can be split in a sum of elemental bilinear forms B_K

$$B(u,v) = \sum_{K\in\mathcal{T}_H} B_K(u,v).$$

For each element K of the 'macro' triangulation $\mathcal{T}_H$ with 'macroscopic' polynomial space $S^p(K) = \mathrm{Span}\,\{\nu_I^{[K]}\}_I$, the elemental bilinear form B_K can be written in terms of the reference element matrix

$$B_K(u,v) = \underline{v}^{\top}\underline{\underline{\mathbf{K}}}^{[K]}\underline{u}, \quad \underline{u} = \{u_{Ii}\}, \quad \underline{v} = \{v_{Ii}\},$$

where $u(x)\big|_K = \sum_{I,i} u_{Ii}\nu_I^{[K]}(x)\Phi_i(x/\varepsilon)$ and $v(x)\big|_K = \sum_{J,j} v_{Jj}\nu_J^{[K]}(x)\Phi_j(x/\varepsilon)$. The entries of the element stiffness matrix $\underline{\underline{\mathbf{K}}}^{[K]}$ are given by

$$\begin{aligned}\underline{\underline{\mathbf{K}}}^{[K]}_{(Ii)\,(Jj)} &= \int_K a\left(\frac{x}{\varepsilon}\right)\left(\nu_I^{[K]}(x)\Phi_i\left(\frac{x}{\varepsilon}\right)\right)'\left(\nu_J^{[K]}(x)\Phi_j\left(\frac{x}{\varepsilon}\right)\right)' dx \\ &+ \int_K a_0\left(\frac{x}{\varepsilon}\right)\nu_I^{[K]}(x)\Phi_i\left(\frac{x}{\varepsilon}\right)\nu_J^{[K]}(x)\Phi_j\left(\frac{x}{\varepsilon}\right) dx,\end{aligned} \tag{26}$$

where a prime denotes $\frac{d}{dx}$. Without loss of generality we assume now that $K = (0, H)$, with $M := H/\varepsilon \in \mathbb{N}$. For simplicity, we consider only the first integral term in (26). Since $K = \cup_{m=0}^{M-1} K_m$, with $K_m = \varepsilon(m + \widehat{Q})$ we obtain that

$$\begin{aligned}\underline{\underline{A}}^{[K]}_{(Ii)\,(Jj)} &= \int_K a\left(\frac{x}{\varepsilon}\right)\left(\nu_I^{[K]}(x)\Phi_i\left(\frac{x}{\varepsilon}\right)\right)'\left(\nu_J^{[K]}(x)\Phi_j\left(\frac{x}{\varepsilon}\right)\right)' dx \\ &= \sum_{\gamma,\delta\le 1}\sum_{\alpha} c^{IJ}_{\gamma\delta\alpha}\,\varepsilon^{n-(\gamma+\delta)+\alpha}\sum_{m=0}^{M-1}\int_{\widehat{Q}} a(\hat{y})\Phi_i^{(\gamma)}(\hat{y})\Phi_j^{(\delta)}(\hat{y})\,(\hat{y}+m)^{\alpha}\,d\hat{y},\end{aligned}$$

with suitable constants $c^{IJ}_{\gamma\delta\alpha} = c^{IJ}_{\gamma\delta\alpha}(K)$ depending only on $I, J, \alpha, \gamma, \delta$ and the element K. We see that for the calculation of the two-scale element stiffness matrices the basic integrals

$$\underline{\underline{\widehat{K}}}^{\gamma\delta\tau}_{\mu} = \left(\int_{\widehat{Q}} a(\hat{y})\Phi_i^{(\gamma)}(\hat{y})\Phi_j^{(\delta)}(\hat{y})\hat{y}^{\tau}\,d\hat{y}\right)_{i,j=1,\dots,\mu} \tag{27}$$

are needed. Let us remark that (27) when $\tau = 0$ and $\delta = \gamma = 1$, corresponds to the global stiffness matrix of the unit cell problem discretized with $\mathcal{M}^{\mu} = \mathrm{Span}\{\Phi_i \,|\, i = 1,\dots,\mu\}$. When $\tau > 0$ we obtain a scale interaction stiffness matrix and a discretization of the unit cell problem with monomial weight functions is generally needed.

The entries $\underline{\underline{A}}^{[K]}_{(Ii)\,(Jj)}$ of the element stiffness matrix are ultimately given by

$$\begin{aligned}&\sum_{\gamma,\delta\le 1}\sum_{\alpha} c^{IJ}_{\gamma\delta\alpha}\sum_{\tau\le\alpha}\left(\underline{\underline{\widehat{K}}}^{\gamma\delta\tau}_{\mu}\right)_{ij}\binom{\alpha}{\tau}\sum_{m=0}^{M-1} m^{\alpha-\tau} \\ &= \sum_{\gamma,\delta\le 1}\sum_{\alpha}\sum_{\tau\le\alpha}\left(\underline{\underline{\widehat{K}}}^{\gamma\delta\tau}_{\mu}\right)_{ij}\sum_{m=0}^{M-1} S^{IJ}_{\gamma\delta\alpha\tau}(m, H, \varepsilon),\end{aligned}$$

with $\sum_{m=0}^{M-1} S^{IJ}_{\gamma\delta\alpha\tau}(m, H, \varepsilon)$ being directly computable.

4.2 Numerical results

We illustrate our error estimates for the two-scale FEM for the model problem

$$-\frac{d}{dx}\left(a\left(\frac{x}{\varepsilon}\right)\frac{du^\varepsilon}{dx}(x)\right) = f(x) \quad \text{in } \Omega = (0,1), \qquad u^\varepsilon|_{\partial\Omega} = 0, \tag{28}$$

where $f(x) = 1$ and

$$a(y) = 2 + \cos(2\pi y).$$

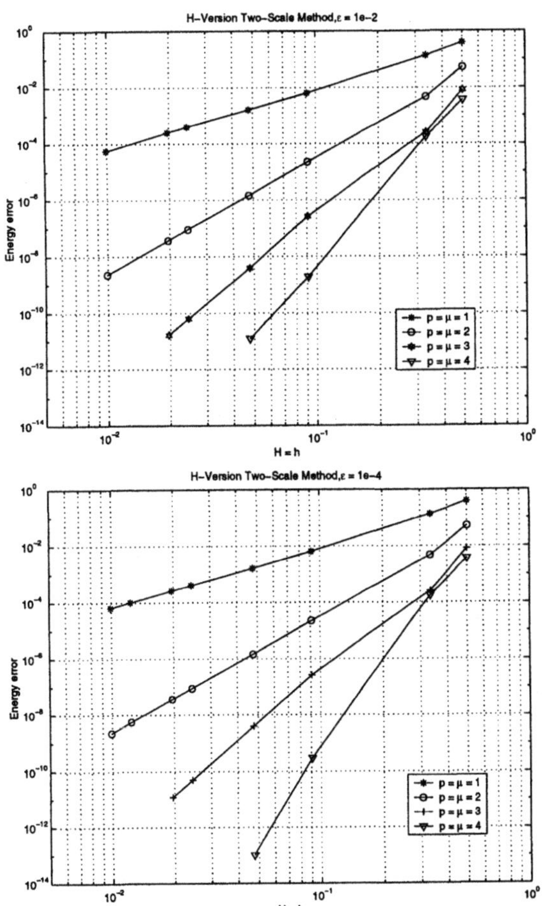

Fig. 3. Energy error in the H-Version of the two-scale FEM

The shift Theorem 1 applies in this case on Ω and the solution does not exhibit boundary layers, since the scales are separated in the following sense: $u^\varepsilon(x) = U^\varepsilon(x, x/\varepsilon)$, with $U^\varepsilon(x, y)$ smooth on $\Omega \times \widehat{Q}$ and 1-periodic in y.

In Figure 3 we plot the energy error versus $H = h$ and for different $p = \mu \in \{1, 2, 3, 4\}$. Computations were performed for two different ε-scales, 10^{-2} and 10^{-4}, respectively. We see that the rate of convergence of $\|u^\varepsilon - u^\varepsilon_{FE}\|^2_{H^1(\Omega)}$ is proportional to H^{2p} as expected from the error estimates in Theorem 11. Moreover, we observe robustness of the convergence rates with respect to the parameter ε.

The next set of numerical experiments shows that simultaneous refinement on both scales is indeed necessary. To that end, calculations for $\varepsilon = 10^{-4}$, $\mu = 1$ and fixed h, p were performed. In Figure 4 we plot the error in energy versus H (for several fixed p). In agreement with our *a-priori* estimates $O(H^{2p} + h^{2\mu})$ we observe a saturation effect.

We remark that for analytic or piecewise analytic $U^\varepsilon(x, y)$, it is possible to obtain even robust exponential convergence rates of the two-scale FEM. For numerical examples, we refer to [6].

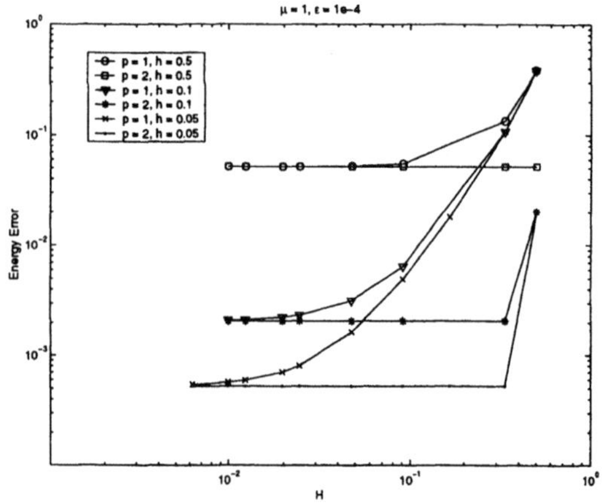

Fig. 4. Energy error versus H for fixed micro scale resolution h ($\mu = 1$, $\varepsilon = 10^{-4}$)

References

1. D. Cioranescu and Jeanine Saint Jean Paulin, "Homogenization of Reticulated Structures", Springer Applied Mathematical Sciences (1999).
2. T. Y. Hou and X. Xin, "Homogenization of Linear Transport Equations with Oscillatory Vector Fields", SIAM J. Appl. Math. **52** (1992), 34–45.
3. A. M. Matache, Spectral- and p-Finite Elements for problems with microstructure, Doctoral Dissertation ETH Zürich (2000).

4. A.M. Matache, I. Babuška and Ch. Schwab, "Generalized p-FEM in Homogenization", Numerische Mathematik **86** (2000) 319-375.
5. A.M. Matache and Ch. Schwab, Homogenization via p-FEM for Problems with Microstructure, Applied Numerical Mathematics, **33** Issue 1–4, May 2000.
6. A.M. Matache and Ch. Schwab, Two-Scale FEM for Homogenization Problems Basic Regularity and Approximation Estimates, Report, SAM, ETH Zürich, Switzerland.
7. R.C. Morgan and I. Babuška, "An approach for constructing families of homogenized solutions for periodic media, I: An integral representation and its consequences, II: Properties of the kernel", SIAM J. Math. Anal. **22** (1991) 1–33.
8. O. A. Oleinik, A.S. Shamaev, G. A. Yosifian, "Mathematical Problems in Elasticity and Homogenization", North-Holland, 1992.
9. Ch. Schwab, "p- and hp- Finite Element Methods", Oxford Science Publications, 1998.
10. Ch. Schwab, A.-M. Matache, High order generalized FEM for lattice materials, Proceedings of the 3rd European Conference on Numerical Mathematics and Advanced Applications, Finland, 1999, ed. by P. Neittaanmäki, T. Tiihonen and P. Tarvainen, World Scientific, Singapore, 2000
11. B. Szabó, I. Babuška, "Finite Element Analysis", John Wiley & Sons, Inc. 1991.

Numerical Analysis of Electromagnetic Problems

Fumio Kikuchi

Graduate School of Mathematical Sciences, University of Tokyo,
Komaba, Meguro, Tokyo, 153-8914 Japan

Abstract. We give theoretical and computational overview of numerical analysis of the finite element methods for electromagnetics. In particular, theoretical comments on the edge and face elements, frequently employed in the finite element discretizations, are given. Moreover, we present some iteration methods which are effective to solve discrete equations arising from finite element methods.

1 Introduction

Numerical analysis of electromagnetic problems is now quite important in wide fields of science and engineering. The application of the finite element method (FEM) to such ends is very effective especially for 3-D problems since FEM is well suited for complex regions with various boundary conditions. By appropriate modeling of the Maxwell equations of electromagnetics, we have a variety of problems describing electromagnetic phenomena in practice.

In FEM, we first derive weak forms to such problems, and then obtain discrete equations by the finite element procedures. In this process, we often utilize mixed formulations based on the Lagrange multiplier techniques. Then the Nedelec type edge elements and the Raviart-Thomas type face ones are very effective to approximate vector fields with their rotations and/or divergences. They are also well suited to the tangential and normal boundary conditions in electromagnetics, and their effectiveness is now widely recognized through practical experiences. Although there have been proposed various approaches without relying on such vector elements, most of them have been unsuccessful as is known the "nightmare" in computational electromagnetics.

The Raviart-Thomas face elements were proposed in [26], and they have been generalized in various fashions as summarized in [7]. On the other hand, basic edge elements of simplex or cube-based shapes were early proposed by Nedelec [23],[24], and then their generalizations such as the covariant interpolation elements have been developed [6], [27].

Once the discrete equations are obtained, they must be solved by means of appropriate computational methods. Since the equations are usually of very large-scale especially in 3-D cases, we often prefer to reliable iterative methods even in linear cases.

The outline of this note is as follows. We first present some basic electromagnetic problems and give them variational formulations, most of them are of mixed type, either explicitly or implicitly, and easy to implement as the finite element methods. Then we show the outline of theoretical numerical analysis of such finite element schemes. Among them, we give comments on the discrete compactness properties, which are discrete analogs of the compactness properties of electromagnetic spaces and difficult to check theoretically [13]. Finally, we present some computational techniques to solve discrete equations arising from the finite element discretizations.

2 Electromagnetic problems

First we will explain typical electromagnetic problems with some variational formulations.

2.1 Maxwell's equations

As is well known, the governing equations of electromagnetics are the Maxwell partial differential equations, which may be expressed as follows when considered in a 3-D domain Ω occupied by the medium:

$$\mathrm{rot}\,\boldsymbol{E} + \frac{\partial \boldsymbol{B}}{\partial t} = \boldsymbol{0}, \quad \mathrm{div}\,\boldsymbol{D} = \rho, \quad \mathrm{rot}\,\boldsymbol{H} - \frac{\partial \boldsymbol{D}}{\partial t} = \boldsymbol{j}, \quad \mathrm{div}\,\boldsymbol{B} = 0 \;\; in\; \Omega, \tag{1}$$

where "rot" and "div" are the usual differential operators, $\boldsymbol{E}$ is the electric field, $\boldsymbol{H}$ the magnetic field, $\boldsymbol{D}$ the electric flux density, $\boldsymbol{B}$ the magnetic flux density, ρ the electric charge density, $\boldsymbol{j}$ the electric current density, and t the time variable. Moreover, the following "constitutive" relations are assumed:

$$\boldsymbol{D} = \varepsilon \boldsymbol{E}, \quad \boldsymbol{B} = \mu \boldsymbol{H}, \tag{2}$$

where ε and μ are the dielectric constant and the magnetic permeability of the medium, respectively. Here, we assume that both ε and μ are positive constants for simplicity.

Of course, appropriate initial and boundary conditions must be imposed on the fields. For example, when the boundary $\partial\Omega$ of Ω corresponds to a perfectly conducting wall, the boundary conditions are given by

$$\boldsymbol{E} \times \boldsymbol{n} = \boldsymbol{0}, \quad \boldsymbol{B} \cdot \boldsymbol{n} = 0 \;\; on\; \partial\Omega, \tag{3}$$

where $\boldsymbol{n}$ is the outward unit normal on $\partial\Omega$, and $\times$ and $\cdot$ denote the operations of vector and scalar products, respectively.

2.2 Basic problems in electromagnetics

By introducing various assumptions and/or simplifications to the original Maxwell equations, we may obtain a number of model problems in electromagnetics. For simplicity, we assume that $\Omega \subset \mathbf{R}^3$ is a simply-connected bounded Lipschitz domain and that $\partial\Omega$ is connected. Then we can assure the existence of both the scalar and vector potentials [1]. We can explain some of the essential features of numerical analysis of electromagnetic problems under such fairly strong conditions on the domain as well as very special boundary conditions (3), although it is possible to relax them considerably.

Cavity resonator problem. A typical model problem is the classical cavity resonator eigenvalue problem, which is essentially to determine non-trivial time-harmonic electromagnetic fields satisfying the Maxwell equations in a bounded vacuum cavity surrounded by a perfectly conducting wall.

In this case, the Maxwell equations reduce to

$$\operatorname{rot}\boldsymbol{E} + \frac{\partial \boldsymbol{B}}{\partial t} = \boldsymbol{0}, \quad \operatorname{div}\boldsymbol{D} = 0, \quad \operatorname{rot}\boldsymbol{H} - \frac{\partial \boldsymbol{D}}{\partial t} = \boldsymbol{0}, \quad \operatorname{div}\boldsymbol{B} = 0 \ \ in\ \Omega, \tag{4}$$

with the boundary conditions (3) and the relations (2) for the vacuum:

$$\boldsymbol{D} = \varepsilon_0 \boldsymbol{E}\,, \quad \boldsymbol{B} = \mu_0 \boldsymbol{H}\,, \tag{5}$$

where $\varepsilon_0 > 0$ and $\mu_0 > 0$ are the (constant) values of ε and μ for the vacuum.

Introducing the time-harmonic assumption to the above and eliminating the common factor $e^{i\omega t}$, we have, for the spatial parts of the fields,

$$\operatorname{rot}\boldsymbol{E} = -i\omega\mu_0\boldsymbol{H}, \ \operatorname{div}\boldsymbol{E} = 0, \ \operatorname{rot}\boldsymbol{H} = i\omega\varepsilon_0\boldsymbol{E}, \ \operatorname{div}\boldsymbol{H} = 0 \ \ in\ \Omega, \tag{6}$$

$$\boldsymbol{E}\times\boldsymbol{n} = \boldsymbol{0}\,, \quad \boldsymbol{H}\cdot\boldsymbol{n} = 0 \quad on\ \partial\Omega, \tag{7}$$

where i is the imaginary unit and ω the angular frequency, and we have also used the notations of original time-dependent fields as their spatial parts.

The above equations contain two unknown vector functions $\boldsymbol{E}$ and $\boldsymbol{H}$. If we eliminate $\boldsymbol{H}$, we have the following eigenvalue problem in $\boldsymbol{E}$ only:

Cavity resonator problem: $\boldsymbol{E}$

$$\operatorname{rot}\operatorname{rot}\boldsymbol{E} = \lambda\boldsymbol{E}\,, \quad \operatorname{div}\boldsymbol{E} = 0 \ \ in\ \Omega; \quad \boldsymbol{E}\times\boldsymbol{n} = \boldsymbol{0} \ \ on\ \partial\Omega\,, \tag{8}$$

where $\lambda = \varepsilon_0\mu_0\omega^2$, which can be shown to be actually real (and positive).

Similarly, by eliminating $\boldsymbol{E}$, we have the following one in terms of $\boldsymbol{H}$:

Cavity resonator problem: $\boldsymbol{H}$

$$\operatorname{rot}\operatorname{rot}\boldsymbol{H} = \lambda\boldsymbol{H}, \ \operatorname{div}\boldsymbol{H} = 0 \ in\ \Omega; \ \boldsymbol{H}\cdot\boldsymbol{n} = 0, \ (\operatorname{rot}\boldsymbol{H})\times\boldsymbol{n} = \boldsymbol{0} \ on\ \partial\Omega. \tag{9}$$

Electrostatic and magnetostatic problems. It is also essential to determine stationary, i. e., static, states of electromagnetic fields. In such cases, all the quantities must be independent of the time variable t. Then the governing equations can be obtained by setting the time derivatives zero in (1):

$$\operatorname{rot}\boldsymbol{E} = \boldsymbol{0}, \quad \operatorname{div}\boldsymbol{D} = \rho, \quad \operatorname{rot}\boldsymbol{H} = \boldsymbol{j}, \quad \operatorname{div}\boldsymbol{B} = 0 \ \ in\ \Omega\,, \tag{10}$$

along with appropriate boundary conditions. If we employ the static version of (3) as the boundary conditions, the present problem may be separated into two, that is, the electrostatic problem and the magnetostatic one:

Electrostatic problem: $\boldsymbol{E}$, $\boldsymbol{D}$

$$\operatorname{rot}\boldsymbol{E} = \boldsymbol{0}\,,\quad \operatorname{div}\boldsymbol{D} = \rho \ \ in\ \Omega\,; \qquad \boldsymbol{E}\times\boldsymbol{n} = \boldsymbol{0}\ \ on\ \partial\Omega\,, \tag{11}$$

Magnetostatic problem: $\boldsymbol{H}$, $\boldsymbol{B}$

$$\operatorname{rot}\boldsymbol{H} = \boldsymbol{j}\,,\quad \operatorname{div}\boldsymbol{B} = 0 \ \ in\ \Omega\,; \qquad \boldsymbol{B}\cdot\boldsymbol{n} = 0\ \ on\ \partial\Omega. \tag{12}$$

In the present case, it is also possible to use the scalar potential ϕ and the vector one $\boldsymbol{A}$ to deal with the linear relations $\operatorname{rot}\boldsymbol{E} = \boldsymbol{0}$ and $\operatorname{div}\boldsymbol{B} = 0$:

Electrostatic problem: ϕ

$$\operatorname{div}(\varepsilon\operatorname{grad}\phi) = \rho \ \ in\ \Omega\,; \qquad \phi = 0\ \ on\ \partial\Omega\,, \tag{13}$$

Magnetostatic problem: $\boldsymbol{A}$

$$\operatorname{rot}(\mu^{-1}\operatorname{rot}\boldsymbol{A}) = \boldsymbol{j},\quad \operatorname{div}\boldsymbol{A} = 0 \ \ in\ \Omega\,; \qquad \boldsymbol{A}\times\boldsymbol{n} = \boldsymbol{0}\ \ on\ \partial\Omega\,. \tag{14}$$

Here the Coulomb gauge is imposed on $\boldsymbol{A}$.

As may be easily seen from the above, the scalar potential formulation leads to a boundary-value problem of a Poisson-like equation, which can be dealt with by the classical FEM. On the other hand, the vector potential formulation gives a fully vector-based system of equations.

There are various other electromagnetic problems such as the eddy current problems, the forced vibration of dielectric media appearing in the design of microwave ovens, etc. Fully time-dependent analysis is of course possible, but it inevitably becomes to be of quite large-scale.

2.3 Function spaces for electromagnetics

Besides the usual spaces $L_2(\Omega)$, $H^1(\Omega)$ and $H_0^1(\Omega)$, we also use some function spaces to express our electromagnetic problems mathematically:

$$H(\mathrm{rot};\Omega) = \{\boldsymbol{u}\in L_2(\Omega)^3;\ \operatorname{rot}\boldsymbol{u}\in L_2(\Omega)^3\}\,, \tag{15}$$

$$H_0(\mathrm{rot};\Omega) = \{\boldsymbol{u}\in H(\mathrm{rot};\Omega);\ \boldsymbol{u}\times\boldsymbol{n} = \boldsymbol{0}\ on\ \partial\Omega\}\,, \tag{16}$$

$$H(\mathrm{rot}^0;\Omega) = \{\boldsymbol{u}\in H(\mathrm{rot};\Omega);\ \operatorname{rot}\boldsymbol{u} = \boldsymbol{0}\}\,, \tag{17}$$

$$H_0(\mathrm{rot}^0;\Omega) = H_0(\mathrm{rot};\Omega)\cap H(\mathrm{rot}^0;\Omega)\,, \tag{18}$$

$$H(\mathrm{div};\Omega) = \{\boldsymbol{u}\in L_2(\Omega)^3;\ \operatorname{div}\boldsymbol{u}\in L_2(\Omega)\}\,, \tag{19}$$

$$H_0(\mathrm{div};\Omega) = \{\boldsymbol{u}\in H(\mathrm{div};\Omega);\ \boldsymbol{u}\cdot\boldsymbol{n} = 0\ on\ \partial\Omega\}\,, \tag{20}$$

$$H(\mathrm{div}^0;\Omega) = \{\boldsymbol{u}\in H(\mathrm{div};\Omega);\ \operatorname{div}\boldsymbol{u} = 0\}\,, \tag{21}$$

$$H_0(\mathrm{div}^0;\Omega) = H_0(\mathrm{div};\Omega)\cap H(\mathrm{div}^0;\Omega)\,. \tag{22}$$

These spaces become Hilbert spaces by equipping them with appropriate inner products, and are effective to describe electromagnetic problems for homogeneous media. Moreover, we will use $(\cdot,\cdot)$ and $\|\cdot\|$ as notations of the inner products and the norms of both $L_2(\Omega)$ and $L_2(\Omega)^3$. For details of the above spaces, especially the definitions of boundary conditions, see e. g. [7],[14].

2.4 Variational formulations

Let us present some weak or variational formulations to the basic electromagnetic problems stated above.

Cavity resonator problem. First we have the following variational formulations by applying the standard approaches to (8) and (9).

$[\mathrm{CR1}]_E$ *Find* $\{\lambda, \boldsymbol{E}\} \in \mathbf{R}\times\{H_0(\mathrm{rot};\Omega)\cap H(\mathrm{div}^0;\Omega)\}$ *such that* $\boldsymbol{E} \neq \mathbf{0}$ *and*

$$(\mathrm{rot}\,\boldsymbol{E}, \mathrm{rot}\,\boldsymbol{E}^*) = \lambda(\boldsymbol{E}, \boldsymbol{E}^*)\,;\ \forall \boldsymbol{E}^* \in H_0(\mathrm{rot};\Omega)\,. \tag{23}$$

$[\mathrm{CR1}]_H$ *Find* $\{\lambda, \boldsymbol{H}\} \in \mathbf{R}\times\{H(\mathrm{rot};\Omega)\cap H_0(\mathrm{div}^0;\Omega)\}$ *such that* $\boldsymbol{H} \neq \mathbf{0}$ *and*

$$(\mathrm{rot}\,\boldsymbol{H}, \mathrm{rot}\,\boldsymbol{H}^*) = \lambda(\boldsymbol{H}, \boldsymbol{H}^*)\,;\ \forall \boldsymbol{H}^* \in H(\mathrm{rot};\Omega)\,. \tag{24}$$

It is noted that the trial spaces are different from the test ones, but we can take the test ones to the trial ones without essentially changing the problems [16]. Sometimes $\boldsymbol{u}$ will be used instead of $\boldsymbol{E}$ and $\boldsymbol{H}$ in what follows.

It is easy to see that the present eigenvalue problems may be considered those of symmetric bounded non-negative operators. Moreover, for the present Ω, we have the compact imbedding properties:

$$V_E\,,\ V_H \subset\ L_2(\Omega)^3\ \ (\textit{compactly})\,, \tag{25}$$

where

$$V_E := H_0(\mathrm{rot};\Omega)\cap H(\mathrm{div};\Omega),\quad V_H := H(\mathrm{rot};\Omega)\cap H_0(\mathrm{div};\Omega). \tag{26}$$

See Amrouche et al. [1] for details.

Based on the above and the spectral theory in Hilbert spaces, we have various nice properties on the eigenvalues and eigenspaces, well-known for symmetric positive compact operators. That is, our model problem is a standard and nice one from purely analytical standpoint under (25).

In $[\mathrm{CR1}]_E$ and $[\mathrm{CR1}]_H$, the divergence-free conditions for $\boldsymbol{u} \in H_0(\mathrm{rot};\Omega)$ ($H(\mathrm{rot};\Omega)$, resp.) can be expressed weakly as

$$(\boldsymbol{u}, \mathrm{grad}\,\varphi) = 0\ ;\ \forall\varphi \in H_0^1(\Omega)\ \left(H^1(\Omega),\ \textit{resp.}\right). \tag{27}$$

More precisely, these are equivalent to $\boldsymbol{u} \in H(\mathrm{div}^0; \Omega)$ ($H_0(\mathrm{div}^0; \Omega)$ resp.) for $\boldsymbol{u} \in H_0(\mathrm{rot}; \Omega)$ ($H(\mathrm{rot}; \Omega)$, resp.). Then we can see that each $\boldsymbol{u}$ of $[\mathrm{CR1}]_E$ or $[\mathrm{CR1}]_H$ for $\lambda \neq 0$ satisfies (27) even when they are not required beforehand. At the same time, the problems then become eigenvalue problems for non-compact operators. In fact, the eigenspaces associated to $\lambda = 0$ without (27) become infinite-dimensional: for the present Ω, they are spanned by $\mathrm{grad}\,\varphi$ for $\varphi \in H_0^1(\Omega)$ or $\varphi \in H^1(\Omega)$. It is also not difficult to show the equivalence between $[\mathrm{CR1}]_E$ and $[\mathrm{CR1}]_H$ under natural correspondence between $\boldsymbol{E}$ and $\boldsymbol{H}$ suggested by (6), cf. [18].

It is now natural to use the Lagrange multipliers to deal with the divergence-free conditions (27) as linear constraints. Choosing the Lagrange multiplier p from $H_0^1(\Omega)$, we have the following mixed variational formulation for $[\mathrm{CR1}]_E$:

$[\mathrm{CR2}]_E$ *Find* $\{\lambda, \boldsymbol{E}, p\} \in \mathbf{R} \times H_0(\mathrm{rot}; \Omega) \times H_0^1(\Omega)$ *such that* $\boldsymbol{E} \neq \mathbf{0}$ *and*

$$(\mathrm{rot}\,\boldsymbol{E}, \mathrm{rot}\,\boldsymbol{E}^*) + (\mathrm{grad}\,p, \boldsymbol{E}^*) = \lambda(\boldsymbol{E}, \boldsymbol{E}^*)\ ; \forall \boldsymbol{E}^* \in H_0(\mathrm{rot}; \Omega), \tag{28}$$

$$(\boldsymbol{E}, \mathrm{grad}\,q) = 0\ ;\ \forall q \in H_0^1(\Omega). \tag{29}$$

Similar formulation may be derived for $[\mathrm{CR1}]_H$. Substituting $\mathrm{grad}\,q$ for $q \in H_0^1(\Omega)$ into $\boldsymbol{E}^*$ of (28), we can see that $\mathrm{grad}\,p = \mathbf{0}$. That is, the multiplier p essentially vanishes, and hence it may be considered a "hidden" variable. Thus $[\mathrm{CD2}]_E$ reduces to the original formulation $[\mathrm{CR1}]_E$.

Another popular approach to deal with linear constraints is the penalty method. To apply it to our problem, we can use the function spaces V_E and V_H defined by (26). Using $s > 0$ as the penalty parameter for the divergence-free condition, we have the following variational formulation for $[\mathrm{CR1}]_E$:

$[\mathrm{CR3}]_E$ *Find* $\{\lambda, \boldsymbol{E}\} \in \mathbf{R} \times V_E$ *such that* $\boldsymbol{E} \neq \mathbf{0}$ *and*

$$(\mathrm{rot}\,\boldsymbol{E}, \mathrm{rot}\,\boldsymbol{E}^*) + s^{-1}(\mathrm{div}\,\boldsymbol{E}, \mathrm{div}\,\boldsymbol{E}^*) = \lambda(\boldsymbol{E}, \boldsymbol{E}^*)\ ;\ \forall \boldsymbol{E}^* \in V_E\ . \tag{30}$$

Similar formulation may be given for $[\mathrm{CR1}]_H$.

The present penalty formulation appears to be quite attractive since we can rely on the compactness (25). Moreover, the original eigenpairs satisfy the corresponding penalized equations, although additional spurious pairs also satisfy them and pollute the original spectrum. Such spurious ones, however, can be made diverge to infinity by letting $s \to +0$, see [18] for the selection of s. Nevertheless, our penalty approach has a serious drawback when implemented numerically as we will see later.

It is possible to give other variational formulations to the present problem, some of which use $H(\mathrm{div}; \Omega)$ or $H_0(\mathrm{div}; \Omega)$ as well, see [3],[5].

Electrostatic and magnetostatic problems. It is also possible to give various variational formulations to the electrostatic and magnetostatic problems. It is quite common to use various mixed formulations in such processes.

The electrostatic equations (11) consist of two parts, one involving the rotation and the other involving the divergence. If we apply the least square technique to the rotation part and deal with the divergence one as the constraint condition, we are naturally lead to a mixed formulation. Essentially the same idea is applicable to the magnetostatic equations (12), and we have:

$[\mathrm{ES1}]_E$ *Given* $\rho \in L_2(\Omega)$, *find* $\{\boldsymbol{E}, p\} \in H_0(\mathrm{rot};\Omega) \times H_0^1(\Omega)$ *such that*

$$(\mathrm{rot}\,\boldsymbol{E}, \mathrm{rot}\,\boldsymbol{E}^*) + (\mathrm{grad}\,p, \epsilon\boldsymbol{E}^*) = 0\,;\ \forall \boldsymbol{E}^* \in H_0(\mathrm{rot};\Omega)\,, \tag{31}$$

$$(\epsilon\boldsymbol{E}, \mathrm{grad}\,q) = -(\rho, q)\,;\ \forall q \in H_0^1(\Omega)\,. \tag{32}$$

$[\mathrm{MS1}]_H$ *Given* $\boldsymbol{j} \in L_2(\Omega)^3$, *find* $\{\boldsymbol{H}, p\} \in H(\mathrm{rot};\Omega) \times H^1(\Omega)$ *such that*

$$(\mathrm{rot}\,\boldsymbol{H}, \mathrm{rot}\,\boldsymbol{H}^*) + (\mathrm{grad}\,p, \mu\boldsymbol{H}^*) = (\boldsymbol{j}, \mathrm{rot}\,\boldsymbol{H}^*)\,;\ \forall \boldsymbol{H}^* \in H(\mathrm{rot};\Omega)\,, \tag{33}$$

$$(\mu\boldsymbol{H}, \mathrm{grad}\,q) = 0\,;\ \forall q \in H^1(\Omega)\,. \tag{34}$$

Here, some of the boundary conditions are implicitly expressed as the natural ones. Moreover, it is easy to see that $\mathrm{grad}\,p = \boldsymbol{0}$ in these formulations. For $\boldsymbol{j} \in H(\mathrm{div}^0;\Omega)$, the solution $\boldsymbol{H}$ of $[\mathrm{MS1}]_H$ is that of (12). Otherwise, it is a kind of generalized inverse solution of (12), cf. [19].

It is also possible to consider the "dual" formulations by dealing with the rotation equations as the constraint conditions:

$[\mathrm{ES2}]_D$ *Given* $\rho \in L_2(\Omega)$, *find* $\{\boldsymbol{D}, \boldsymbol{p}\} \in H(\mathrm{div};\Omega) \times H(\mathrm{rot};\Omega)$ *such that*

$$(\mathrm{div}\,\boldsymbol{D}, \mathrm{div}\,\boldsymbol{D}^*) + (\mathrm{rot}\,\boldsymbol{p}, \epsilon^{-1}\boldsymbol{D}^*) = (\rho, \mathrm{div}\,\boldsymbol{D}^*)\,;\ \forall \boldsymbol{D}^* \in H(\mathrm{div};\Omega)\,, \tag{35}$$

$$(\epsilon^{-1}\boldsymbol{D}, \mathrm{rot}\,\boldsymbol{q}) = 0\,;\ \forall \boldsymbol{q} \in H(\mathrm{rot};\Omega)\,. \tag{36}$$

$[\mathrm{MS2}]_B$ *Given* $\boldsymbol{j} \in L_2(\Omega)^3$, *find* $\{\boldsymbol{B}, \boldsymbol{p}\} \in H_0(\mathrm{div};\Omega) \times H_0(\mathrm{rot};\Omega)$ *such that*

$$(\mathrm{div}\,\boldsymbol{B}, \mathrm{div}\,\boldsymbol{B}^*) + (\mathrm{rot}\,\boldsymbol{p}, \mu^{-1}\boldsymbol{B}^*) = 0\,;\ \forall \boldsymbol{B}^* \in H_0(\mathrm{div};\Omega)\,, \tag{37}$$

$$(\mu^{-1}\boldsymbol{B}, \mathrm{rot}\,\boldsymbol{q}) = (\boldsymbol{j}, \boldsymbol{q})\,;\ \forall \boldsymbol{q} \in H_0(\mathrm{rot};\Omega)\,. \tag{38}$$

Again, some of the boundary conditions above are implicitly expressed as the natural ones, and $\mathrm{rot}\,\boldsymbol{p} = \boldsymbol{0}$ in these formulations.

We can also consider variational formulations in terms of the scalar and vecotor potentials. Since it is rather trivial to give a scalar potential formulation for electrostatics, we here present only a mixed formulation for magnetostatics in terms of the vector potential:

$[\mathrm{MS3}]_A$ *Given* $\boldsymbol{j} \in L_2(\Omega)^3$, *find* $\{\boldsymbol{A}, p\} \in H_0(\mathrm{rot};\Omega) \times H_0^1(\Omega)$ *such that*

$$(\mu^{-1}\mathrm{rot}\,\boldsymbol{A}, \mathrm{rot}\,\boldsymbol{A}^*) + (\mathrm{grad}\,p, \boldsymbol{A}^*) = (\boldsymbol{j}, \boldsymbol{A}^*)\,;\ \forall \boldsymbol{A}^* \in H_0(\mathrm{rot};\Omega)\,, \tag{39}$$

$$(\boldsymbol{A}, \mathrm{grad}\,q) = 0\,;\ \forall q \in H_0^1(\Omega)\,. \tag{40}$$

Here, $\mathrm{grad}\,p = \boldsymbol{0}$ if $\boldsymbol{j}$ satisfies the range condition $\boldsymbol{j} \in H(\mathrm{div}^0;\Omega)$ for the rotation operator over $H(\mathrm{rot};\Omega)$.

For some other variational formulations for magnetostatics, see e.g. [25].

3 Finite element approximations

Based on the variational formulations in Sec. 2, we can obtain finite element schemes by means of the Galerkin principle and triangulations of Ω.

3.1 Finite element spaces

As we have seen in various variational formulations, it appears to be quite natural to consider finite dimensional subspaces of $H^1(\Omega)$, $H(\mathrm{rot};\Omega)$ and $H(\mathrm{div};\Omega)$. So let us prepare appropriate finite element spaces $G^h \subset H^1(\Omega)$, $R^h \subset H(\mathrm{rot};\Omega)$, $D^h \subset H(\mathrm{div};\Omega)$, and also

$$G_0^h := G^h \cap H_0^1(\Omega), \quad R_0^h := R^h \cap H_0(\mathrm{rot};\Omega), \quad D_0^h := D^h \cap H_0(\mathrm{div};\Omega). \tag{41}$$

For these, we assume the existence of the both scalar and vector potentials inside the finite element spaces so that

$$\begin{aligned} &\mathrm{grad}\, G^h = R^h \cap H(\mathrm{rot}^0, \Omega), \quad \mathrm{grad}\, G_0^h = R_0^h \cap H_0(\mathrm{rot}^0, \Omega), \\ &\mathrm{rot}\, R^h = D^h \cap H(\mathrm{div}^0, \Omega), \quad \mathrm{rot}\, R_0^h = D_0^h \cap H_0(\mathrm{div}^0, \Omega). \end{aligned} \tag{42}$$

It is now well known that such desirable situation can be actually realized for appropriate combination of the nodal, edge, and face elements, cf. [4],[6].

Let us now show two simplest pairs of examples for R^h and D^h presented by Nedelec [23] and Raviart-Thomas [26] that satisfy (42) for appropriate G^h. The associated finite elements are typical edge (face, resp.) elements where edge (normal to face, resp.) values of approximate vector fields are used as the degrees of freedom. Then it is easy to assure the interelement continuity of the tangential (normal, resp.) components of the approximate vector fields, which is required for the fields to belong to $H(\mathrm{rot},\Omega)$ ($H(\mathrm{div},\Omega)$, resp.).

1) Tetrahedral element: $\boldsymbol{u}_h = (u_1, u_2, u_3) \in R^h$ in each element is given by

$$\boldsymbol{u}_h = \boldsymbol{\alpha} + \boldsymbol{\beta} \times \boldsymbol{x}\,, \tag{43}$$

where $\boldsymbol{x} = (x_1, x_2, x_3)$, and $\boldsymbol{\alpha} \in \mathbf{R}^3$ and $\boldsymbol{\beta} \in \mathbf{R}^3$ are coefficient vectors. Similarly, $\boldsymbol{p}_h = (p_1, p_2, p_3) \in D^h$ in each element is of the form

$$\boldsymbol{p}_h = \boldsymbol{\alpha}' + \beta \boldsymbol{x}\,, \tag{44}$$

where $\boldsymbol{\alpha}' \in \mathbf{R}^3$ is a coefficient vector and $\beta \in \mathbf{R}$ is a scalar coefficient. The associated G^h is the classical piecewise linear space of tetrahedral nodal element.

2) Rectangular parallelepiped element: $\boldsymbol{u}_h \in R^h$ in each element is of the form

$$\begin{aligned} &u_1 = \alpha_1 + \beta_1 x_2 + \beta_2 x_3 + \gamma_1 x_2 x_3, \quad u_2 = \alpha_2 + \beta_3 x_3 + \beta_4 x_1 + \gamma_2 x_3 x_1, \\ &u_3 = \alpha_3 + \beta_5 x_1 + \beta_6 x_2 + \gamma_3 x_1 x_2, \end{aligned} \tag{45}$$

while $\boldsymbol{p}_h = (p_1, p_2, p_3) \in D^h$ in each element is of the form

$$p_1 = \alpha_1' + \beta_1' x_1\,,\ p_2 = \alpha_2' + \beta_2' x_2\,,\ p_3 = \alpha_3' + \beta_3' x_3\,. \tag{46}$$

The associated G^h is the nodal finite element space for the popular 8-node trilinear element.

3.2 Finite element schemes

Let us present some examples of finite element schemes based on the variational formulations in 2.4 and the finite element spaces in 3.1.

Cavity resonator problem. We can give a finite element scheme for $[\mathrm{CR1}]_E$ as

$[\mathrm{CR1}]_E^h$ *Find* $\{\lambda_h, \boldsymbol{E}_h\} \in \mathbf{R} \times R_0^h$ *such that* $\boldsymbol{E}_h \neq \mathbf{0}$ *and*

$$(\mathrm{rot}\, \boldsymbol{E}_h, \mathrm{rot}\, \boldsymbol{E}_h^*) = \lambda_h (\boldsymbol{E}_h, \boldsymbol{E}_h^*)\,;\ \forall \boldsymbol{E}_h^* \in R_0^h\,. \tag{47}$$

We can easily check that $\boldsymbol{E}_h$ of $[\mathrm{CR1}]_E^h$ for $\lambda_h \neq 0$ satisfies

$$(\boldsymbol{E}_h, \mathrm{grad}\, \varphi_h) = 0 \quad ; \ \forall \varphi_h \in G_0^h\,. \tag{48}$$

Clearly this is a discrete analogs of (27), but the divergence-free condition is not satisfied strictly so that we cannot rely on the compactness (25). A similar scheme may be derived for $[\mathrm{CR1}]_H$ with similar observations.

We can also construct a mixed finite element scheme based on $[\mathrm{CR2}]_E$ by using G_0^h as the space for the approximate Lagrange multiplier p_h :

$[\mathrm{CR2}]_E^h$ *Find* $\{\lambda_h, \boldsymbol{E}_h, p_h\} \in \mathbf{R} \times R_0^h \times G_0^h$ *such that* $\boldsymbol{E}_h \neq \mathbf{0}$ *and*

$$(\mathrm{rot}\, \boldsymbol{E}_h, \mathrm{rot}\, \boldsymbol{E}_h^*) + (\mathrm{grad}\, p_h, \boldsymbol{E}_h^*) = \lambda_h (\boldsymbol{E}_h, \boldsymbol{E}_h^*)\,;\ \forall \boldsymbol{E}_h^* \in R_0^h\,, \tag{49}$$

$$(\boldsymbol{E}_h, \mathrm{grad}\, q_h) = 0\,;\ \forall q_h \in G_0^h\,. \tag{50}$$

Similarly to the continuous case, we have $\mathrm{grad}\, p_h = \mathbf{0}$ by taking $\boldsymbol{E}_h^*$ in (49) as $\mathrm{grad}\, q_h$ for $q_h \in G_0^h$, and the present mixed scheme reduces to $[\mathrm{CR1}]_E^h$.

We can also consider penalty finite element schemes based on the formulation $[\mathrm{CR3}]_E$, if we can find a finite-dimensional space V_E^h such that

$$V_E^h \subset V_E\,. \tag{51}$$

Such a space is obtainable if we use the popular "nodal" finite elements with the tangential boundary conditions approximated adequately. Then our approximation is an "internal" one and we can take full advantage of the compactness (25) for V_E. That is, each eigenpair of (30) can be well approximated by assuring appropriate approximation capabilities on V_E^h. Moreover, polluting eigenpairs may be deleted by choosing s sufficiently small.

Actually it is not easy to assure approximation capabilities. That is, if we use usual piecewise polynomial spaces, then condition (51) requires that

$$V_E^h \subset H^1(\Omega)^3. \tag{52}$$

Such a condition is satisfied when we use usual nodal elements. The serious fact is that, for general (in particular, non-convex, non-smooth) domain Ω,

$$V_E \cap H^1(\Omega)^3 \textit{ may be a proper subspace of } V_E\,, \tag{53}$$

cf. [10]. If such being the case, approximation is impossible when nodal type elements are used as above [11],[18]. Of course, when the singular parts of V_E are known and finite-dimensional, we can include them to V_E^h to obtain reasonable results. Various other attempts have been also made to use nodal elements, but the use of edge elements appears to be more effective at present.

In [18], numerical results based on the penalty approach by the piecewise linear triangular element are given for various domains. In particular, we consider pentagonal domains which can be either convex or non-convex. When it is non-convex, it has at least one reentrant corner.

The results are generally reasonable when Ω is convex or $\partial\Omega$ is smooth. In particular, we can observe typical spectral pollution when s^{-1} is close to 0, but polluting eigenvalues can be made diverge to ∞ by letting $s \to +0$ with physical eigenvalues being almost insensitive to variation of s.

On the other hand, when Ω is non-convex with non-smooth $\partial\Omega$, there are eigenvalues which do not approximate exact ones. Even in such cases, finite element schemes based on edge elements are usually robust to such geometrical singularities, and also free from the spectral pollution [16].

Electrostatic and magnetostatic problems. First we present a few finite element schemes for magnetostatics.

$[\mathrm{MS1}]_H^h$ *Given* $\boldsymbol{j} \in L_2(\Omega)^3$, *find* $\{\boldsymbol{H}_h, p_h\} \in R^h \times G^h$ *such that*

$$(\mathrm{rot}\,\boldsymbol{H}_h, \mathrm{rot}\,\boldsymbol{H}_h^*) + (\mathrm{grad}\,p_h, \mu \boldsymbol{H}_h^*) = (\boldsymbol{j}, \mathrm{rot}\,\boldsymbol{H}_h^*)\ ;\ \forall \boldsymbol{H}_h^* \in R^h, \tag{54}$$

$$(\mu \boldsymbol{H}_h, \mathrm{grad}\,q_h) = 0\ ;\ \forall q_h \in G^h. \tag{55}$$

$[\mathrm{MS2}]_B^h$ *Given* $\boldsymbol{j} \in L_2(\Omega)^3$, *find* $\{\boldsymbol{B}_h, \boldsymbol{p}_h\} \in D_0^h \times R_0^h$ *such that*

$$(\mathrm{div}\,\boldsymbol{B}_h, \mathrm{div}\,\boldsymbol{B}_h^*) + (\mathrm{rot}\,\boldsymbol{p}_h, \mu^{-1}\boldsymbol{B}_h^*) = 0\ ;\ \forall \boldsymbol{B}_h^* \in D_0^h, \tag{56}$$

$$(\mu^{-1}\boldsymbol{B}_h, \mathrm{rot}\,\boldsymbol{q}_h) = (\boldsymbol{j}, \boldsymbol{q}_h)\ ;\ \forall \boldsymbol{q}_h \in R_0^h\,. \tag{57}$$

$[\mathrm{MS3}]_A^h$ *Given* $\boldsymbol{j} \in L_2(\Omega)^3$, *find* $\{\boldsymbol{A}_h, p_h\} \in R_0^h \times G_0^h$ *such that*

$$(\mu^{-1}\mathrm{rot}\,\boldsymbol{A}_h, \mathrm{rot}\,\boldsymbol{A}_h^*) + (\mathrm{grad}\,p_h, \boldsymbol{A}_h^*) = (\boldsymbol{j}, \boldsymbol{A}_h^*)\ ;\ \forall \boldsymbol{A}_h^* \in R_0^h, \tag{58}$$

$$(\boldsymbol{A}_h, \mathrm{grad}\,q_h) = 0\ ;\ \forall q_h \in G_0^h. \tag{59}$$

For electrostatics, we present here the following one based on $[\mathrm{ES2}]_D$:

$[\mathrm{ES2}]_D^h$ *Given* $\rho \in L_2(\Omega)$, *find* $\{\boldsymbol{D}_h, \boldsymbol{p}_h\} \in D^h \times R^h$ *such that*

$$(\operatorname{div} \boldsymbol{D}_h, \operatorname{div} \boldsymbol{D}_h^*) + (\operatorname{rot} \boldsymbol{p}_h, \epsilon^{-1}\boldsymbol{D}_h^*) = (\rho, \operatorname{div} \boldsymbol{D}_h^*) \ ; \ \forall \boldsymbol{D}_h^* \in D^h, \tag{60}$$

$$(\epsilon^{-1}\boldsymbol{D}_h, \operatorname{rot} \boldsymbol{q}_h) = 0 \ ; \ \forall \boldsymbol{q}_h \in R^h \, . \tag{61}$$

4 Analysis of finite element schemes

4.1 General

In order to analyze the above mixed finite element schemes, we should check various conditions such as (cf. [7],[12],[14])

1) approximation properties for R^h, R_0^h, D^h, D_0^h, G^h and G_0^h,
2) uniform coerciveness for some bilinear forms,
3) uniform lifting properties (inf-sup conditions),
4) discrete compactness properties.

The first condition is clearly necessary and has been shown for various concrete finite element spaces. The second and the third ones are also required since our schemes are essentially of mixed type, but closely related to 4) in the present cases. On the other hand, the fourth one has been quite difficult to show until recently. We will discuss it in a little more detail later.

Once these conditions are established, it is rather a standard process to show the validity of finite element schemes given in the preceding section. Of course, we actually consider an appropriate h-family of finite element spaces associated with a family of triangulations of Ω, where h is the discretization parameter of a representative element size for each triangulation.

4.2 Discrete compactness properties

As was already noted, the discrete compactness is quite delicate. To give examples of such definitions, let us first introduce some discrete operators :

$$\operatorname{div}_h : R_0^h \to G_0^h; \ (\operatorname{div}_h \boldsymbol{v}_h, q_h) = -(\boldsymbol{v}_h, \operatorname{grad} q_h) \ (\forall \boldsymbol{v}_h \in R_0^h, \ \forall q_h \in G_0^h), \tag{62}$$
$$\operatorname{div}'_h : R^h \to G^h; \ (\operatorname{div}'_h \boldsymbol{v}_h, q_h) = -(\boldsymbol{v}_h, \operatorname{grad} q_h) \ (\forall \boldsymbol{v}_h \in R^h, \ \forall q_h \in G^h), \tag{63}$$
$$\operatorname{rot}_h : D_0^h \to R_0^h; \ (\operatorname{rot}_h \boldsymbol{q}_h, \boldsymbol{v}_h) = (\boldsymbol{q}_h, \operatorname{rot} \boldsymbol{v}_h) \ (\forall \boldsymbol{q}_h \in D_0^h, \ \forall \boldsymbol{q}_h \in R_0^h), \tag{64}$$
$$\operatorname{rot}'_h : D^h \to R^h; \ (\operatorname{rot}'_h \boldsymbol{q}_h, \boldsymbol{v}_h) = (\boldsymbol{q}_h, \operatorname{rot} \boldsymbol{v}_h) \ (\forall \boldsymbol{q}_h \in D^h, \forall \boldsymbol{v}_h \in R^h). \tag{65}$$

These operators are well-defined thanks to assumptions (42).

Now we can give two examples of discrete versions of (25) in terms of the above discrete divergence and/or rotation operators.

[DC1] *Let* $\{\boldsymbol{u}_h\}_{h>0}$ *be an arbitrary h-family such that*

$$\boldsymbol{u}_h \in R_0^h\,, \quad \|\boldsymbol{u}_h\|^2_{H(\mathrm{rot};\Omega)} + \|\mathrm{div}_h \boldsymbol{u}_h\|^2 \le 1\,. \tag{66}$$

Then there exist a subfamily, again denoted by $\{\boldsymbol{u}_h\}_{h>0}$*, and an element* $\boldsymbol{u}_0 \in H_0(\mathrm{rot};\Omega) \cap H(\mathrm{div};\Omega)$ *such that* $\boldsymbol{u}_h \to \boldsymbol{u}_0$ *weakly in* $H_0(\mathrm{rot};\Omega)$ *and strongly in* $\{L_2(\Omega)\}^3$*, and* $\mathrm{div}_h \boldsymbol{u}_h \to \mathrm{div}\, \boldsymbol{u}_0$ *weakly in* $L_2(\Omega)$*, as* $h \downarrow 0$.

[DC2] *Let* $\{\boldsymbol{p}_h\}_{h>0}$ *be an arbitrary h-family such that*

$$\boldsymbol{p}_h \in D_0^h\,, \quad \|\boldsymbol{p}_h\|^2_{H(\mathrm{div};\Omega)} + \|\mathrm{rot}_h\, \boldsymbol{p}_h\|^2 \le 1\,. \tag{67}$$

Then there exist a subfamily, again denoted by $\{\boldsymbol{p}_h\}_{h>0}$*, and an element* $\boldsymbol{p}_0 \in H(\mathrm{rot};\Omega) \cap H_0(\mathrm{div};\Omega)$ *such that* $\boldsymbol{p}_h \to \boldsymbol{p}_0$ *weakly in* $H_0(\mathrm{div};\Omega)$ *and strongly in* $L_2(\Omega)^3$*, and* $\mathrm{rot}_h\, \boldsymbol{p}_h \to \mathrm{rot}\, \boldsymbol{p}_0$ *weakly in* $L_2(\Omega)^3$*, as* $h \downarrow 0$.

In the above, the "strong-convergence" parts are essential since the "weak" ones follow from the boundedness of the families and the approximation conditions for finite element spaces.

We can give two more examples of discrete compactness. First, using R^h and div'_h in place of R_0^h and div_h in [DC1], we have a discrete analog of compact imbedding of $H(\mathrm{rot};\Omega) \cap H_0(\mathrm{div};\Omega)$ to $L_2(\Omega)^3$. Similarly, using D^h and rot'_h in place of D_0^h and rot_h in [DC2], we have a discrete analog of compact imbedding of $H_0(\mathrm{rot};\Omega) \cap H(\mathrm{div};\Omega)$ to $L_2(\Omega)^3$.

The property [DC1] was shown by the present author [17] for the simplest Nedelec elements of triangular and tetrahedral shapes in [23]. Notice here that the case $\mathrm{div}_h \boldsymbol{u}_h = 0$ is essential. Recently, Boffi [3] presented more general results for fundamental Nedelec's elements of simplex and cube-based types presented in [23] by using the Fortin operator and the results in [1],[23]. See also [8],[9] and [20] for some related results and generalizations . Thus it will be important to analyze the discrete compactness for more general edge elements including the covariant interpolation ones.

If we consider an appropriate pair of edge and face elements, the analysis of [DC2] is quite similar to that of [DC1], and follows from [DC1] in the case where $\mathrm{div}\, \boldsymbol{p}_h = 0$, cf. [22]. Furthermore, the case $\mathrm{rot}_h\, \boldsymbol{p}_h = \boldsymbol{0}$ is essential.

5 Computational techniques

There are at least two or three important factors for implementation of the proposed finite element schemes for electromagnetics. The first one is the special data structures for the edge- and face-type elements compared with the conventional node-type ones. The second is how to solve the algebraic equations arising from the discretization, which may contain indefinite matrices and are of very large-scale especially in 3-D cases. Moreover, we need to specify tangential or normal boundary conditions, but such conditions are rather easy to deal with in edge- and face-type elements.

The first factor may be coped with preparing directional segment or arc data as well as face ones, besides the usual element and node data.

5.1 Cavity resonator problem

For this problem, we should solve the matrix eigenvalue problem associated with (47), which has zero eigenvalue with very large multiplicity since condition (48) is not a priori imposed. To cope with such a difficulty, we may essentially use the subspace iteration method with shift techniques, which appears to be effective to separate non-zero eigenvalues from the zero one [2]. However, especially in 3-D analysis, we need good solvers for large-scale linear simultaneous equations in the inverse iteration step. For such a purpose, we can for example use the CG-type methods combined with Iwashita's zero filtering technique, which enables us to deal with symmetric positive-definite matrices in the inverse iteration process [15]. His idea is to consider the following form of a matrix eigenvalue problem:

$$([K] + \lambda^*[M])\{u\} = (\lambda + \lambda^*)[M]\{u\}, \tag{68}$$

where λ is the eigenvalue, $\{u\}$ the eigenvector, λ^* the shift > 0 ($\lambda^* \leq 0$ in the usual inverse iteration method), $[K]$ the stiffness matrix, and $[M]$ the mass matrix. Usually $[K] + \lambda^* [M]$ is symmetric positive definite, and hence the CG-type method is applicable to it. Furthermore, the eigenspace associated with the zero eigenvalue can be removed by the zero-filtering, which is a simple subtraction process for iteration vectors.

An alternative to be tested as an eigensolver is the Lanczos method.

5.2 Electrostatic and magnetostatic problems

The finite element schemes proposed in Sec. 3 for magneto- and electrostatics are in general of large-scale, and the coefficient matrices associated with the discrete linear equations are indefinite. Moreover, the Lagrange multipliers are also contained as unknowns although they are essentially zero. To cope with such difficulties, some iteration methods have been proposed, cf. [21].

Let us consider $[\mathrm{MS1}]_H^h$ as an example. Let τ be a positive parameter, and consider the following perturbation problem [21].

$[\mathrm{MS1}]_H^{h;\tau}$ *Given* $\boldsymbol{j} \in L_2(\Omega)^3$ *and* $\tau > 0$, *find* $\{\boldsymbol{H}_h^\tau, p_h\} \in R^h \times G^h$ *such that*

$$(\mathrm{rot}\,\boldsymbol{H}_h^\tau, \mathrm{rot}\,\boldsymbol{H}_h^*) + \tau(\mu\boldsymbol{H}_h^\tau, \boldsymbol{H}_h^*) + (\mathrm{grad}\,p_h, \mu\boldsymbol{H}_h^*) = (\boldsymbol{j}, \mathrm{rot}\,\boldsymbol{H}_h^*)\,; \quad \forall \boldsymbol{H}_h^* \in R^h, \tag{69}$$

$$(\mu\boldsymbol{H}_h^\tau, \mathrm{grad}\,q_h) = 0\ ;\ \forall q_h \in G^h. \tag{70}$$

Then we find that $\mathrm{grad}\,p_h = \mathbf{0}$, so that we have the equation in $\boldsymbol{H}_h^\tau$ only:

$$(\mathrm{rot}\,\boldsymbol{H}_h^\tau, \mathrm{rot}\,\boldsymbol{H}_h^*) + \tau(\mu\boldsymbol{H}_h^\tau, \boldsymbol{H}_h^*) = (\boldsymbol{j}, \mathrm{rot}\,\boldsymbol{H}_h^*)\ ;\ \forall \boldsymbol{H}_h^* \in R^h. \tag{71}$$

If τ is small (but not too small to cause numerical instability), we can show that $\boldsymbol{H}_h^\tau$ is close to the unperturbed $\boldsymbol{H}_h$ under some conditions on the finite element spaces. Clearly, the associated coefficient matrix is now positive definite. Moreover, we may consider an iteration method to correct $\boldsymbol{H}_h^\tau$:

$$(\mathrm{rot}\,\boldsymbol{H}_h^{(k)}, \mathrm{rot}\,\boldsymbol{H}_h^*) + \tau(\mu\boldsymbol{H}_h^{(k)}, \boldsymbol{H}_h^*) = (\boldsymbol{j}, \mathrm{rot}\,\boldsymbol{H}_h^*) + \tau(\mu\boldsymbol{H}_h^{(k-1)}, \boldsymbol{H}_h^*)\,;$$
$$\forall \boldsymbol{H}_h^* \in R^h \quad (k = 1, 2, 3, \ldots). \quad (72)$$

For the analysis of such a perturbation problem and its iterative version, the discrete compactness properties again play essential roles [21].

Essentially the same approach is available to solve $[\mathrm{MS3}]_A^h$ and $[\mathrm{ES2}]_D^h$. To deal with $[\mathrm{MS3}]_A^h$ for general $\boldsymbol{j} \in L_2(\Omega)^3$, we first obtain p_h and modify $\boldsymbol{j}$ so that the new p_h becomes zero. More specifically, we obtain for p_h that

$$(\mathrm{grad}\, p_h, \mathrm{grad}\, q_h) = (\boldsymbol{j}, \mathrm{grad}\, q_h)\ ;\ \forall q_h \in G_0^h, \quad (73)$$

which can be solved by the standard methods based on nodal elements. Then we modify $\boldsymbol{j}$ as $\boldsymbol{j} - \mathrm{grad}\, p_h$, for which the new Lagrange multiplier becomes zero and hence the afore-mentioned approach is available.

6 Concluding remarks

We have given overview of numerical analysis of electromagnetics by the finite element method. It appears that theoretical and computational basis of such approaches become considerably firm, but further study appears still necessary. Theoretically, we should establish discrete compactness for general (curved, covariant) edge and face elements. Computationally, development of appropriate iteration methods is desirable although it is not easy because the arising matrices are frequently indefinite.

We have been mainly concerned with finite element schemes developed around the present author, but there may exist various other effective approaches for computational electromagnetics. One important subject is to develop effective finite elements well suited to electromagnetic fields and based on sound theoretical basis, and another is to obtain nice variational formulations convenient for large-scale computations and adaptive error controls. Of course these two are closely related to each other, and it seems that the role of numerical analysts will become increasingly important in the new century.

References

1. Amrouche, C., Bernardi, C., Dauge, M., Girault, V.: Vector potentials in three-dimensional non-smooth domains. Math. Meth. Appl. Sci. **21** (1998) 823-864
2. Bathe, K.-J.: Finite Element Procedures. Prentice-Hall (1996)
3. Boffi, D.: Fortin operator and discrete compactness for edge elements. Numer. Math. **87** (2000) 229-246

4. Boffi, D.: A note on the de Rahm complex and a discrete compactness property. Appl. Math. Lett. **14** (2001) 33-38
5. Boffi, D., Fernandes, P., Gastaldi, L., Perugia, I.: Computational models of electromagnetic resonators : analysis of edge element approximation. SIAM J. Numer. Anal. **36** (1999) 1264-1290
6. Bossavit, A.: Computational Electromagnetism : Variational Formulations, Complementary, Edge Elements. Academic Press (1998)
7. Brezzi, F., Fortin, M.: Mixed and Hybrid Finite Element Methods. Springer-Verlag (1991)
8. Caorsi, S., Fernandes, P., Raffetto, M.: On the convergence of Galerkin finite element approximations of electromagnetic eigenproblems. SIAM J. Num. Anal. **38** (2000) 580-607
9. Caorsi, S., Fernandes, P., Raffetto, M.: Approximation of electromagnetic eigenproblems: a general proof of convergence for edge finite elements of any order of both Nedelec's families. CNR-IMA, Genova, Italy, Technical Report No. 16/99 (1999)
10. Costabel, M.: A coercive bilinear form for Maxwell's equations. J. Math. Anal. Appl. **157** (1991) 527-541
11. Costabel, M., Dauge, M.: Maxwell and Lamé eigenvalues on polyhedra. Math. Meth. Appl. Sci. **22** (1999) 243-258
12. Descloux, J., Nassif, N., Rappaz, J.: On spectral approximation. Part 2. Error estimates for the Galerkin method. RAIRO, Analyse Numerique **12** (1978) 113-119
13. Fernandes, P., Raffetto, M.: The question of spurious modes revisited. Int. Compumag Soc. Newsletter **7**(1) (2000) 5-8
14. Girault, V., Raviart, P.-A.: Finite Element Methods for Navier-Stokes Equations. Springer-Verlag (1986)
15. Iwashita, Y.: General eigenvalue solver for large sparse symmetric matrix with zero filtering. Bull. Inst. Chem. Res., Kyoto Univ. **67** (1989) 32-39
16. Kikuchi, F.: Mixed and penalty formulations for finite element analysis of an eigenvalue problem in electromagnetism. Computer Meth. Appl. Mech. Engng **64** (1987) 509-521
17. Kikuchi, F.: On a discrete compactness property for the Nedelec finite elements. J. Fac. Sci., Univ. Tokyo, Sect. IA Math. **36** (1989) 479-490
18. Kikuchi, F.: Weak formulations for finite element analysis of an electromagnetic eigenvalue problem. Sci. Papers of Coll. Arts & Sci., The Univ. Tokyo **38** (1989) 43-67
19. Kikuchi, F.: Numerical analysis of electrostatic and magnetostatic problems. Sugaku Expositions **6** (1993) 33-51
20. Kikuchi, F.: Theoretical analysis of Nedelec's edge elements. (to appear in Jap. J. Indust. Appl. Math.)
21. Kikuchi, F., Fukuhara, M.: An iterative method for finite element analysis of magnetostatic problems. Advances in Numerical Mathematics (Eds.: Ushijima, T., Shi, Z.-C., Kako, T.). Kinokuniya (1995) 93-105
22. Kikuchi, F., Yamamoto, M., Fujio, H.: Theoretical and computational aspects of Nedelec's edge elements for electromagnetics. Computational Mechanics – New Trends and Applications (Eds.: Oñate, E., Idelsohn, S. R.). ©CIMNE, Barcelona, Spain (1998)
23. Nedelec, J.-C.: Mixed finite elements in $\mathbf{R}^3$. Numer. Math. **35** (1980) 315-341

24. Nedelec, J.-C.: A new family of mixed finite elements in $\mathbf{R}^3$. Numer. Math. **50** (1986) 57-81
25. Perugia, I.: A mixed formulation for 3D magnetostatic problems: theoretical analysis and face-edge finite element approximation. Numer. Math. **84** (1999) 305-326
26. Raviart, P. A., Thomas, J. M.: A mixed finite element method for second order elliptic problems. Mathematical Aspects of the Finite Element Method (Eds.: Galligani, I., Magenes, E.). Lecture Notes in Math. **606** Springer-Verlag (1977)
27. Silvester, P. P., Ferrari, R. L.: Finite Elements for Electrical Engineers. 3rd edn. Cambridge Univ. Press (1996)

A-priori Domain Decomposition of PDE Systems and Applications

S. Delpino[1], J.L. Lions[2], and O. Pironneau[3]

[1] UPMC, Analyse Numérique
[2] Collège de France, 3 rue d'Ulm, 75005 Paris
[3] UPMC, Analyse Numérique, 75252 Paris. email:pironneau@ann.jussieu.fr

Abstract. Domain Decomposition has been extensively studied as a tool for parallel computing. But in many cases the problem posed includes domain decomposition in its statement. For these the necessary numerical analysis is different because domain decomposition is not only at the discrete level but also on the continuous problem. Therefore non-matching grids for their numerical solutions is more natural, but requires new error estimates.
Our main purpose is to compute with the data of Virtual Reality. In this paper we shall review earlier works, including our own[9][10][11] and we shall present the project **freefem3d**.

1 A-Priori Domain Decomposition

The Domain Decomposition Method (DDM) has been invented mostly as an acceleration technique for parallel computations on multi-processors machines or networks. For example let $(\cdot,\cdot)$ denote the scalar product in $L^2(\Omega)$ and $a(u,v) = (\nabla u, \nabla v)$. In [15] for instance, it is seen that a model problem such as

$$a(u,\hat{u}) = (f,\hat{u}) \quad \forall \hat{u} \in V \equiv H_0^1(\Omega), \quad u - g \in V, \tag{1}$$

is first discretized, and then decomposed into sub-problems. So DDM appears as a solution method.
With the P^1 finite element method[3], (1) is discretized as

$$\int_{\Omega_h} (\nabla u \nabla v - fv) = 0 \;\; \forall v \in V_{0h} \qquad u - g_h \in V_{0h} \tag{2}$$

where V_h is the space of continuous P^1 functions on a triangulation Ω_h of Ω, and V_{0h} is the subset of such functions which are zero on the boundary.
The Schwarz algorithm rely on a decomposition of Ω_h into $\Omega_h = \Omega_{1_h} \cup \Omega_{2_h}$ with $\Omega_{1_h} \cap \Omega_{2_h} \neq \emptyset$ where all sets are unions of triangles of the triangulation of Ω. Then the solution of (2) is the limit of the following loop which starts with $u^0 = g_h$

$$\int_{\Omega_{i_h}} (\nabla u_i^{n+1} \nabla v - fv) = 0 \;\; \forall v \in V_{i_{0h}}$$

$$u_i^{n+1} - u_j^n \in V_{i0h}, \qquad i = 1,2, \quad j = i+1 \bmod 2 \tag{3}$$

where V_{i_h} is the space of continuous P^1 functions on the triangulation of Ω_{i_h}.

Note that the same method works on the continuous problem[13]:

$$-\Delta u_i^{n+1} = f \quad \text{in } \Omega_i \qquad u_i^{n+1} - u_j^n \in H_0^1(\Omega_i), \ i = 1,2, \ j = i+1 \bmod 2 \tag{4}$$

and that it can be discretized on non-compatible meshes, but then the convergence is more difficult to establish. We recall a result established in [7] on a modified version of the Schwarz algorithm.

2 A Modified Schwarz Algorithm

To solve (1) when $g = 0$, we choose a coercive bilinear form on $L^2(\Omega)$, $b(\cdot,\cdot)$, and find $u_i^{n+1} \in V_{i0h}$, $i = 1,2$ by solving

$$b(u_i^{n+1} - u_i^n, \hat{u}_i) + a(u_i^{n+1} + u_j^n, \hat{u}_i) = (f, \hat{u}_i) \quad \forall \hat{u}_i \in V_{i0h} \tag{5}$$

We consider the case where Ω_1 and Ω_2 are triangularized independently. Integrals of piecewise constant functions g are computed exactly by

$$\int_{\Omega_i} g = \sum_{k=1}^{n_i} |T_k^i| g(\xi_k^i) \tag{6}$$

where n_i is the number of triangles of the triangulation of Ω_i and ξ_k^i is the chosen quadrature point in triangle T_k^i (its center for instance).
To compute integrals involving products of functions on two triangulations like $\int \nabla u_{1h} \nabla v_{2h}$ we propose the following formula

$$\int_{\Omega_1 \cap \Omega_2} g \approx \frac{1}{2} \sum_{\{k:\xi_k^1 \in \Omega_1 \cap \Omega_2\}} |T_k^1| g(\xi_k^1) + \frac{1}{2} \sum_{\{k:\xi_k^2 \in \Omega_1 \cap \Omega_2\}} |T_k^2| g(\xi_k^2) \tag{7}$$

This can be summarized by saying that when $u \in V_{ih}$ and $v \in V_{jh}$, $i \neq j$, then $a(\cdot,\cdot)$ is replaced by $a_h(\cdot,\cdot)$ with

$$a_h(u,v) = \sum_{k=1}^{n_1} \left(\frac{|T_k^1| \nabla u \cdot \nabla v}{I_{\Omega_1}(x) + I_{\Omega_2}(x)} \right) |_{x=\xi_k^1} + \sum_{k=1}^{n_2} \left(\frac{|T_k^2| \nabla u \cdot \nabla v}{I_{\Omega_1}(x) + I_{\Omega_2}(x)} \right) |_{x=\xi_k^2}. \tag{8}$$

In the above formula $I_{\Omega_i}(x)$ is one if $x \in \Omega_i$ and zero otherwise.

With such definitions the discrete version of (5) is :
- Find $u_{ih}^{n+1} \in V_{ih}$ such that $\forall v_{ih} \in V_{ih}$

$$b(u_{1h}^{n+1} - u_{1h}^n, \hat{u}_{1h}) + a_h(u_{1h}^{n+1} + u_{2h}^n, \hat{u}_{1h}) = (f, \hat{u}_{1h}) \quad \forall \hat{u}_{1h} \in V_{1h}$$

$$b(u_{2h}^{n+1} - u_{2h}^n, \hat{u}_{2h}) + a_h(u_{1h}^n + u_{2h}^{n+1}, \hat{u}_{2h}) = (f, \hat{u}_{2h}) \quad \forall \hat{u}_{2h} \in V_{2h} \quad (9)$$

These equations define u_{ih}^{n+1} uniquely. At convergence the problem solved is

- Find $u_{ih} \in V_{ih}$ such that $\forall \hat{u}_{ih} \in V_{ih}$

$$a_h(u_{1h} + u_{2h}, \hat{u}_{1h} + \hat{u}_{2h}) = (f, \hat{u}_{1h} + \hat{u}_{2h}). \quad (10)$$

The bilinear form is symmetric but this discrete problem may not have a solution because the form may not be coercive. For this there is a compatibility condition between the triangulation which is that the bilinear form must be coercive: $a_h(u_{1h} + u_{2h}, u_{1h} + u_{2h}) \geq c\|u_{1h} + u_{2h}\|^2$ for all $u_{ih} \in V_{ih}$

Proposition

Assume that the triangulations of Ω_1 and Ω_2 are compatible in the sense that they give a coercive bilinear form. Then the error between the approximate problem (10) and the continuous problem is

$$\|u - u_h\| < Ch(\|u_1\|_{2,\Omega_1} + \|u_2\|_{2,\Omega_2})$$

In the case of the Laplace operator, if the mesh is uniform in $\Omega_1 \cap \Omega_2$ and if each triangle of one mesh contains one and only one quadrature point of the other mesh then a_h is positive and coercive .

3 The Chimera Method

Chimera as found in Stegger[16] is a method to solve problems for which the construction of a triangulation of the complete domain Ω is impossible or undesirable. This happens in CFD where the geometries are complicated. For instance one could compute a full aircraft using DDM by computing first the wings then the fuselage, provided that the meshes for each overlap in a suitable manner.

The previous theory applies there too. Assume that $\Omega = \Omega' \setminus C$ where $C \subset \Omega'$.

We take a larger set Ω'_2 containing C and inside Ω'. For Ω_1 we choose a set Ω'_1 containing C but inside Ω_2:

$$C \subset \Omega'_1 \subset \Omega'_2 \subset \Omega' \quad (11)$$

Then we take

$$\Omega_1 = \Omega' \setminus \Omega'_1, \qquad \Omega_2 = \Omega'_2 \setminus C \quad (12)$$

Obviously we have $\Omega = \Omega_1 \cup \Omega_2$.

In the discrete case, the domains Ω_i are found automatically by finding first all the triangles of Ω_{1h} which are touching C then taking one or two layers of triangles around it; this determines the boundary S_1. Then surrounding C with a boundary S_2 of the same type as ∂C which contains S_1 in its interior and is contained in Ω_1. This may not be possible if the triangles of Ω' are too large.

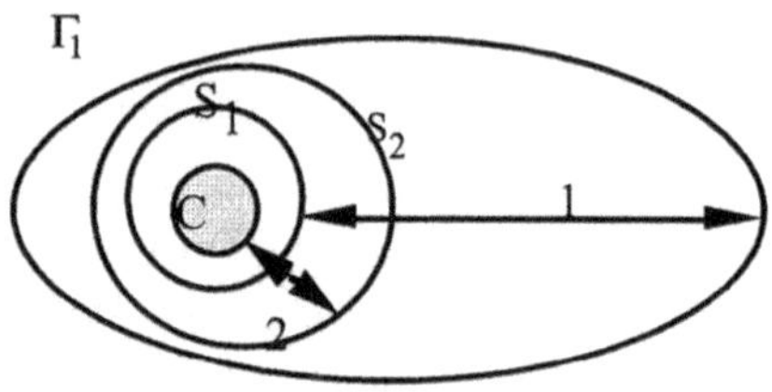

Fig. 1.

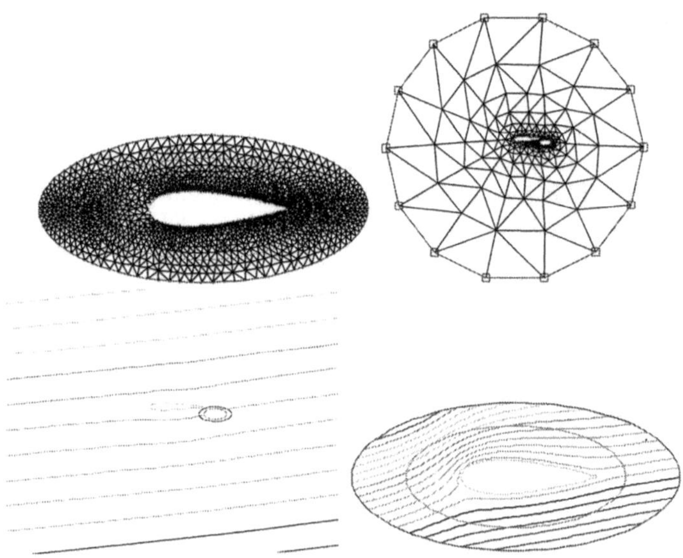

Fig. 2. *Stream function around a two-pieces airfoil, namely solution of $\Delta\psi = 0$ with Dirichlet data by the Chimera method (i.e. Schwarz algorithm). The method allows a finer mesh around the smaller airfoil without the penalty of a fine mesh for the subproblem on the larger domain. The convergence is obtained after 4 iterations.*

4 Constructive Solid Geometry

The algorithms of Virtual Reality (VR)[2][6][18] have to compromise between realistic rendering and speed. We consider the case where a Partial Differential Equation has to be solved in the framework of VR.
Constructive Solid Geometry is quite popular in VR and consequently the domain of integration of the PDE (if any) is not given by its boundary but by set operations on simple elementary volumes, typically unions and intersections and sometimes extruding of elementary shapes. Consider for example the following scene (see figure 3) described in the POV-Ray language:

```
#declare altere = union{ cylinder {<-1.5,0,0> <1.5,0,0>, .35}
                sphere {<-1.5,0,0>, 0.5} sphere {<1.5,0,0>, .5}}
union { object { altere rotate z*90 } object {altere scale 1}
        object { altere rotate y*90 } sphere{<0,0,0>, 0.6 } }
```

It is a set of spheres and cylinders with some realistic steel-like texture; notice that there is no information about their intersection. This is seen only in the rendering phase which is based on the z-buffer algorithm (zbuff[] is initially filled with large values):

```
for(t=0;tmax;t++) if(z[t] <= zbuff[x[t]][y[t]])
         { putpixel(x[t],y[t],z[t]); zbuff[x[t]][y[t]]=z[t];}
```

This C-program displays 3D objects $\{t \to (x[t], y[t], z[t])\}$ on a screen of size hmax × vmax. By displaying all objects this way each new voxel (3D point) of an object lights a screen-pixel (2D point) only if it is in front of all previously displayed voxels that fall on the same pixel.
Engineering data on the other hand use Bezier or B–spline patches from which it is easier to derive a triangulation of the surfaces in the scenes, a necessary step for the generation of three-dimensional meshes.

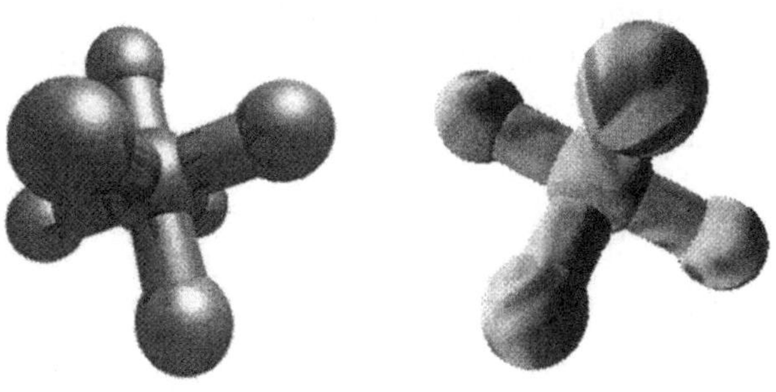

Fig. 3. Left *A scene displayed by POV-Ray. The objects are never intersected, it is the graphic rendering that takes care of the problem.* **Right** *The trace of the real part of the scattered acoustic field on the surface of the geometry.*

5 Geological Flows

Geological layers such as clay and chalk have very different diffusion and Darcy coefficients with the consequence that flows through porous media could benefit from DDM. Implicit solutions of the convection-diffusion equation with largely varying coefficients is a challenge. By concentrating the difficulties at the interfaces between the layers, and use DDM, we may build more efficient numerical methods.
The method described in [12] is well suited and will allow different time steps in the different regions.

We consider the convection-diffusion equation

$$\partial_t u + \mathbf{v} \cdot \nabla u - \nabla \cdot (\nu \nabla u) = 0 \text{ in } \Omega \times (0,T)$$

with zero initial and boundary conditions, except at the right boundary where an homogeneous Neumann condition is applied.
The problem is in $\Omega = (0,1) \times (0,2)$. It is an academic example of the dissipation of a pollutant from an enclosure C into a medium Ω_1 (rock) with low diffusion but cracked (the vertical boundary next to the circle on figure 4). Furthermore below Ω_1 there is another medium Ω_2 (sand) with large diffusion constant ν_2; there the pollutant is also convected (water in sand) at velocity $\mathbf{v}$. The velocity derives from a potential ϕ solution of (see figure 7)

$$-\nabla \cdot (\mu \nabla \phi) = 0 \text{ in } \Omega_2 \quad \phi|_{x=0} = 0 \quad \phi|_{x=1} = 1 \text{ and } \frac{\partial \phi}{\partial n} = 0 \text{ elsewhere}$$

and $\mathbf{v} = -\mu \nabla \phi$. The equations are discretized in time by an implicit Euler scheme and in space by the finite element method of order one on triangles. The convection term is treated by the Galerkin-Characteristic method[14] and $X(x)$ denotes an approximation of $x - \mathbf{v}\delta t$.
The following Domain Decomposition Method is fully described in [12]; it relies on Lagrange multipliers for the constraint $u_1 = u_2$ on the interface of the (non-overlapping) subdomains.

$$\begin{aligned}
&\frac{1}{\delta t}(u_i^{n+1} - u_i^n o X, \hat{u}_i) + (\nu_i \nabla u_i^{n+1}, \nabla \hat{u}_i) \\
&+d(u_i - u_j, \hat{u}_i) - (-1)^i \int_S \lambda \hat{u}_i = 0 \quad \forall \hat{u}_i \in V_i \quad i,j = 1,2, \; j \neq i \\
&\frac{\epsilon}{\delta t} \int_S (\lambda^{n+1} - \lambda^n)\hat{\lambda} + \eta \int_S \frac{d\lambda}{ds}\frac{d\hat{\lambda}}{ds} = \int_S (u_1 - u_2)\hat{\lambda} \quad \forall \hat{\lambda} \in L_h \qquad (13)
\end{aligned}$$

where V_i and L_h are the finite element spaces of piecewise linear continuous functions on Ω_i and S respectively, and

$$d(u,v) = \int_S (\alpha uv + \beta \frac{\partial u}{\partial s}\frac{\partial v}{\partial s}) \quad S = \overline{\Omega}_1 \cap \overline{\Omega}_2$$

The parameters chosen are:

$$\mu = 10,\ \nu_2 = 0.1,\ \delta t = 0.1,\ T = 0.6,\ \alpha = 1,\ \beta = 0,\ \epsilon = 1,\ \eta = 0.1,\ \nu_2/\nu_1 = 100.$$

The results are shown on figure 4. The mesh has some 1500 vertices (2/3 in the top region Ω_1). Sensitivity with respect to numerical coefficients is mild except for ϵ and α. The size of λ is proportional to $1/\epsilon$ and its smoothness (or localization) is controlled by η; a good choice of α improves the quality of the results but β does not seem to play a major role. If ϵ is too small the method is unstable.
The numerical tests show that the method is feasible and well adapted to discontinuous coefficients. Different time steps in different sub-domains (with rendez-vous) improve the computing time and do not affect the precision.

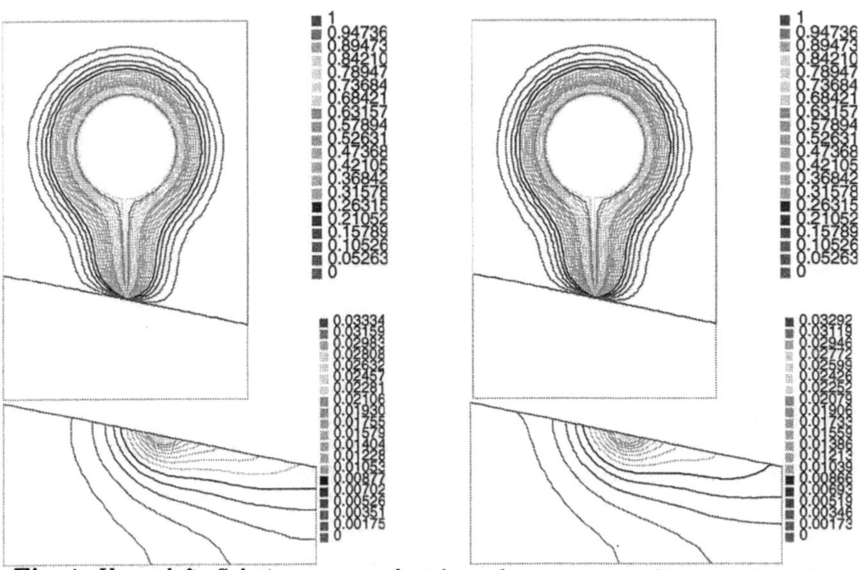

Fig. 4. *Upper left: Solution computed without decomposition. Upper right: Solution computed with decomposition. No level lines are visible in the bottom regions because very little diffusion has occurred. Below is a zoom of both on subdomain Ω_2 where the convection and diffusion are large.*

6 The freefem Project

Started in 1995 as an educational multi-platform software `freefem`[1] has received a larger audience than expected and many users have requested a three dimensional version of this public domain software.
Presently freefem+ is a language driven 2D-PDE solver based on the finite

element method of order 1. It is written in C++ using template and generic programming so that both scalar equations and systems can be solved. It has two solvers in its kernel: a general elliptic PDE solvers based on a direct Choleski factorization of the linear systems and a convection solver based on the Galerkin-Characteristic method [14]. It can handle several meshes in one program thanks to its powerful interpolator[8]. Therefore it is a convenient tool for DDM.

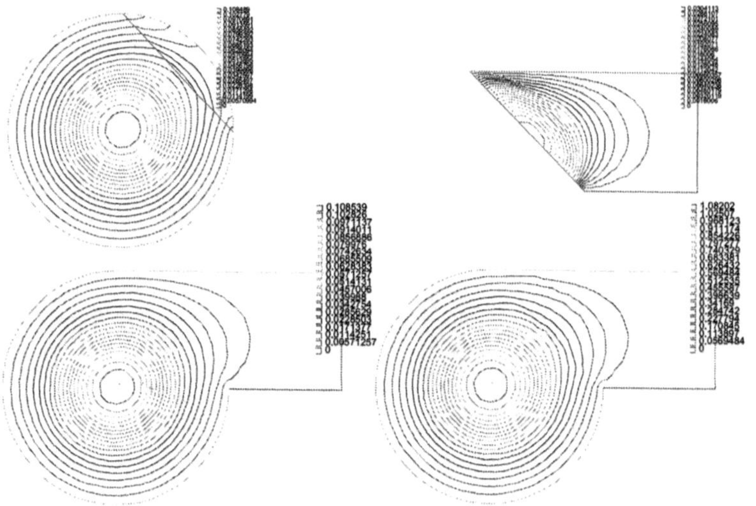

Fig. 5. Solution of (1) with $f = 1$ in a circle of radius a third of the larger one, by solving (9) on non-matching meshes with `freefem+`. The bottom left figure shows the direct solution of the same problem without DDM.

In `freefem3d` we delegate the description of the domain to `POV-Ray` or `vrml`. Similarly the display of result is done either in `POV-Ray` or in `dx` (ibm's data-explorer[4]). The language to describe the partial differential equations is similar to `freefem+`. The following program is the source code of Figure 6.

```
    vector a =(-1,-1,-1);   // Lower left  domain corner
    vector b = (1,1,1);     // upper right corner
    vector n = (38,38,38);  // nb of mesh points

structmesh Mesh(n,a,b);

scene S("fic6.pov",Mesh);   // POV-Ray file

double i=0; double dt=0.1;
do{
```

```
  solve(u) { - div (dt * grad(u))+u = u;
              u=1  on <1,0,0>;
              dnu(u) = 0 on d Mesh;
            };
  i=i+1;
  dxplot("u.dat",u,Mesh);
}while(i<=5);
```

However the solver is quite different from `freefem+`: it is a fictitious domain method with an iterative solution of the linear systems by the preconditioned conjugate gradient method (see [5] for more details).

To use the *virtual reality* description of the computational domain (Figure 3 and the program of Section 2), one has to generate a characteristic function for the object D.
The construction of this function was done in two steps, first by writing an interpreter for the language describing the scene and then by constructing the characteristic functions of all the elementary objects of the scene.
The *virtual reality* language chosen was `POV-Ray`[18] because it is stable and portable but it can also be done with `vrml` as easily. Also `vrml` just implements the set union but no set intersection. Moreover, one can use it as a visualization tool thanks to the iso-surface patch of Suzuki[17].
To parse the program of Section 1 is easy but `POV-Ray` has also conditional statements, loops, symbolic variables, functions, etc.; so we have used `lex` and `yacc` because both of them generate `C` or `C++` parsers.
In the `C++` produced, an abstract `Shape` class is defined; it provides a standard interface to every kind of shapes. Its children are specialization such as `Sphere` or `Cube` (primitives), but also boolean operations such as `Union` and `Intersection`. The same kind of design is used for geometrical transformations of `POV-Ray`, (`scale`, `rotate` or `translate`).
As every primitive shape is simple (sphere, cube, cone,...), one can easily associate to it its characteristic function. This is done by specializing the virtual function of the class `Shape` for each of them. To know if a vertex is in an *object* one has to compute the inverse of each transformation associated to it and then check if the antecedent is in the primitive `Shape`.
This implementation makes the formulation very close to the mathematics, and as the results of boolean operations on shapes are shapes, complex objects descriptions and their characteristic function evaluation may be recursive. There is a problem however with two dimensional structures like polygonal facets.

7 Conclusion

We have presented a few examples where Domain Decomposition is part of the definition of the problems.

We have recalled a modified Domain Decomposition Method which is well suited to these problems and for which we have some convergence results for an arbitrary number of subdomains.
We have given some results in two dimensions and presented the work plan for three dimensional results. These are preliminary results which will be followed by more realistic ones in the near future.

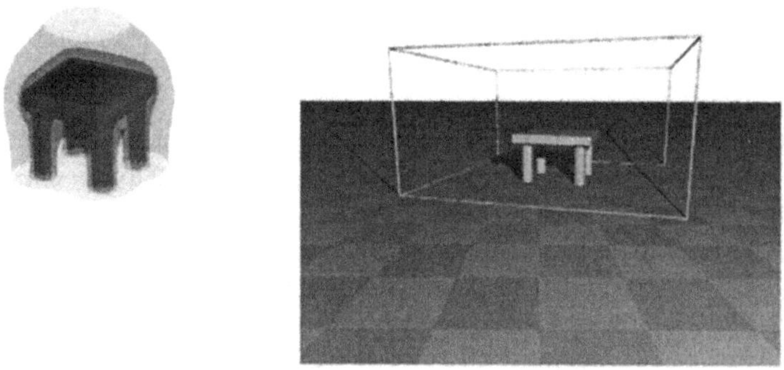

Fig. 6. *Displayed are 4 iso-temperature surfaces (0.95, 0.5, 0.25, 0.05) for a transient solution of the heat equation at time 0.1, around a table-shaped object at temperature 1 with Neumann conditions on the boundary of the computational domain, (shown on the right with POV-Ray) and initial temperature zero; the program in* `freefem3d` *language is given above.*

References

1. Bernardi D., F. Hecht, K. Otsuka, O. Pironneau : freefem+, a finite element software to handle several meshes.Dowloadable from ftp://ftp.ann.jussieu.fr/pub/soft/pironneau/, 1999.
2. Burden G., Coiffe Ph. : Virtual Reality Technology, New York, Wiley 1994.
3. Ciarlet P.G : The Finite Element Method,Prentice Hall, 1977.
4. Data Exploreur : IBM Corporation Thomas J. Watson Research Center/Hawthorne from http://www.dx.com, 1997.
5. Del Pino S., Heikkola E., Pironneau O., Toivanen J. : A finite element method for virtual reality data. C.R.A.S., June 2000.
6. Hartman J., Wernecke J. : "The VRML 2.0 Handbook" Addison-Wesley 1996.
7. Hecht F, Lions J.L., Pironneau O. : Domain Decomposition Algorithm for Computed Aided Design. (To appear in the anniversary book of Necas)
8. Hecht F., Pironneau O. : Multiple meshes and the implementation of freefem+, INRIA report March, 1999. Also on the web at ftp://ftp.ann.jussieu.fr/pub/soft/pironneau.
9. Lions J.L., Pironneau O. : Algorithmes parallèles pour la solution de problèmes aux limites, C.R.A.S., 327, pp 947-352, Paris 1998.

10. Lions J.L., Pironneau O. : Domain decomposition methods for CAD. C.R.A.S., 328, pp 73-80, Paris 1999.
11. Lions J.L., Pironneau O. : Domain decomposition methods for CAD. C.R.A.S., 328, pp 73-80, Paris 1999.
12. Lions J.L., Pironneau O. : Non-Overlapping Domain Decomposition of Evolution Operators C.R.A.S. Paris June 2000.
13. Lions P.L. : On the Schwarz alternating method. I,II,III. Int Symposium on Domain decomposition Methods for Partial Differential Equations. SIAM, Philadelphia, 1988,89,90.
14. Pironneau O. : *Finite Element Methods for Fluids* Wiley, Chichester 1987.
15. Smith B., Bjørstad P., Gropp W. : *Domain decomposition. Parallel multilevel methods for elliptic partial differential equations.* Cambridge University Press, Cambridge, 1996.
16. Steger J.L. : The Chimera method of flow simulation,Workshop on applied CFD, Univ of Tennessee Space Institute, August 1991.
17. R. Suzuki R. : A patch to **POV-Ray** for iso-surfaces. In http://www.public.usit.net/rsuzuki/e/povray/iso/index.html.
18. Wardley A. : Persistence of Vision, **POV-Ray** in http://www.povray.org/.

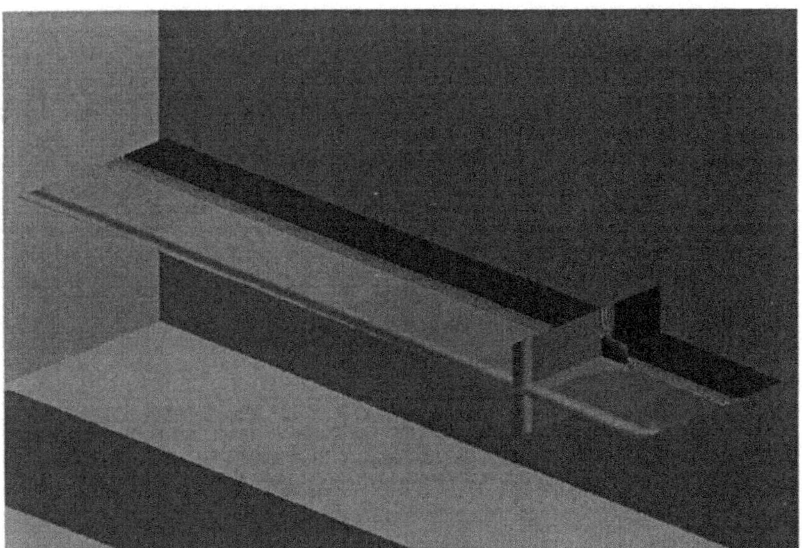

Fig. 7. Two cuts across the parallelipedic 3D domain showing the levels of the hydrostatic potential for a 4 layer medium with a man made impermeable cave (in grey), computed with `freefem3d`. The deepest layer (in violet) is of dogger, the lower middle one is clay the next one is lime stone and the nearest to us is made of marne.

A One Dimensional Model for Blood Flow: Application to Vascular Prosthesis

Luca Formaggia[1], Fabio Nobile[1], Alfio Quarteroni[1,2]

[1] Mathematics Department, EPFL, Lausanne, CH-1015, Switzerland
[2] Mathematics Department, Politecnico di Milano, I-20133 Milano, Italy

Abstract. We investigate a one dimensional model of blood flow in human arteries. In particular we consider the case when an abrupt variation of the mechanical characteristic of an artery is caused by the presence of a vascular prosthesis (e.g. a stent). The derivation of the model and the numerical scheme adopted for its solution are detailed. Numerical experiments show the effectiveness of the model for the problem at hand.

1 Introduction

The growing interest in the use of mathematical modelling and numerical simulations for the investigation of biomedical issues and, in particular, the human cardiovascular system, is testified by the numerous works which have appeared on the subject in recent years, among which we mention [1,6,10] and the references therein. In this context, often simple models are already able to provide good indications for the practitioners, at a reasonable computational cost.

Here, we will focus on the application of a one-dimensional model of blood flow in a compliant vessel to study the effect on the flow pattern caused by the local stiffening of an artery. This can be due to a stent implantation, or to the presence of a vascular prosthesis. A common pathology in the human circulatory system is the on-rise of atherosclerotic plaques which cause a restriction of the arterial lumen called a stenosis; in the most severe cases this may hinder, or even stop, the flow of blood. One of the techniques nowadays used for curing this problem is the implantation of a *stent* (an expandable metal mesh) into the affected region which has the purpose of returning the artery lumen to approximately its original shape. Whenever possible, this procedure is preferred to more invasive ones, such as surgical by-pass.

However, besides other effects, the presence of a stent causes an abrupt variation in the elastic properties of the vessel wall, since the stent is usually far more rigid than the rather soft arterial tissue. This fact may cause a disturbance in the blood flow pattern (and in particular in the pressure) with the appearance of reflected waves. Indeed, the so–called pressure pulse is generated by the interaction between the blood flow and the compliance of the circulatory system and is intrinsically related to the elastic properties of the arteries. The alteration in the pressure pattern is even more important in the case of vascular prosthesis in large arteries, such as the ones used, for instance, to cure aneurisms of the aorta or the femoral artery. Indeed, the superposition of the waves reflected by the prosthesis or the stent with the pressure pulse coming from the heart may generate localised pressure peaks, which may have dangerous effects on the heart. In the case of an aortic or femoral prosthesis, should these peaks occur at the region of the suture, they could even break it apart.

Here, we carry out a numerical investigation of the modification of pressure and flow pattern caused by an abrupt variation of the vessel elastic characteristic, by employing a one dimensional model of blood flow. This model is derived from the one described in [8] and already used for simulations in arteries in normal conditions.

In section 2 we will present the model and in particular the modifications which have been necessary to account for variable elastic properties. In section 3 we detail the numerical algorithm proposed for its solution, which is based on a Taylor-Galerkin finite element scheme, and we briefly comment on the application of boundary conditions for the simulations at hand. Section 4 presents some numerical results.

2 One dimensional models of blood flow in arteries

A one dimensional model for blood flow in arteries may be derived from the following Navier Stokes equations, under a certain number of simplifying hypotheses,

$$\begin{cases} \dfrac{\partial \mathbf{u}}{\partial t} + (\mathbf{u} \cdot \nabla)\,\mathbf{u} + \dfrac{1}{\rho}\nabla P + \mathrm{div}\left(\nu \dfrac{\nabla \mathbf{u} + (\nabla \mathbf{u})^T}{2}\right) = \mathbf{0} & \text{in } \Omega_t,\ t > 0 \\ \mathrm{div}\ \mathbf{u} = 0 \end{cases} \quad (1)$$

The domain Ω_t is depicted in Fig. (1) and is a cylinder which exemplifies a portion of an artery. We have used the subscript t to put into evidence that

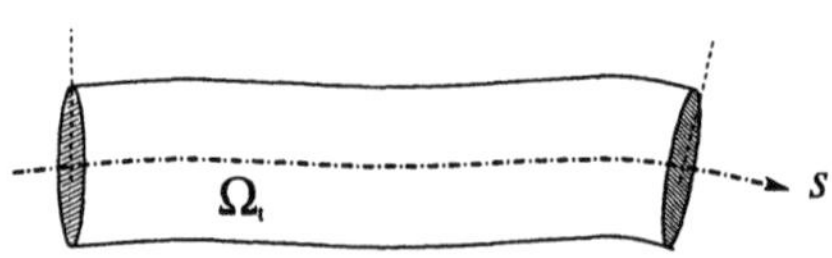

Fig. 1. The domain Ω_t representing the portion of an artery.

the domain is in fact moving with time, because of the compliance of the vessel walls. Therefore, the equation should be coupled with an appropriate model of the vessel wall dynamics, as for instance one of those that are proposed in [10,3]. The unknowns are the fluid velocity $\mathbf{u}$, the pressure P

and the wall displacement. The two positive constants ν and ρ denote the kinematic viscosity and the fluid density, respectively.

A simplified model is derived by assuming a cylindrical geometry with circular cross section, like the one depicted in Fig. (2). Without loss of generality,

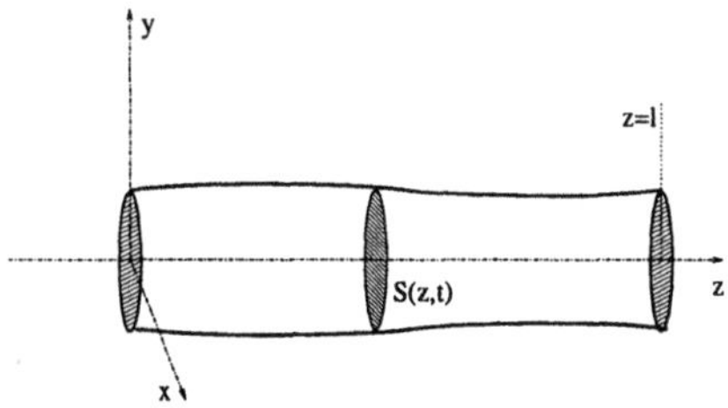

Fig. 2. Simplified geometry. The vessel is assumed to by a straight cylinder with circular cross section.

we assume that the z coordinate is aligned with the axis of the cylinder. By integrating the Navier-Stokes equations on a generic axial section $S = S(z,t)$ one obtains the following set of two partial differential equations

$$\begin{cases} \frac{\partial A}{\partial t} + \frac{\partial Q}{\partial z} = 0 \\ \frac{\partial Q}{\partial t} + \frac{\partial}{\partial z}\left(\alpha \frac{Q^2}{A}\right) + \frac{A}{\rho}\frac{\partial p}{\partial z} + K_R \frac{Q}{A} = 0, \end{cases} \qquad z \in (0,l),\ t > 0, \tag{2}$$

where A, Q and p denote the section area, average volumic flux and mean pressure, and are defined as

$$\begin{gathered} A(z,t) = \int_{S(z,t)} d\gamma, \quad Q(z,t) = \int_{S(z,t)} u_z(\mathbf{x},t) d\gamma \\ p(z,t) = (A(z,t))^{-1} \int_{S(z,t)} P(\mathbf{x},t) d\gamma, \end{gathered} \tag{3}$$

where $\mathbf{x} = (x,y,z)$ and u_z denotes the z-component of the velocity. The coefficient α is a function of the velocity profile on each section, and accounts for the fact that the actual momentum flux differs from that obtained using the averaged quantities. Here, it has been taken constant and equal to one. It will be therefore ignored in the rest of the paper. The coefficient K_R is a resistance parameter linked to the blood viscosity.

We wish to point out that in order to derive (2) from (1) we have made the approximation of considering the viscous effects to be important only in the wall boundary layer. Consequently, we have neglected all second derivative terms along the z coordinate coming from the viscous stress tensor in the Navier-Stokes equations.

The number of unknowns (p, A and Q) exceeds the number of equations in (2). A possible way to close the system is to explicitly provide a relation linking the pressure to the vessel area A and possibly its time derivatives. This relation is derived from the model of the dynamics of the vessel wall, and some possible expressions have been given in [8]. Here we consider the case of an elastic wall with varying elastic Young's modulus E to simulate the presence of a vascular prosthesis or of a stent. We have neglected the inertia effect and used the relation of static equilibrium in the radial direction for a cylindrical tube, in the hypothesis of linear elastic material and small deformations, to derive the following pressure relation

$$p = p_{\text{ext}} + \beta\left(\sqrt{A} - \sqrt{A_0}\right), \tag{4}$$

where p_{ext} is the external pressure, here taken constant and equal to zero. A_0 is the reference vessel section area which is a function of z since an artery is usually tapered. However, for the sake of simplicity we have here considered A_0 constant, the steps required to deal with tapering have been described in [8]. Finally, $\beta = \beta(z)$ is a parameter linked to the wall elastic properties, which, under the assumption of incompressible material, takes the form

$$\beta(z) = \frac{4\sqrt{\pi}h_0 E(z)}{3A_0} \ . \tag{5}$$

The fact that β varies with z will affect the derivation of the conservation form for equations (2), as we will detail in the following section.

Other possible relationships between p and A can be found in [5],[12],[7]. The methodology here presented may be extended to those formulations as well.

2.1 The conservation form

We aim at writing (2) in the form

$$\frac{\partial \mathbf{U}}{\partial t} + \frac{\partial \mathbf{F}}{\partial z}(\mathbf{U}) = \mathbf{B}(U), \tag{6}$$

where $\mathbf{U} = [A, Q]^T$ are the conservation variables, $\mathbf{F} = [F_A, F_Q]^T$ the corresponding fluxes and $\mathbf{B} = [B_A, B_Q]^T$ a source term which is of zeroth order in $\mathbf{U}$.
To that purpose, we note that

$$\frac{\partial p}{\partial z} = \frac{d\beta}{dz}A^{\frac{1}{2}} + \frac{\beta}{2}A^{-\frac{1}{2}}\frac{\partial A}{\partial z} - \frac{d\beta}{dz}A_0^{\frac{1}{2}},$$

and, consequently,

$$\frac{A}{\rho}\frac{\partial p}{\partial z} = \frac{\partial}{\partial z}\left(\frac{\beta A^{\frac{3}{2}}}{3\rho}\right) + \frac{A}{\rho}\frac{d\beta}{dz}\left(\frac{2}{3}\sqrt{A} - \sqrt{A_0}\right) \ . \tag{7}$$

Comparing with the case when β is constant, we may note that an additional source term has appeared which depends on the derivative of β. We are now in the position of writing (2) under the conservative form (6) by setting

$$\mathbf{F} = \begin{bmatrix} Q \\ \frac{Q^2}{A} + \frac{\beta}{3\rho} A^{\frac{3}{2}} \end{bmatrix}, \mathbf{B} = \begin{bmatrix} 0 \\ -K_R \frac{Q}{A} + \frac{A}{\rho}\frac{d\beta}{dz}\left(\sqrt{A_0} - \frac{2}{3}\sqrt{A}\right) \end{bmatrix} . \tag{8}$$

Alternative one-dimensional models for blood flow in artery, which adopt the non-conservative variables p and Q as unknowns, have been used in the literature (see e.g. [9]).

System (6) may be written in *quasi-linear* form as

$$\frac{\partial \mathbf{U}}{\partial t} + \mathbf{H}\frac{\partial \mathbf{U}}{\partial z} = \mathbf{B}(\mathbf{U}), \text{ where } \mathbf{H} = \begin{bmatrix} 0 & 1 \\ c^2 - \left(\frac{Q}{A}\right)^2 & 2\frac{Q}{A} \end{bmatrix} . \tag{9}$$

Here,

$$c = \sqrt{\frac{A}{\rho}\frac{\partial p}{\partial A}}$$

is the "sound speed". Indeed, it may be shown that for all allowable values of $\mathbf{U}$, the matrix $\mathbf{H}$ possesses two real and distinct eigenvalues,

$$\lambda_{1,2} = \bar{u} \pm c,$$

and a complete set of eigenvectors. We have put $\bar{u} = Q/A$ to indicate the average value of the component u_z of the velocity.

System (2) is then hyperbolic. Although a non-linear hyperbolic system may develop discontinuous solutions even if the initial and boundary data are smooth, for the values attained by the mechanical parameters and blood velocities in physiologic (sub-critical) conditions, one has $c > \bar{u}$, then the two eigenvalues have opposite sign. A recent study [2] shows that in this situation the pulsatile nature of blood flow may inhibit the formation of discontinuities. Therefore, we will assume in the following that the solution is continuous.

We now indicate by $\mathbf{L}$ and $\boldsymbol{\Lambda} = \mathrm{diag}(\lambda_1, \lambda_2)$ the matrix of left eigenvectors and that of the eigenvalues of $\mathbf{H}$, respectively. Then, relation (9) may be written as

$$\mathbf{L}\frac{\partial}{\partial t}\mathbf{U} + \boldsymbol{\Lambda}\mathbf{L}\frac{\partial \mathbf{U}}{\partial z} = \mathbf{L}\mathbf{B}(\mathbf{U}) . \tag{10}$$

It is possible to find a vector function $\mathbf{W} = \mathbf{W}(\mathbf{U}) = [W_1(\mathbf{U}), W_2(\mathbf{U})]$ such that $\partial_{\mathbf{U}}\mathbf{W} = \mathbf{L}$. The quantities W_1 and W_2 are the *characteristic variables* of problem (2) and their expression may be given as function of the average velocity $\bar{u}$ and the pressure p (or of A and Q) as follows

$$W_{1,2} = \bar{u} \pm 2\sqrt{\frac{2}{\rho}}\sqrt{p + \beta\sqrt{A_0}} = \left(\frac{Q}{A}\right) \pm 2\sqrt{\frac{2}{\rho}}\beta^{\frac{1}{2}}A^{\frac{1}{4}} \ . \tag{11}$$

By inverting these relations we may express the conserved variables in terms of W_1 and W_2,

$$A = \left(\frac{\rho}{\beta}\right)^2 \frac{(W_1 - W_2)^4}{4^5}, \quad Q = A\frac{W_1 + W_2}{2} \ . \tag{12}$$

In the case $\mathbf{B} = \mathbf{0}$, equations (10) decouples and may be written componentwise as

$$\begin{cases} \dfrac{\partial}{\partial t}W_1 + \lambda_1\dfrac{\partial W_1}{\partial z} = 0, \\ \dfrac{\partial}{\partial t}W_2 + \lambda_2\dfrac{\partial W_2}{\partial z} = 0. \end{cases} \tag{13}$$

Equation (6) must be supplemented by appropriate boundary conditions at $z = 0$ (proximal section) and $z = l$ (distal section). The medical terms "proximal" and "distal" correspond to the sections nearest and farthest from the heart.

Since the eigenvalues have opposite sign, a single boundary condition must be specified at both ends. In particular, we may impose the entering characteristic variable, i.e., $\forall t > 0$

$$W_1(t) = g_1(t) \ \text{ at } z = 0, \text{ and } W_2(t) = g_2(t) \ \text{ at } z = l, \tag{14}$$

where g_1 and g_2 are given functions of t. Alternative boundary conditions applied to the primary variables Q and A (or p) can be devised as well, under suitable restrictions [4]. For the numerical discretisation of (6), the two boundary conditions above need to be supplemented by two additional equations, one at each side, in order to allow the computation of two vector unknowns $\mathbf{U}(0)$ and $\mathbf{U}(L)$. We will address this issue in more detail in section 3.1.

We recall here an a-priori estimate which has been derived in [3] for the case of constant Young's modulus using slightly different, yet equivalent, formulation for the characteristic variables. If

$$g_1(t) > -4\sqrt{\frac{\beta}{2\rho}}A_0^{\frac{1}{4}} \ \text{ and } g_2(t) < 4\sqrt{\frac{\beta}{2\rho}}A_0^{\frac{1}{4}}, \quad \forall t > 0,$$

then there exists a positive function $G = G(g_1, g_2)$ such that the following inequality holds, for all $T \geq 0$,

$$E(T) + \rho K_R \int_0^T \int_0^l \bar{u}^2 dz dt \leq E(0) + \int_0^T G(g_1(t), g_2(t)) dt,$$

where

$$E(t) = \frac{1}{2}\rho \int_0^l A(z,t)\bar{u}^2(z,t)dz + \beta \int_0^l \int_{A_0}^{A(z,t)} \left(\sqrt{A} - \sqrt{A_0}\right) dA\, dz \ .$$

Clearly, $E(0)$ depends only on the initial data.

3 Numerical Discretisation

The equations in conservation form (6) have been discretised by adopting a second order Taylor-Galerkin scheme, which is the finite element counterpart of the well known Lax-Wendroff scheme.

The presence of a non-constant source term and the explicit dependence of the momentum flux F_Q on the variable z due to the varying β has made the derivation of the scheme slightly more complex. In the rest of this paper we will use the abridged notation

$$\mathbf{F_U} = \frac{\partial \mathbf{F}}{\partial \mathbf{U}}, \quad \mathbf{B_U} = \frac{\partial \mathbf{B}}{\partial \mathbf{U}}.$$

We now follow the usual route to derive the Lax-Wendroff scheme, by writing

$$\frac{\partial \mathbf{U}}{\partial t} = \mathbf{B} - \frac{\partial \mathbf{F}}{\partial z} \tag{15}$$

$$\frac{\partial^2 \mathbf{U}}{\partial t^2} = \mathbf{B_U}\frac{\partial \mathbf{U}}{\partial t} - \frac{\partial^2 \mathbf{F}}{\partial t \partial z} = \mathbf{B_U}\frac{\partial \mathbf{U}}{\partial t} - \frac{\partial}{\partial z}\left(\mathbf{F_U}\frac{\partial \mathbf{U}}{\partial t}\right) =$$
$$\mathbf{B_U}\left(\mathbf{B} - \frac{\partial \mathbf{F}}{\partial z}\right) - \frac{\partial\left(\mathbf{F_U}\mathbf{B}\right)}{\partial z} + \frac{\partial}{\partial z}\left(\mathbf{F_U}\frac{\partial \mathbf{F}}{\partial z}\right) \tag{16}$$

We note that, in contrast with what is usually done for the derivation of a Lax-Wendroff scheme in standard cases, here we have not further developed the z derivative of the fluxes, since in our case it is not anymore true that

$$\frac{\partial \mathbf{F}}{\partial z} = \mathbf{F_U}\frac{\partial \mathbf{U}}{\partial z},$$

because of the dependence of $\mathbf{F}$ on z through β.

For the time discretisation, we consider a time step Δt and indicate with the superscript n quantities computed at time $t^n = n\Delta t$. Using, as in a standard Lax-Wendroff procedure, a truncated Taylor expansion in time around

t^n and exploiting (15) and (16) we finally obtain the following time-marching scheme

$$\mathbf{U}^{n+1} = \mathbf{U}^n - \Delta t \frac{\partial}{\partial z}\left[\mathbf{F}^n + \frac{\Delta t}{2}\mathbf{F}_{\mathbf{U}}^n \mathbf{B}^n\right] - \frac{\Delta t^2}{2}\left[\mathbf{B}_{\mathbf{U}}^n \frac{\partial \mathbf{F}^n}{\partial z} - \frac{\partial}{\partial z}\left(\mathbf{F}_{\mathbf{U}}^n \frac{\partial \mathbf{F}^n}{\partial z}\right)\right] + \Delta t\left(\mathbf{B}^n + \frac{\Delta t}{2}\mathbf{B}_{\mathbf{U}}^n \mathbf{B}^n\right) . \quad (17)$$

Space discretisation is carried out by using linear finite elements. To that purpose, let us subdivide the interval $[0,l]$ into N finite elements $[z_i, z_{i+1}]$, with $i = 0, \cdots, N$ and $z_i = ih$ being h the constant element size. We indicate by $\mathbf{V}_h = [V_h]^2$ the space of continuous vector functions defined on $[0,l]$, linear on each element, and with $\mathbf{V}_h^0$ the set formed by functions of $\mathbf{V}_h$ which are zero at $z = 0$ and $z = l$. Furthermore, we indicate by

$$(\mathbf{u}, \mathbf{v}) = \int_0^l \mathbf{u} \cdot \mathbf{v}\, dz$$

the L^2 vector product.

Let us put $\mathbf{F}_{LW} = \mathbf{F} + (\Delta t/2)\mathbf{F}_{\mathbf{U}}\mathbf{B}$ and $\mathbf{B}_{LW} = \mathbf{B} + (\Delta t/2)\mathbf{B}_{\mathbf{U}}\mathbf{B}$. A finite element formulation of (17) is: for $n \geq 0$, find $\mathbf{U}_h^{n+1} \in \mathbf{V}_h$ which satisfies

$$(\mathbf{U}_h^{n+1}, \psi_h) = (\mathbf{U}_h^n, \psi_h) + \Delta t(\mathbf{F}_{LW}^n, \frac{\partial \psi_h}{\partial z}) - \frac{\Delta t^2}{2}(\mathbf{B}_{\mathbf{U}}^n \frac{\partial \mathbf{F}^n}{\partial z}, \psi_h) - \frac{\Delta t^2}{2}(\mathbf{F}_{\mathbf{U}}^n \frac{\partial \mathbf{F}^n}{\partial z}, \frac{\partial \psi_h}{\partial z}) + \Delta t(\mathbf{B}_{LW}^n, \psi_h), \quad \forall \psi_h \in \mathbf{V}_h^0 . \quad (18)$$

The boundary values of $\mathbf{U}_h^{n+1}$ will be calculated using the presented boundary conditions and following the technique described in section 3.1. $\mathbf{U}_h^0$ will be taken as the finite element interpolant of the given initial data $\mathbf{U}_0$.

Remark 1. We have integrated by parts the term $(\frac{\partial \mathbf{F}_{LW}}{\partial z}, \psi_h)$ in order to make a discontinuous flux $\mathbf{F}_{LW}$ admissible. Such event may be possible if $\beta \notin C^1(0,l)$. However, the formulation here adopted does not allow for discontinuities in β, because of the presence of $d\beta/dz$ in the source term $\mathbf{B}$ and in its gradient $\mathbf{B}_{\mathbf{U}}$.

Remark 2. Since we have ruled out the appearance of discontinuous solutions, there is no need to adopt the limiting techniques normally implemented in the context of the numerical simulation of non-linear hyperbolic equations by explicit schemes.

A Von Neumann linear stability analysis for the proposed finite element scheme on a grid with uniform spacing h, gives a stability condition of the type

$$\frac{\Delta t \sup_{0<z<l}(\max_{i=1,2} |\lambda_i|)}{h} < \frac{1}{\sqrt{3}},$$

which is more restrictive than the classical CFL condition for finite difference of finite volume Lax Wendroff schemes. This has been confirmed by our numerical experiments.

In (18) there is the need of integrate numerically the terms containing the fluxes and the sources. As for the terms involving $\mathbf{F}^n$ and $\mathbf{F}^n_{\mathbf{U}}$ we have used the technique of projecting each component on the finite element function space V_h by interpolation. The same applies for the other vector products which involve only $\mathbf{F}^n$ and $\mathbf{F}^n_{\mathbf{U}}$.

As for the terms containing $\mathbf{B}^n$ and $\mathbf{B}^n_{\mathbf{U}}$, we have adopted a slightly different approach in order to assure the strong consistency of the numerical scheme. More precisely, we wanted our numerical scheme to be able to represent exactly constant solutions of the differential problem. In view of that the term $d\beta/dz$ has to be approximated by piecewise constants. Precisely, on each element $[z_i, z_{i+1}]$ we have approximated $d\beta/dz$ by $[\beta(z_{i+1}) - \beta(z_i)]/h$. For the remaining terms we have used the same technique adopted for the fluxes. This gives rise to a piecewise linear discontinuous representation for the source terms.

3.1 Computing the boundary data for the numerical scheme

The numerical scheme (18) needs to have a complete boundary data $\mathbf{U}^{n+1}$ at the boundary nodes. We may note that the knowledge of W_1^{n+1} and W_2^{n+1} at the boundary would in principle enable us to compute the corresponding $\mathbf{U}^{n+1}$, thanks to relation (12). However, for the well posedness of the differential problem only one condition at each end, for instance those in (14), has to be assigned.

In order to compute the complete boundary data we need an additional equation, which must be compatible with the original differential problem. Here, we have adopted a technique based on the extrapolation of the outgoing characteristics. We remind that we are interested in simulating the effect on the flow (and in particular on the pressure) of a vascular prosthesis, which is accounted for by a variation in the parameter β. The prosthesis will occupy a portion of the model of the artery, let us say the interval $z \in [a_1, a_2]$, with $a_1 > 0$ and $a_2 < l$. Therefore, at the boundary points $z = 0$ and $z = l$ we have $\dfrac{d\beta}{dz} = 0$; furthermore the term $K_R(Q/A)$ is normally of small size. We feel then legitimated to assume that in the vicinity of the boundary the flow is essentially governed by equations (13). Let us consider now the proximal boundary $z = 0$ (the case of the distal boundary will be treated similarly), and a generic time step of our numerical procedure. We assume that $\mathbf{U}^n$ is known and we linearise λ_2 in the second equation in (13) by taking its value at time t^n and at $z = 0$. The solution corresponding to this linearised problem at the time level t^{n+1} gives

$$W_2^{n+1}(0) = W_2^n(-\lambda_2^n(0)\Delta t) \ ,$$

and is in fact a first order extrapolation of the outgoing characteristic variable W_2 from the previous time level. Higher order extrapolations can be used as well. By using this information together with the value of W_1 provided by the boundary condition, $W_1^{n+1} = g_1(t^{n+1})$, we are able to compute, using (12), the required boundary data, $\mathbf{U}^{n+1}(0)$.

This technique may be extended to the case when the boundary conditions are not given in terms of the characteristic variable, for instance when one wants to impose a given law for the pressure $p(0,t) = \psi(t)$ at the proximal boundary. However, we wish to point out that a homogeneous condition on the incoming characteristics corresponds to a non-reflecting boundary condition and prevents unwanted spurious wave reflections at the boundary.

Remark 3. The methodology just illustrated is not the only possibility to provide boundary data for the discrete problem. Another possible approach is to adopt the so-called compatibility conditions [4,11], namely

$$\begin{aligned} \mathbf{l}_2^T\left(\frac{\partial \mathbf{U}}{\partial t} + \mathbf{H}\frac{\partial \mathbf{U}}{\partial z} - \mathbf{B}(\mathbf{U})\right) &= 0, \quad z = 0, t > 0, \\ \mathbf{l}_1^T\left(\frac{\partial \mathbf{U}}{\partial t} + \mathbf{H}\frac{\partial \mathbf{U}}{\partial z} - \mathbf{B}(\mathbf{U})\right) &= 0, \quad z = l, t > 0\ . \end{aligned} \tag{19}$$

A way to apply these conditions in a finite element scheme is to rewrite the weak form (18) by letting $\psi_h \in V_h$(beware that in this case we cannot drop the boundary terms which originate from the integration by parts). Now using as test function the finite element functions associated to the two boundary nodes we obtain two relations per boundary node. We may then form a linear combination of the equations associated to node $z = 0$ (resp. $z = l$) with coefficients $\mathbf{l}_2$ (resp. $\mathbf{l}_1$), thus producing one equation for each boundary node. They effectively represent a discretisation of (19) which conforms to the adopted numerical scheme. In fact, a linearisation is usually required to this purpose, for instance by evaluating the eigenvectors $\mathbf{l}_i$, $i = 1, 2$, in (19) at the previous time step t^n. These two extra equations are then added to the ones produced by (18) and, together with the assigned boundary conditions, allow to solve the discrete problem.

This technique has been tested as well, yet in this work we preferred the method based on the extrapolation of the characteristics, for its simplicity.

4 Numerical Tests

In order to assess the effect of the changes in vessel wall elastic characteristic on the pressure pattern, we have devised several numerical experiments. Three types of pressure input have been imposed at $z = 0$, namely an impulse input, that is a single sine wave with a small time period, a single sine wave with a more realistic time period and a periodic sine wave (see Fig. 4).

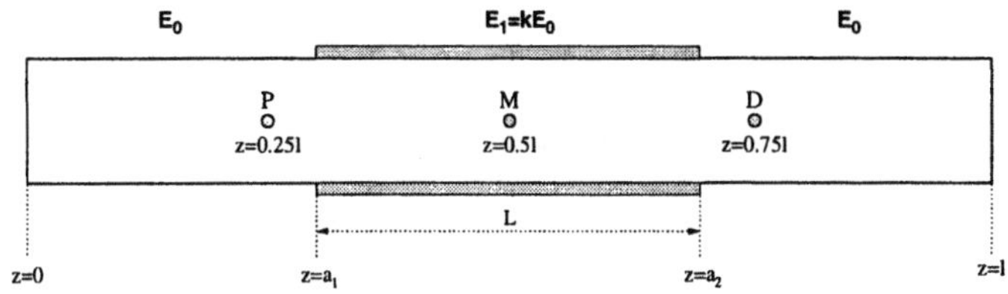

Fig. 3. The layout of our numerical experiment.

The impulse have been used to better highlight the reflections induced by the vascular prosthesis.

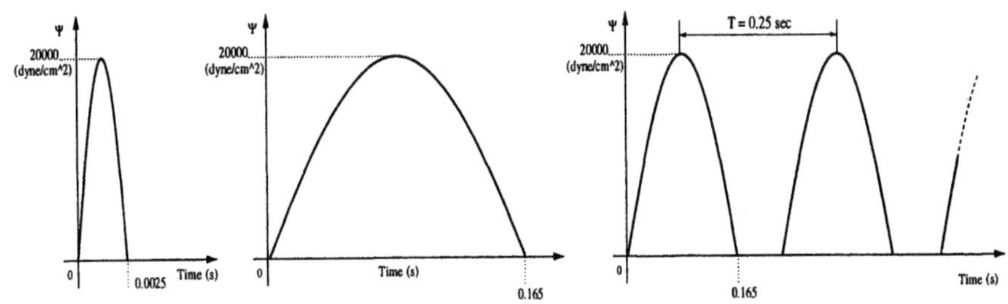

Fig. 4. The three types of pressure input profiles used in the numerical experiments: an impulse (left) a more realistic sine wave (middle) and a periodic sine wave (right).

Fig. 3 shows the layout of the numerical experiment. The part that simulates the presence of the prosthesis or stent of length L is comprised between coordinates a_1 and a_2. The corresponding Young's modulus has been taken as a multiple of the basis Young's modulus E_0 associated to the physiological tissue.

Three locations along the vessel have been identified and indicated by the letters D (distal), M (medium) and P proximal. They will be taken as monitoring point for the pressure variation. Different prosthesis length L have been considered; in all cases points P and D are located outside the region occupied by the prosthesis. Table 1 indicates the basic data which have been used in all numerical experiments. A time step $\Delta t = 2 \times 10^{-6}$s and the initial values $A = A_0$ and $Q = 0$ have been used throughout.

The boundary data for this numerical tests have been taken as follows. At the distal boundary $z = l$ we leave W_2 constant and equal to its initial value. This simulates a tube of "infinite" length. At the proximal boundary, we would like to impose the chosen pressure input $p(0,t) = \psi(t)$. Yet, imposing directly the pressure, as already noted in section 3.1, will produce a reflecting boundary condition. Therefore, we wish to transform the pressure condition

Table 1. Data used in the numerical experiments.

	Parameter	
FLUID	Input Pressure Amplitude	$20\text{x}10^3$dyne/cm^2
	Viscosity, ν	0.035poise
	Density, ρ	1g/cm^3
STRUCTURE	Young's Modulus, E_0	$3\text{x}10^6$dyne/cm^2
	Wall Thickness, h	0.05cm
	Reference Radius, R_0	0.5cm

in a condition on W_1, i.e. $W_1(0,t) = g_1(t)$, for an appropriate function g_1. To that purpose we note that W_1 may be written in terms of W_2 and p

$$W_1 = W_2 + 4\sqrt{\frac{2}{\rho}}\left(\sqrt{p + \beta\sqrt{A_0}}\right). \tag{20}$$

We have then chosen the boundary data g_1 by setting the pressure at the desired level while keeping W_2 in (20) equal to the its initial value $W_2^0(0)$ at the proximal boundary, i.e. we put

$$g_1(t) = W_2^0 + 4\sqrt{\frac{2}{\rho}}\left(\sqrt{\psi(t) + \beta(0)\sqrt{A_0}}\right). \tag{21}$$

Although this relation imposes the pressure only implicitly and not in exact terms, it has been proved very effective and enjoys good non-reflecting properties. Furthermore, sufficiently small t, when no W_2-waves generated at the prosthesis location have jet reached the proximal boundary, (21) is effectively equivalent to impose the desired value for p and indeed this is confirmed by the numerical experiments.

As previously pointed out, the formulation here adopted does not allow for a discontinuous variation of the Young modulus E. Therefore, we smoothed

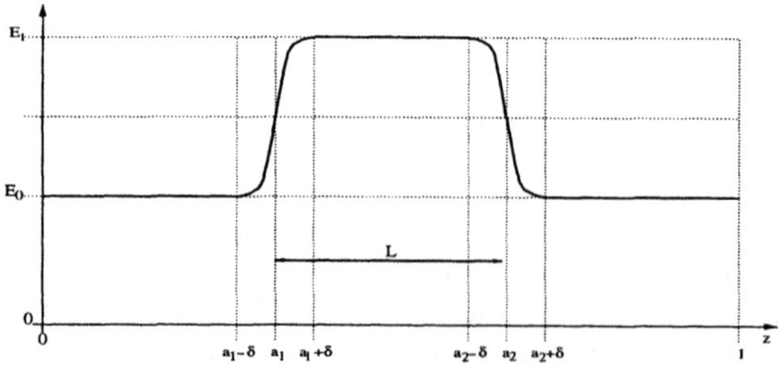

Fig. 5. Variation of Young's modulus.

out the transition between E_0 and E_1, as depicted in Fig. 5. A transition zone of thickness 2δ has been set around the point $z = a_1$ and $z = a_2$. In that region the Young modulus varies between E_0 and E_1 with a fifth order polynomial law.

4.1 Case of an impulsive pressure wave

In Fig. 6 we show the results obtained for the case of a pressure impulse. We compare the results obtained with uniform Young's modulus E_0 and the corresponding solution when $E_1 = 100E_0$, $L = a_2 - a_1 = 5$cm and $\delta = 0.5$cm. We have taken $l = 15$cm and a non uniform mesh of 105 finite elements, refined around the points a_1 and a_2. When the Young modulus is uniform, the impulse travels along the tube undisturbed. The numerical solution shows a little dissipation and dispersion due to the numerical scheme. In the case of varying E the situation changes dramatically. Indeed, as soon as the wave enters the region at higher Young's modulus it gets partially reflected (the reflection is registered by the positive pressure value at point P and $t \approx 0.015$s) and it accelerates. Another reflection occurs at the exit of the 'prosthesis', when E returns to its reference value E_0. The point M indeed registers an oscillatory pressure which corresponds to the waves that are reflected back and forth between the two ends of the prosthesis. The wave at point D is much weaker, because part of the energy has been reflected back and part of it has been 'captured' inside the prosthesis itself.

4.2 Case of a sine wave

Now, we present the case of the pressure input given by the sine wave with a larger period shown in Fig. 4, which describes a situation closer to reality than the impulse. We present again the results for both cases of a constant and a variable E. All other problem data have been left unchanged from the previous simulation. Now, the interaction among the reflected waves is more complex and eventually results in a less oscillatory solution (see Fig. 7). The major effect of the presence of the stent is a pressure build-up at the proximal point P, where the maximum pressure is approximately 2500dynes/cm^2 higher than in the constant case. By a closer inspection one may note that the interaction between the incoming and reflected waves shows up in discontinuities in the slope, particularly for the pressure history at point P. In addition, the wave is clearly accelerated inside the region where E is larger.

In table 2 we show the effect of a change in the length of the prosthesis by comparing the maximum pressure value recorded for a prosthesis of 4, 14 and 24 cm, respectively. The values shown are the maximal values in the whole vessel, over one period. Here, we have taken $l = 60$cm, $\delta = 1$cm, a mesh of 240 elements and we have positioned in the three cases the prosthesis in the middle of the model. The maximum value is always reached at a point upstream the prosthesis. In the table we give the normalised distance between

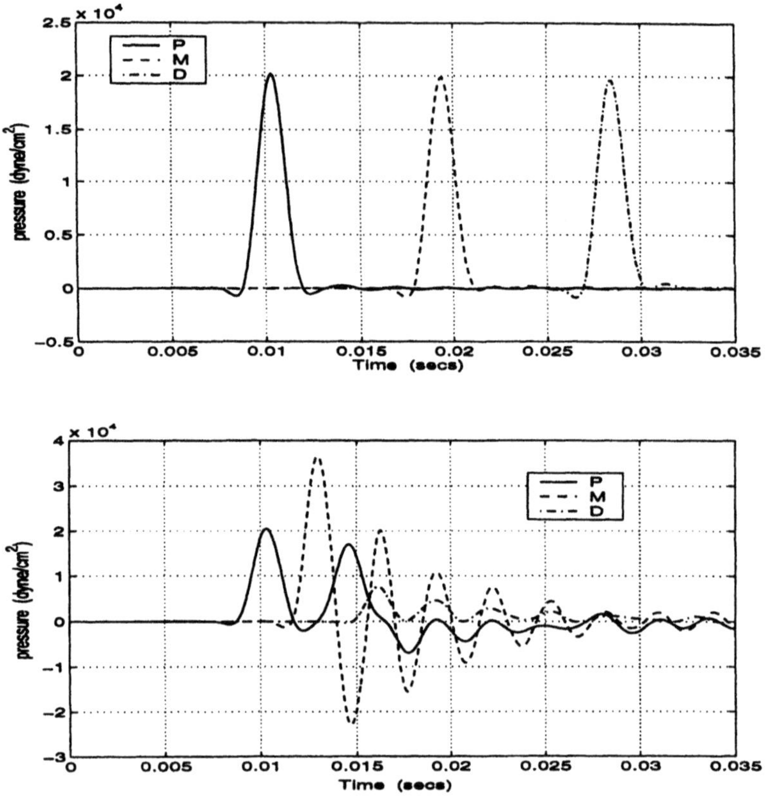

Fig. 6. Pressure history at points P, M and D of figure 3, for an impulsive input pressure, in the case of constant (upper) and variable (lower) E.

the upstream prosthesis section and of the point where the pressure attains its maximum.

Finally, we have investigated the variation of the pressure pattern due to an increase of $k = E/E_0$. Fig. 8 shows the result corresponding to $L = 20$cm and $\delta = 1$cm and various values for k. The numerical result confirms the fact that a stiffer prosthesis causes a higher excess pressure in the proximal region.

4.3 Case of a periodic sine wave

We consider here the case where the sine wave of the previous test case is repeated periodically with a period $T = 0.25$sec as illustrated in Fig. 4. We have taken $l = 120$cm and a prosthesis of 10cm between the points $a_1 = 70$cm and $a_2 = 80$cm. All other problem data have been left unchanged.

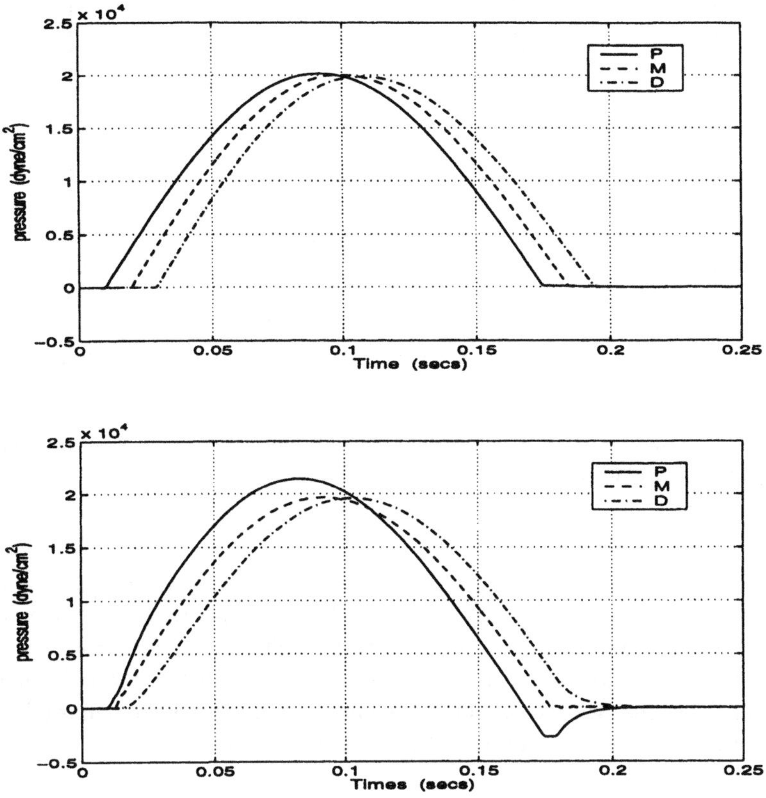

Fig. 7. Pressure history at points P, M and D of figure 3, for a sine wave input pressure, in the case of constant (upper) and variable (lower) E.

Table 2. Maximum pressure value for prosthesis of different length.

Prosthesis length (cm)	*Maximal pressure* (dyne/cm^2)	*Maximum location* $z_{\max}/l$
4	23.5×10^3	0.16
14	27.8×10^3	0.11
24	30.0×10^3	0.09

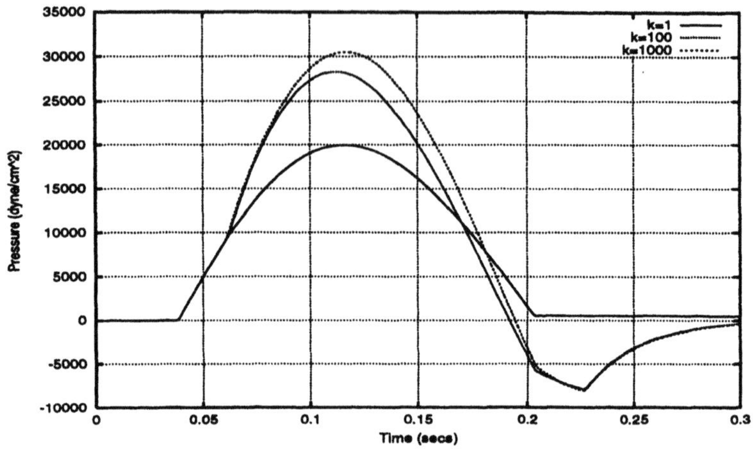

Fig. 8. Pressure history at point P of figure 3, for a sine wave input pressure and different Young's moduli $E = kE_0$.

We have simulated six periods. Fig. 9 shows the pressure at the proximal position $z = 40$cm, i.e. a point which is 30cm far from the prosthesis. In that position, the incoming pressure wave adds to the reflected one and the result is a build-up of the maximum pressure of approximately 2650dyne/cm^2. This simulation shows that the effects of the presence of a prosthesis are remarkable even far away form the prosthesis in the proximal region.

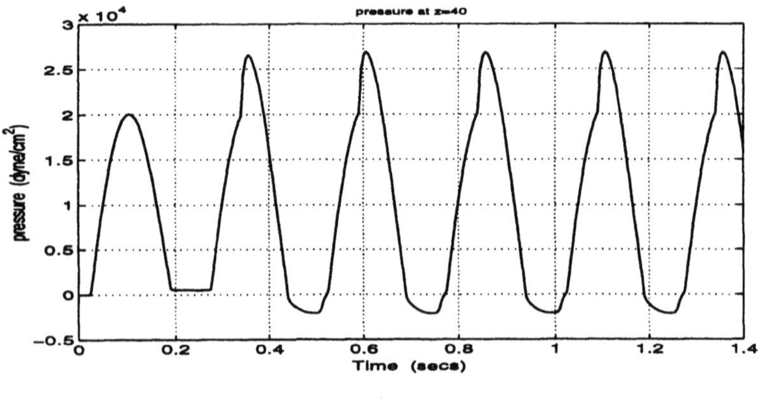

Fig. 9. Pressure history at point $z = 40$cm, for a periodic sine wave input, in the case of a prosthesis positioned between $a_1 = 70$cm and $a_2 = 80$cm.

Acknowledgements

We wish to thank Mr. Simon Tweddle from Imperial College, who wrote part of the software and carried out the numerical tests during a four month stay at the Mathematics Department of EPFL. We also gratefully acknowledge the enlightening discussions with Dr. M. Tuveri, who has motivated us to investigate stented arteries.

This research has been partially supported by the Swiss National Research Foundation, under the research grant N.21-54139.98.

References

1. R. Botnar, G. Rappitsch, M.B. Scheidegger, D. Liepsch, K. Perktold, and P. Boesiger. Hemodynamics in the carotid artery bifurcation: A comparison between numerical simulations and in-vitro measurements. *J. of Biomech.*, 33:137–144, 2000.
2. S. Canic. On the shock formation in pulsaltile blood flow through a "stented" aorta modeled by a system of hyperbolic conservation laws with discontinuous coefficients. (In Preparation), 2000.
3. L. Formaggia, J.-F. Gerbeau, F. Nobile, and A.Quarteroni. On the coupling of 3D and 1D Navier-Stokes equations for flow problems in compliant vessels. Technical Report 3862, INRIA, 2000. to appear in Comp. Methods in Appl. Mech. Engng.
4. E. Godlewski and P.-A. Raviart. *Numerical approximation of hyperbolic systems of conservation laws*, volume 118 of *Applied Mathematical Sciences*. Springer, New York, 1996.
5. K. Hayashi, K. Handa, S. Nagasawa, and A. Okumura. Stiffness and elastic behaviour of human intracranial and extracranial arteries. *J. Biomech.*, 13:175–184, 1980.
6. T.H. Hughes, C. Taylor, and C. Zarins. Finite element modeling of blood flow in arteries. *Comp. Meth. Appl. Mech. Eng.*, 158:155–196, 1998.
7. G.L. Langewouters, K.H. Wesseling, and W.J.A. Goedhard. The elastic properties of 45 human thoracic and 20 abdominal aortas *in vitro* and the parameters of a new model. *J. Biomech.*, 17:425–435, 1984.
8. L.Formaggia, F. Nobile, A. Quarteroni, and A. Veneziani. Multiscale modelling of the circulatory system: a preliminary analysis. *Computing and Visualisation in Science*, 2:75–83, 1999.
9. F. Phythoud, N. Stergiopulos, and J.-J. Meister. Forward and backward waves in the arterial system: nonlinear separation using Riemann invariants. *Technology and Health Care*, 3:201–207, 1995.
10. A. Quarteroni, M. Tuveri, and A. Veneziani. Computational vascular fluid dynamics: Problems, models and methods. *Computing and Visualisation in Science*, 2:163–197, 2000.
11. A. Quarteroni and A. Valli. *Domain decomposition methods for partial differential equations.* Oxford University Press, Oxford, 1999.
12. A. Tozeren. Elastic properties of arteries and their influence on the cardiovascular system. *ASME J. Biomech. Eng.*, 106:182–185, 1984.

Essential Spectrum and Mixed Type Finite Element Method

Takashi Kako and Haniffa M. Nasir

The University of Electro-Communications, Chofu, Tokyo 182-8585, Japan

Abstract. In the error analysis of finite element method, the inf-sup condition or the uniform lifting property plays an important role. In this paper, we discuss the relationship between the uniform inf-sup condition and the essential spectrum of the operator that appears in the problem. In general, one can not expect the convergence of the finite element approximation due to the spectral pollution that stems from the inappropriate mixing of the eigen-subspaces that correspond to two distinct components of the essential spectrum. As examples of our consideration, we treat the Stokes problem, mixed approximations of elliptic problems and a structural-acoustic coupling problem. In these problems, two distinct components might appear in the essential spectrum of the corresponding operators.

1 Introduction

In the finite element method, to assure the convergence of a sequence of approximate solutions to the exact solution, we have to choose appropriate approximation subspaces. In the case of conforming finite element method, it is well known that the inf-sup condition gives an abstract sufficient condition for the convergence (see Babuška [1]). On the other hand, Descloux-Nassif-Rappaz [9] introduced a necessary and sufficient condition for the non-pollution of the approximate spectra in the finite element approximation for a bounded linear operator. One of the main topics of this paper is to point out a relation between these conditions with some typical examples.

As the application of these considerations, we investigate the mixed method for the Poisson equation and a simple 1-D Stokes problem. The essential spectrum of an operator plays an important role in the analysis.

We also consider a fictitious domain method applied to a structural-acoustic coupling problem as an example of a mixed finite element method and show some numerical results without pollution although its exact error analysis has not yet been done.

2 Finite Element Method and Abstract Error Estimation

The error analysis for the conforming finite element method has been investigated for a long time since the works of Babuška [1] and Brezzi [5]. In the following, we review the basic formulation for the finite element method and the fundamental abstract error analysis of the problem in the framework of operator theory.

Let X be a Hilbert space with inner product $(\cdot,\cdot)$ and X_h, $0 < h \le h_0$, be a family of its finite dimensional subspaces, and let P_h be the orthogonal projections from X onto X_h. Then we consider the following linear equation for the self-adjoint operator A in X :

$$Ax = f \tag{1}$$

and its approximation problem:

$$A_h x_h = f_h, \quad \text{with} \quad A_h = P_h A|_{X_h} \quad \text{and} \quad f_h = P_h f. \tag{2}$$

As for the weak formulation of this problem, we introduce the bounded bilinear form $a(\cdot,\cdot)$ associated with A as follows:

$$a(x,y) = (Ax,y) \quad \text{for all } x,y \in X,$$

and get equivalent problems to (1) and (2): To find $x \in X$ such that

$$a(x,y) = (f,y) \quad \text{for all } y \in X, \tag{3}$$

and to find $x_h \in X_h$ such that

$$a(x_h,y_h) = (f,y_h) \quad \text{for all } y_h \in X_h. \tag{4}$$

In the following, we assume the unique existence of the solution of the equation (1) and hence a bounded inverse of A. This is equivalent to assume that the range of A is dense:

$$a(x,y) = 0 \quad \text{for all } x \in X \quad \text{implies} \quad y = 0,$$

and the following inf-sup condition is satisfied:

$$\inf_{x\in X} \sup_{y\in X} \frac{a(x,y)}{||x||\,||y||} > 0.$$

As for the approximation problem, the uniform inf-sup condition:

$$c_{IS} \equiv \inf_{h\in(0,h_0]} \inf_{x_h\in X_h} \sup_{y_h\in X_h} \frac{a(x_h,y_h)}{||x_h||\,||y_h||} > 0,$$

which is equivalent to the uniform invertibility of the approximate problems:

$$\sup_{h\in(0,h_0]} ||A_h^{-1}|| \le c_{IS}^{-1} < \infty$$

plays an essential role. Namely, under this uniform invertibility condition, we obtain the abstract basic error estimate due to Babuška[1]:

$$\|x - x_h\| \le c\,\|(1-P_h)x\|.$$

The proof of this fact is given as follows. For any y_h in X_h, we have

$$\begin{aligned} x - x_h &= x - y_h + y_h - x_h = x - y_h + (A_h)^{-1}(A_h y_h - f_h) \\ &= x - y_h + (A_h)^{-1}(A_h y_h - P_h f) \\ &= x - y_h + (A_h)^{-1}(A_h y_h - P_h A x) \\ &= x - y_h + (A_h)^{-1} P_h A(y_h - x) \\ &= \{1 - (A_h)^{-1} P_h A\}(x - y_h). \end{aligned}$$

Hence using the uniform boundedness of $\|(A_h)^{-1}\|$, we get

$$\begin{aligned} \|x - x_h\| &\leq (1 + \|(A_h)^{-1}\| \cdot \|P_h A\|) \inf_{y_h \in X_h} \|x - y_h\| \\ &\leq (1 + \{\sup_h \|(A_h)^{-1}\|\} \|A\|) \|(1 - P_h)x\| \\ &\leq c\, \|(1 - P_h)x\|. \end{aligned} \tag{5}$$

Combining this estimate with the approximation property of X_h (see (P-1) condition (31) in Section 4) we have the convergence of the approximate solution, and higher regularity of the solution implies the corresponding higher order of convergence of the approximate solution (see for example [4], [3] and [7]).

3 Application to coupled systems

Various problems such as the mixed formulation of the Poisson equation, the Stokes problem, the electro-magnetic problem and the domain decomposition method with interfacial conditions can be written in the following form of a coupled system:

$$T \begin{bmatrix} u \\ p \end{bmatrix} = \begin{bmatrix} f \\ g \end{bmatrix} \quad \text{with} \quad T \equiv \begin{pmatrix} A & B^* \\ B & 0 \end{pmatrix}. \tag{6}$$

Here, A is a selfadjoint operator in some Hilbert space V and B is an operator from V to another Hilbert space W with its adjoint B^*.

We introduce a direct product X of Hilbert spaces V and W: $X \equiv V \times W$. Then the operator T becomes selfadjoint in X. The weak formulation of the equation (6) which is used for the finite element method is written as to find $u \in V$ and $p \in W$ such that

$$a(u, v) + b^*(p, v) = (f, v)_V, \quad \forall v \in V, \tag{7}$$

$$b(u, q) = (g, q)_W, \quad \forall q \in W, \tag{8}$$

where $a(u, v) = (Au, v)_V$, $b^*(p, v) = (B^* p, v)_V$ and $b(u, q) = (Bu, q)_W$. We note the relation:

$$b(v, q) = (Bv, q)_W = (v, B^* q)_V = \overline{b^*(q, v)}. \tag{9}$$

In the following we show some typical examples which can be written in this form.

Example 1 (Mixed method for the Poisson equation): Consider the following Poisson equation:

$$\Delta p = f, \quad p \in H^2(\Omega) \cap H_0^1(\Omega), \tag{10}$$

where Ω is a bounded region in the N-dimensional Euclidean space R^N, and $H^2(\Omega)$ and $H_0^1(\Omega)$ are the usual Sobolev spaces on Ω. We assume that the inhomogeneous term f belongs to $L^2(\Omega)$. For the mixed formulation of the problem, we introduce u as $u \equiv \text{grad}\, p$, and put

$$V = H(\text{div}; \Omega) \equiv \{u | u \in (L^2(\Omega))^n,\ \text{div}\, u \in L^2(\Omega)\} \quad \text{and} \quad W = L^2(\Omega), \tag{11}$$

and define

$$A \equiv (I - \text{grad} \cdot \text{div})^{-1}, \quad B \equiv \text{div} \quad \text{and} \quad B^* \equiv -A\, \text{grad}. \tag{12}$$

Then (10) can be written in the form (6). The corresponding weak formulation is given as

$$(u, v)_{L^2} + (p, \text{div}\, v)_W = 0, \tag{13}$$

$$(\text{div}\, u, q)_W = (f, q)_W, \tag{14}$$

with $(u, v)_V = (u, v)_{L^2} + (\text{div}\, u,\ \text{div}\, v)_{L^2}$ and $(p, q)_W = (p, q)_{L^2}$. We note that the relation $a(u, v) = (Au, v)_V = (u, v)_W$ holds.

Example 2 (The Stokes problem): Let v be a velocity and p a pressure for the incompressible fluid under the external force f, then the stationary Stokes problem is written as

$$-\Delta v + \text{grad}\, p = f, \tag{15}$$

$$-\text{div}\, v = 0. \tag{16}$$

Introduce function spaces $V \equiv H_0^1(\Omega)^3$, $W = L_0^2(\Omega) \equiv L^2(\Omega) \cup \{p | \int_\Omega p\, dx = 0\}$ and let $A \equiv I$, $B \equiv -\text{div}$ and $B^* \equiv (-\Delta)^{-1}\text{grad}$. Then the problem takes the form (6) with $(u, v)_V \equiv (\text{grad}\, u,\ \text{grad}\, v)_{L^2}$ and $(p, q)_W \equiv (p, q)_{L^2}$.

4 Eigenvalue problem for coupled systems and finite element method

The spectrum of the selfadjoint operator T which appears in the equation (6) is a closed subset of the real line, and the essential spectrum of the operator T is defined as

$$\sigma_{ess}(T) \equiv \{\lambda \in \mathbf{C} \mid \exists \{f_j\},\ (f_j, f_k) = \delta_{jk},\ \|Tf_j - \lambda f_j\| \to 0,\ j \to \infty\}. \tag{17}$$

If this set includes two mutually distinct non-empty subsets, the finite element approximation of its spectrum could be suffered from the so-called spectral pollution in the sense that there could be a point in the resolvent set of T in the convex hull of two spectral subsets to which a sequence of eigenvalues of the approximate operator converges. The simplest example of the spectral pollution can be found in Kako [13] (see also Kako [14]) and is given as follows.

4.1 A simple example of the spectral pollution

Let $\mathcal{H}$ be a Hilbert space and let $\{\phi_j\}_{j=1}^{\infty}$ be a set of orthonormal vectors in $\mathcal{H}$. We consider the following eigenvalue problem in the direct product space $\mathcal{X} \equiv \mathcal{H} \times \mathcal{H}$:

$$T \begin{bmatrix} u \\ v \end{bmatrix} \equiv \begin{pmatrix} \alpha & 0 \\ 0 & \beta \end{pmatrix} \begin{bmatrix} u \\ v \end{bmatrix} = \lambda \begin{bmatrix} u \\ v \end{bmatrix}. \tag{18}$$

Here α and β are two distinct complex numbers. The operator T in the left-hand side has the essential spectrum which consists of two points α and β:

$$\sigma_{ess}(T) = \{\alpha\} \cup \{\beta\}. \tag{19}$$

Now introduce a special sequence of approximation subspaces $\mathcal{X}_n \subset \mathcal{X}$, and consider the Ritz-Galerkin approximation. Let s_n and t_n be sequences of real numbers which satisfy $s_n^2 + t_n^2 = 1$ for each integer $n > 0$. We define $\mathcal{X}_n$ as

$$\mathcal{X}_n \equiv \{ \begin{bmatrix} u \\ v \end{bmatrix} \mid \begin{bmatrix} u \\ v \end{bmatrix} = \Sigma_{k=1}^{n} \begin{bmatrix} c_k\phi_k \\ d_k\phi_k \end{bmatrix} \text{ with arbitrary complex numbers}$$

$$c_k, d_k, 1 \le k \le n-1, \text{and the } n\text{-th coefficients} \begin{bmatrix} c_n \\ d_n \end{bmatrix} = b_n \begin{bmatrix} s_n \\ t_n \end{bmatrix}$$

$$\text{with an arbitrary complex number } b_n \}.$$

Then the Ritz-Galerkin approximation of the eigenvalues of T consists of α, β and $s_n^2\alpha + t_n^2\beta$. The last value is situated on the interval between α and β. From this observation, choosing s_n and t_n appropriately, we can construct a sequence of subspaces for which the approximate eigenvalue converges to any point on the interval between α and β. Furthermore, changing $\mathcal{X}_n$ appropriately we can construct an example where any closed subset of the interval $[\alpha, \beta]$ becomes a set of accumulation points of approximate eigenvalues. This is a typical simple(est) example of pollution phenomena which is sometimes observed in the finite element approximation of the saddle point problem.

4.2 Examples of essential spectrum

In the following, we compute the essential spectrum of the operators appearing in the problems introduced in Section 3, and we can conclude that the spectral pollution might occur for the problems involving these operators.

Example 1 (Mixed formulation for the Poisson equation): The operator

$$T = T_M \equiv \begin{pmatrix} (I - \text{grad} \cdot \text{div})^{-1} & -(I - \text{grad} \cdot \text{div})^{-1} \text{ grad} \\ \text{div} & 0 \end{pmatrix} \tag{20}$$

has the essential spectrum given as follows.

$$\sigma_{ess}(T_M) = \{1\} \cup \{-1\}. \tag{21}$$

Hence the origin $0 \in [-1, 1]$ can be polluted for a certain special finite element approximation. The fact (21) can be shown as follows. The concrete expression of the eigenvalue problem: $T_M(u,p)^t = \lambda(u,p)^t$ is given as

$$\begin{aligned} Au + B^*p &= \lambda u \\ Bu &= \lambda p. \end{aligned} \tag{22}$$

The corresponding weak formulation is written as

$$\begin{aligned} (u, v)_{L^2} + (p, \text{div } v)_{L^2} &= \lambda\{(u, v)_{L^2} + (\text{div } u, \text{div } v)_{L^2}\} \\ (\text{div } u, q) &= \lambda(p, q)_{L^2}. \end{aligned} \tag{23}$$

Eliminating u from this system of equations, we can deduce a condition for p:

$$(\lambda^2 - 1)(-\Delta)_D p = -\lambda(\lambda - 1)p, \tag{24}$$

where $-\Delta_D$ denotes the Laplacian with homogeneous zero Dirichlet boundary condition. Hence $\lambda = 1$ is always the point spectrum with infinite multiplicity which belongs to the essential spectrum. A sequence of eigenvalues $\{\mu_j\}_{J=1,2,\ldots}$ of $-\Delta_D$ which tends to infinity gives a sequence of eigenvalues $\{\lambda_j\}_{j=1,2,\ldots}$ of T_M:

$$\lambda_j = -1 + \frac{1}{1 + \mu_j} \tag{25}$$

which tends to -1. Hence -1 also belongs to the essential spectrum of T_M.

Example 2 (The Stokes problem): Let us denote by $[\,\cdot\,]$ the closure of an operator. The operator $T = T_S$ in this case is written as

$$T = T_S \equiv \begin{pmatrix} I & [(-\Delta_D)^{-1} \text{ grad}] \\ -\text{div} & 0 \end{pmatrix}. \tag{26}$$

The essential spectrum of T_S then becomes

$$\sigma_{ess}(T_S) = \{1\} \cup \{\frac{1 + \sqrt{5}}{2}\} \cup \{\frac{1 - \sqrt{5}}{2}\}. \tag{27}$$

Hence also in this case, the origin 0 can be polluted. The proof of (27) is given as follows.

After some calculation we see that $\lambda = 1$ is an eigenvalue with infinite multiplicity whose eigen-functions consist of those pairs of functions satisfying the condition:

$$\operatorname{div} u = 0, \quad p = 0. \tag{28}$$

In case that $\Delta_D\, p \neq 0$ with non-zero p, we can deduce the following non-trivial condition

$$(\lambda^2 - \lambda - 1)(-\Delta_D)p = 0 \tag{29}$$

which leads to the second order algebraic equation in λ:

$$\lambda^2 - \lambda - 1 = 0. \tag{30}$$

Hence the roots of this equation $(1 \pm \sqrt{5})/2$ are eigenvalues of T_S with infinite multiplicity and hence belong to the essential spectrum.

As for the non-pollutive spectral approximation, we mention the following results by Descloux-Nassif-Rappaz ([9]) under the conditions (P-1) and (P-2) below. Let us introduce a family of subspaces of X as $X \supset X_h, \quad h \in (0, h_0]$, and define the orthogonal projection P_h from X onto X_h: $(P_h x, y_h) = (x, y_h), \quad \forall y_h \in X_h$. We introduce conditions (P-1) and (P-2) as:

$$\text{(P-1)} \quad \lim_{h\to 0} P_h x = x, \quad \forall x \in X, \tag{31}$$

$$\text{(P-2)} \quad \lim_{h\to 0} \|(I - P_h)TP_h\| = 0. \tag{32}$$

Then we have the following

Theorem 1 ([9]). *The conditions (P-1) and (P-2) are the necessary and sufficient conditions for that the spectrum and eigenspaces of the operator T can be approximated by those of T_h without pollution.*

5 Inf-sup condition and spectral pollution

In this section, we first briefly discuss the relations among the conditions (P-1) and (P-2), the spectral pollution and the inf-sup condition. Next we show a typical example of spectral pollution phenomena related to the 1-D Stokes problem.

5.1 Abstract results

We have the following results (see [14]).

Theorem 2. *Consider the finite element approximation of a uniquely solvable problem: $Tu = f$ with a bounded selfadjoint operator T in a Hilbert space*

X. Let $X_h, 0 < h \le h_0$ be a family of finite dimensional subspaces with associated orthogonal projections P_h and approximation operators $T_h = P_h T|_{X_h}$. Then the uniform inf-sup condition:

$$0 < \inf_{h\in(0,h_0]} \inf_{x_h\in X_h} \frac{\|T_h x_h\|}{\|x_h\|} = \inf_{h\in(0,h_0]} \inf_{x_h\in X_h} \sup_{y_h\in X_h} \frac{|(T_h x_h, y_h)|}{\|x_h\| \, \|y_h\|} \tag{33}$$

is satisfied if and only if the origin 0 is not polluted.

Proof. If the origin is not polluted, there exists a neighborhood of the origin:

$$B_\epsilon \equiv \{z| \ |z| < \epsilon\} \tag{34}$$

such that, for sufficiently small h, B_ϵ is included in the resolvent set of $T_h, h \in (0,\ h_0]$:

$$B_\epsilon \subset \rho(T_h), \quad h \in (0,\ h_0], \tag{35}$$

which implies the uniform inf-sup condition.

On the other hand, if the pollution occurs in the sense that there exists a sequence of approximate eigenvalues which converges to the origin which belongs to the resolvent set of T by the solvability of the original problem:

$$\begin{aligned} T_h x_h &= \lambda_h x_h, \quad \|x_h\| = 1, \\ \lambda_h &\to 0, \quad \text{as} \quad h \to 0. \end{aligned} \tag{36}$$

If we consider (33) for the eigenvectors x_h which causes the pollution, it is easily seen that the uniform inf-sup condition is violated.

For the finite element approximation of the inhomogeneous problem (6), we have the following results (see [14]).

Theorem 3. *Let us assume that the problem (6) is solvable and also assume that $0 \notin \sigma(T)$. Then, under the conditions (P-1) and (P-2), the solution u_h, p_h of the finite element approximation problem:*

$$\begin{bmatrix} P_h^V A & P_h^V B^* \\ P_h^W B & 0 \end{bmatrix} \begin{bmatrix} u_h \\ p_h \end{bmatrix} = \begin{bmatrix} P_h^V f \\ P_h^W g \end{bmatrix} \tag{37}$$

converges to the solution u, p of the original equation. Here P_h^V and P_h^W are the orthogonal projections from V and W onto their respective subspaces V_h and W_h.

5.2 One dimensional Stokes problem

We consider the 1-D Stokes problem as one of the simplest typical examples for the possible spectral pollution. The problem is given as to find functions

$v(x)$ and $p(x)$ on $[0,1]$ which satisfy the system of equations

$$-\frac{d^2}{dx^2}v + \frac{d}{dx}p = f_0, \tag{38}$$

$$-\frac{d}{dx}v = g \tag{39}$$

with boundary condition $v(0) = v(1) = 0$. Applying the operator $(-d^2/dx^2)^{-1}$ to the first equation, we obtain the following equation in $X \equiv V \times W$ with $V = H_0^1(0,1)$ and $W = L_0^2(0,1) \equiv L^2(0,1) \cap \{p \mid \int_0^1 p(x)dx = 0\}$:

$$T\begin{bmatrix} f \\ g \end{bmatrix} = \lambda \begin{bmatrix} f \\ g \end{bmatrix} \quad \text{with} \quad T = \begin{pmatrix} 1 & [(-\frac{d^2}{dx^2})_D^{-1}\frac{d}{dx}] \\ -\frac{d}{dx} & 0 \end{pmatrix}. \tag{40}$$

We choose as the finite element subspace $X_h \subset X$ the set of pairs of functions v_h and p_h which are continuous and piecewise linear with respect to the equi-partition of $[0,1]$ and $v_h(0) = v_h(1) = 0$. Let n be the number of sub-intervals in the partition and let $T_h \equiv P_h T|_{X_h}$. Then a special pair of functions v_{0h} and p_{0h}, which are given as

$$v_{0h}(x) \equiv 0$$

$$p_{0h}(x) = \begin{cases} 1, & \text{at} \quad x = kh/n \text{ with } k = \text{even integer}, \\ -1, & \text{at} \quad x = kh/n \text{ with } k = \text{odd integer}, \\ & \text{linear function for rest } x, \end{cases} \tag{41}$$

belongs to X_h and satisfies the condition: $T_h(v_h, p_h)^t = 0 \cdot (v_h, p_h)^t$. Essential point to prove this is the fact that $(-d^2/dx^2)^{-1}(d/dx)p_h$ is orthogonal to every function in V_h.

Namely the origin 0 is always the eigenvalue of T_h although 0 is not in the spectrum of T. This is a typical pollution phenomena for the Stokes problem which violates the uniform inf-sup condition.

On the other hand, we can explain this phenomena as the spectral pollution of the origin due to the fact that the essential spectra of T is given as $\sigma_{ess}(T)(= \sigma(T)) = \{\lambda_+ = (1+\sqrt{5})/2\} \cup \{\lambda_- = (1-\sqrt{5})/2\}$ and hence the origin belongs to the convex hull of the essential spectra. The pair of functions v_{0h} and p_{0h} are expressed as the sum of eigenfunctions of T with eigenvalue λ_+ and λ_- as

$$\begin{bmatrix} v_{0h} \\ p_{0h} \end{bmatrix} = \begin{bmatrix} v_{0h}^+ \\ p_{0h}^+ \end{bmatrix} + \begin{bmatrix} v_{0h}^- \\ p_{0h}^- \end{bmatrix} \tag{42}$$

with

$$p_{0h}^{\pm}(x) = \mp \frac{\lambda_\mp}{\lambda_+ - \lambda_-} p_{0h}(x), \qquad v_{0h}^{\pm}(x) = \mp \frac{\lambda_+\lambda_-}{\lambda_+ - \lambda_-} \int_0^x p_{0h}(y)dy. \tag{43}$$

From this expression, we can compute the approximate eigenvalue by finite element method as

$$\lambda_h = \lambda_+(||v_{0h}^+||^2 + ||p_{0h}^+||^2) + \lambda_-((||v_{0h}^-||^2 + ||p_{0h}^-||^2)) \tag{44}$$

and after some calculation, we obtain once more the fact that $\lambda_h = 0$. This elementary observation confirms the argument in the previous section that the spectral pollution due to distinct two essential spectra is the origin of the violation of the inf-sup condition.

6 Fictitious domain method as an example of mixed method

In this section we briefly study a structural-acoustic coupling problem applying the fictitious domain method. The problem can be formulated as a mixed type and hence there is a possibility of pollution phenomena. Nevertheless, in the following numerical examples, we do not see any pollution. Its mathematical justification is a future work.

This type of coupling problems is important for the noise reduction inside motorcars, trains, airplanes and others. The mathematical model of the problem is written as to find the inside and outside pressure deviations $p_i, i = 1, 2$ and the shell deformations $\mathbf{u} = (u_1, u_2)$ satisfying

$$
\begin{aligned}
\frac{\partial^2 p_i}{\partial t^2} - c_i^2 \Delta p_i = f_i \qquad & \text{in } \Omega_i, \quad i = 1, 2, \\
\frac{\partial p_1}{\partial n} = -\rho_1 \frac{\partial^2 u_2}{\partial t^2} \quad & \text{on } S, \\
\frac{\partial p_2}{\partial n} = -\rho_2 \frac{\partial^2 u_2}{\partial t^2} \quad & \text{on } S, \\
\rho_0 \frac{\partial^2 \mathbf{u}}{\partial t^2} + \mathbf{A}\mathbf{u} = (p_1 - p_2)|_S \mathbf{n}, &
\end{aligned} \tag{45}
$$

where ρ_0 is the surface mass density of the shell, $\rho_i, i = 1, 2$ are the densities and $c_i, i = 1, 2$ are sound velocities of inside and outside acoustic media respectively, Δ is the Laplacian, A is the shell force operator, $\mathbf{n}$ is the unit normal on the shell towards outside and $f_i, i = 1, 2$ are sound sources situated inside and outside the shell.

The problem can be formulated mathematically in a simular way as was developed in Deng-Kako [8]. In computing the acoustic fields, we use the fictitious domain method following the idea of Heikkola et al [11] and Kuznetsov-Lipnikov [17]. The treatment of outer infinite domain where the Sommerfeld radiation condition is assumed at infinity is based on the methods in [12], [18] and [19] . As for the finite element subspaces for acoustic pressure fields, we use piecewise linear continuous functions on almost uniform mesh with respect to polar coordinates except the vicinity of the shell where the acoustic mesh is modified to adapt the shell. On the other hand, for the shell deformation variables, one dimensional Fourier spectral method is used.

6.1 Mixd type formulation

Time stationary problem for this type of coupled vibration is formulated as a similar system of equations to the mixed formulation of the Poisson equation or to the Stokes problem. The shell deformation variable plays the role of the Lagrange multiplier. Introducing a family of finite demensional function spaces, we obatin the following matrix equation:

$$\begin{bmatrix} M_1 & 0 & 0 & -k^2L_1^T \\ 0 & M_2 & 0 & k^2L_2^T \\ 0 & 0 & A & B^T \\ -L_1 & L_2 & B & C \end{bmatrix} \begin{bmatrix} P_1 \\ P_2 \\ U_1 \\ U_2 \end{bmatrix} = \begin{bmatrix} 0 \\ \mathbf{F} \\ 0 \\ \mathbf{G} \end{bmatrix}, \tag{46}$$

where k is a given wave number and P_1, P_2 and $\mathbf{U} = (U_1, U_2)^t$ are the unknown nodal values of the corresponding continuous quantities p_1, p_2 and $\mathbf{u} = (u_1, u_2)^t$ and

$M_1 : (\nabla\phi_j^{(1)}, \nabla\phi_i^{(1)})_{\Omega_1} - \rho_1 k^2(\phi_j^{(1)}, \phi_i^{(1)})_{\Omega_1}$
; energy form for inside acoustic field,

$M_2 : (\nabla\phi_j^{(2)}, \nabla\phi_i^{(2)})_{\Omega_2} - \rho_2 k^2(\phi_j^{(2)}, \phi_i^{(2)})_{\Omega_2} - (\mathcal{M}\phi_j^{(2)}, \phi_i^{(2)})_{\Gamma_\infty}$
; energy form for outside acoustic field,

$L_1 : (\xi_j, \phi_i^{(1)})_S$
; coupling term between shell and inner pressure field,

$L_2 : (\xi_j, \phi_i^{(2)})_S$
; coupling term between shell and outer pressure field,

$A : (A_1\psi_j, \psi_i)_S - \rho_0 k^2(\psi_j, \psi_i)_S$
; energy form for tangential deformation of shell,

$B : (B_1\xi_j, , \psi_i)_S$
; coupling term between tangential and normal deformations,

$C : (C_0\xi_j, \xi_i)_S - \rho_0 k^2(\xi_j, \xi_i)_S$
; energy form for normal deformation of shell,

$\mathcal{M}$: Dirichlet to Neumann operator or its approximation on the outer artificial bounady used in the boundary condition: $\partial_r u = \mathcal{M}u$.

This resultant linear equation can be solved efficiency by the iteration method of the conjugate gradient or the minimal residual type.

6.2 Numerical examples

Figures 1 and 2 show two numerical examples of our numerical method where the shell is of elliptic shape with major radius 3.0 and minor raidus 1.0. The incident plane wave with wave numbers k comes from the lefthand side and we computeded the scattered wave for the cases with wave numbers 2π and 3π. The artificial boundary is the circle with radius 2.0. The detailed explanation of the method as well as various numerical results will be published elsewhere.

Major axis = 3.0,
Minor axis = 1.0,
Radius of artificial boundary = 2.0,
Incident wave : Plane wave
Wave Number = 2π,
proc-mscom.tex Shell density = 1.0

Fig. 1. Elliptic shell: Case 1

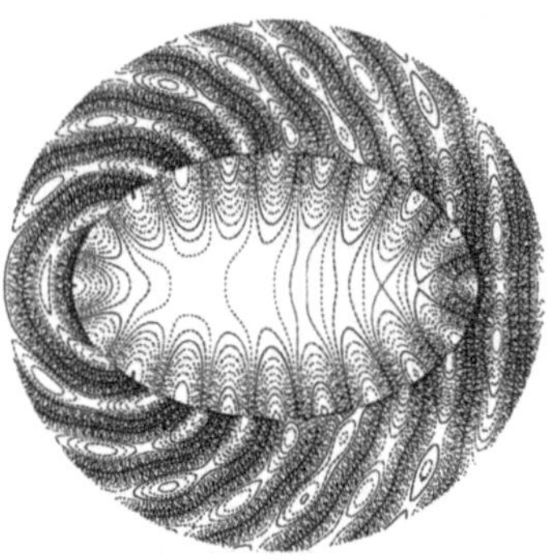

Major axis = 3.0,
Minor axis = 1.0,
Radius of artificial boundary = 2.0,
Incident wave : Plane wave
Wave Number = 3π,
Shell density = 1.0

Fig. 2. Elliptic shell: Case 2

References

1. Babuška, I., Error-bounds for finite element method, Numer. Math., **16**, 322–333 (1971).

2. Babuška, I., The finite element method with Lagrangian multipliers, Numer. Math., **20**, 179–192 (1973).
3. Braess, D., *Finite Elements, Theory, Fast Solvers and Applications in Solid Mechanics*, Cambridge University Press,1997.
4. Bramble, J.H. and Hilbert, S.R., Bounds for a class of linear functionals with applications to Hermite interpolation, Numer. Math., **16**, 362–369 (1971).
5. Brezzi, F., On the existence, uniqueness and approximation of saddle- point problems arising from Lagrangian multipliers, RAIRO Anal. Numér., **8**, 129–151 (1974).
6. Brezzi, F. and Fortin, M., *Mixed and Hybrid Finite Element Methods*, Springer-Verlag (1991).
7. Ciarlet, P. G. and Lions, J. L., *Handbook of Numerical Analysis, Vol II*, Finite element methods, North Holland, 1991.
8. Deng, L. and Kako, T., Finite element approximation of eigenvalue problem for a coupled vibration between acoustic field and plate. J. Compu. Math., **15**, no. 3, 265–278 (1997).
9. Descloux, J., Nassif, N. and Rappaz, J., On spectral approximation. Part 1. The problem of convergence, RAIRO Anal. Numér., **12**, 87–112 (1978).
10. Griffiths, D.F., Discrete eigenvalue problems, LBB constants and stabilization, Numerical Analysis 1995, D F Griffiths and G A Watson eds., 57–75 (1996).
11. Heikkola, E., Kusnetsov, Y. A., Neittaanmaki, P. and Toivanen J., Fictitious domain methods for the numerical solution of two-dimensional scattering problems, J. Comput. Phys., **145**, pp. 89–109 (1998).
12. Kako, T., Approximation of scattering state by means of the radiation boundary condition, Math. Meth. in the Appl. Sci., **3**, pp. 506–515 (1981).
13. Kako, T., MHD numerical simulation and numerical analysis, Journal of the Japan Society for Simulation Technology, Vol.12, 91-98 (1993) (in Japanese).
14. Kako, T., Remark on the relation between spectral pollution and inf-sup condition, GAKUTO International Series Math. Sci. and Appli., **11**, Recent Developments in Domain Decomposition Methods and Flow Problems, 252–258 (1998).
15. Kako, T. and Descloux, J., Spectral approximation for the linearized MHD operator in cylindrical region, Japan J. Indust. Appl. Math., Vol.8, 221-244 (1991).
16. Kikuchi, F., *Mathematical Foundation of Finite Element Method*, Baifukan (1994) (in Japanese).
17. Kuznetsov, Yu. A. and Lipnikov, K. N. 3D Helmholtz wave equation by fictitious domain method, Russ. J. Numer. Anal. Math. Modeling, **13** No. **5**, pp. 371–387 (1998).
18. Liu, X., *Study on Approximation method for the Helmholtz equation in unbounded region*, PhD. thesis, The University of Electro-Communications, Japan, 1999.
19. Liu, X. and Kako, T., Higher order radiation boundary condition and finite element method for scattering problem, Advances in Mathematical Sciences and Applications, **8**, No. 2, pp. 801–819(1998).

Can We Trust the Computational Analysis of Engineering Problems?

I. Babuška[1] and T. Strouboulis[2]

[1] Texas Institute for Computational and Applied Mathematics, University of Texas at Austin, Austin, TX 78712, U.S.A.
[2] Department of Aerospace Engineering, Texas A&M University, College Station, TX 77843-3141, U.S.A.

1 Introduction

Computational science in general and computational mechanics in particular addresses physical and engineering reality with respect to some particular goals. These goals must be clearly specified. They are usually to get good qualitative or quantitative information about reality. The admissible quality of required information should be characterized.

The quality of the information relates directly to *validation* and *verification* of the problem. Validation deals with the question of how well a mathematical model describes reality. Verification deals with the question of the error of the approximate numerical solution in comparison with the exact solution of the mathematical problem (see e.g. [1]).

The usual mathematical problems are deterministic i.e. it is assumed that all data are known perfectly. In practical computations, often the formulation can be assumed to be known (for example the type of differential equation is known) but there is uncertainty in the input data such as the values of the coefficients, right hand side and the domain definition.

For practical purposes, computations make sense only if the uncertainties in the input data have small effect on the output of interest. In other words, that the problem is well posed. By a well posed problem we mean that both the physical one (for example we need reasonable reproducibility of the experiments) and the mathematical one (where we need continuous dependence on the perturbation of the data so that it is relevant for the applications).

Let us show an example when the domain of a partial differential equation is obtained by scanning. Consider the following model problem. Let $D = \left\{x_1, x_2 \middle| \; \frac{1}{2} < x_1^2 + x_2^2 < 1\right\}$, $\Gamma_1 = \left\{x_1, x_2 \middle| \; x_1^2 + x_2^2 = 1\right\}$, and $\Gamma_2 = \left\{x_1, x_2 \middle| \; x_1^2 + x_2^2 = \frac{1}{2}\right\}$ and let us be interested in the problem:

$$\Delta u_0 = 0 \quad \text{on } D, \tag{1.1a}$$

$$\frac{\partial u_0}{\partial n} = 1 \quad \text{on } \Gamma_1, \tag{1.1b}$$

$$u_0 = 0 \text{ on } \Gamma_2. \tag{1.1c}$$

The solution $u \in H_0^1 = \left\{ u \middle| \, u \in H^1(D), u = 0 \text{ on } \Gamma_2 \right\}$ obviously exists, is unique and satisfies for any $v \in H_0^1(\Omega)$.

$$B(u_0, v) = \int_D \nabla u_0 \cdot \nabla v \; dx_1 dx_2 = \int_{\Gamma_1} v \; ds = F(v). \tag{1.2}$$

Let us now assume that the domain D is given by a digital image with pixels of size of ϵ. Let us denote this "pixel" domain by D_ϵ with the boundary $\Gamma_1^\epsilon \cup \Gamma_2^\epsilon$. Let us denote $u_\epsilon \in H_0^1(D_\epsilon) = \left\{ u \in H^1(D_\epsilon), \; u = 0 \text{ on } \Gamma_2^\epsilon \right\}$ which solves the problem (1.1) on D_ϵ in the weak form

$$B_\epsilon(u_\epsilon, v) = \int_{D_\epsilon} \nabla u_\epsilon \cdot \nabla v \; dx_1 dx_2 = \int_{\Gamma_1^\epsilon} v \; ds = F_\epsilon(v) \quad \forall v \in H^1(D_\epsilon). \tag{1.3}$$

Obviously u_ϵ exists for any ϵ. Now we have:

Theorem 1.1 $u_\epsilon(0,0) \not\to u(0,0)$ as $\epsilon \to 0$.

This shows that the classical formulation is not well posed. Nevertheless from physical consideration we can conclude that the physical problem is well posed. The source of the paradox in Theorem 1.1 is that the length of Γ_1^ϵ does not converge to the length of Γ_1. Of course assuming stronger convergence of Γ_1^ϵ to Γ_1 we will get continuous dependence. It is worthwhile to note that this paradox is related to the nonhomogenous Neumann boundary condition. There is no paradox for the homogenous Neumann condition as well as for the homogenous or nonhomogenous Dirichlet boundary condition. The paradox mentioned in Theorem 1.1 can be resolved by different weak formulation. Denoting $G(x_1, x_2) = \Theta$ where Θ is the angle (multivalued function). Then we replace F_ϵ by F_ϵ^* where

$$F_\epsilon^* = \int_{\Gamma_0^\epsilon} v \; dG(x_1, x_2), \tag{1.4}$$

i.e. by Stieltjes integral, which is properly defined in the obvious way for the multivalued G. Denoting u_ϵ^* the weak solution of (1.3) when $F_\epsilon(v)$ is replaced by (1.4) we get

Theorem 1.2

$$u_\epsilon^*(0,0) \to u_0(0,0) \text{ as } \epsilon \to 0. \tag{1.5}$$

The proof follows from the results in [2].

Assuming that the problem is well posed it is important to assess the influence of the uncertainties in the input data on the computed data. We can consider two approaches:

a) The worst scenario approach. Here we assume that a set of admissible data is known and we wish to obtain the range of the ouput when the input ranges over the admissible set. In the example above we can consider the set of admissible domain $D_\epsilon : \left\{ x_1, x_2 \middle| \, \frac{1}{2} + \epsilon \leq x_1^2 + x_2^2 \leq 1 - \epsilon \right\} \subset$ $D_\epsilon \subset \left\{ x_1, x_2 \middle| \, \frac{1}{2} - \epsilon < x_1^2 + x_2^2 < 1 + \epsilon \right\}$. Then we can consider $\sup u_\epsilon(0,0)$ and $\inf u_\epsilon(0,0)$ when D_ϵ is ranging over the admissible set.
b) The stochastic approach. Here we assume that a set D with a specified probability space is given. We are interested in the probability field of the solution. For example we can assume that $D = \left\{ x_1, x_2 \middle| \frac{1}{2} < x_1 + x_2 = 1 + \phi(\theta) \right\}$ where $\phi(\theta)$ is a stochastic function with known probability field. Then the goal is to obtain the probability of $u_\epsilon(0)$.

Problem of this type are related to the validation. The problem of verification is related to the a-posteriori estimate of the error in the output of interest. It is essential to measure the error in an application relevant sense.

2 A-posteriori error estimation for the deterministic problems

2.1 Introduction

Here we summarize the basic results on a-posteriori error estimators. For details we refer to [3].

Let us consider the following model problem:

$$-\Delta u = f \in L_2(\Omega) \text{ on } D \subset \mathbf{R}^2, \tag{2.1a}$$

$$u = 0 \text{ on } \Gamma_D, \tag{2.1b}$$

$$\frac{\partial u}{\partial n} = g \text{ on } \Gamma_N. \tag{2.1c}$$

Let u_{EX} be the exact solution and $u_{S^p_\Delta}$ be the finite element solution. Using the mesh Δ and element of degree p. The exact solution $u_{EX} \in \mathcal{U}$ satisfies

$$B(u_{EX}, v) = F(v) \qquad \forall v \in \mathcal{U} \tag{2.2a}$$

where

$$B(u,v) = \int_D \nabla u \cdot \nabla v \; dx_1 dx_2, \tag{2.2b}$$

$$F(v) = \int_D fv \, dx_1 dx_2 + \int_{\Gamma_N} gv \, ds, \tag{2.2c}$$

and

$$\mathcal{U} = \left\{ u \middle| B(u,u) = \|u\|_{\mathcal{U}}^2 < \infty, \; u = 0 \text{ on } \Gamma_D \right\}. \tag{2.2d}$$

Given the mesh $\Delta = \{\tau\}$ we denote

$$S_\Delta^p = \left\{ u \in \mathcal{U} \middle| \; u|_\tau \text{ is polynomial of degree } p \right\}. \tag{2.3a}$$

Then the finite element solution $u_{S_\Delta^p}$ satisfies

$$B(u_{S_\Delta^p}, v) = F(v) \qquad \forall v \in S_\Delta^p. \tag{2.3b}$$

The elements τ can be triangles or quadrilaterals, straight or curved. Let us denote $\tilde{\tau}$ the master elements which is mapped on the "physical" element τ. If $\tilde{\tau}$ is a triangle then the shape functions are polymomials of degree p. If $\tilde{\tau}$ is rectangle then the shape functions are the polynomials of degree p in both variables. If the mapping of $\tilde{\tau}$ on τ is not linear, then the shape functions on τ are the "pullback" polynomials. By $e_{S_\Delta^p} = u_{EX} - u_{S_\Delta^p}$ we denote the the error of the finite element solution and we are interested in estimating $\|e_{S_\Delta^p}\|_{\mathcal{U}}$. More precisely, we are interested in the construction of a computable upper (resp. lower) estimator $\mathcal{E}^U$ (resp. $\mathcal{E}^L$) so that

$$\mathcal{E}^L \le \|e_{S_\Delta^p}\|_{\mathcal{U}} \le \mathcal{E}^U. \tag{2.4}$$

The upper estimator $\mathcal{E}^U$ is used for the acceptance of the computed data and $\mathcal{E}^L$ is used as the quality measure of $\mathcal{E}^U$. The interval $(\mathcal{E}^U, \mathcal{E}^L)$ is called the reliability interval. $\mathcal{E}^U$ and $\mathcal{E}^L$ have to be computable and utilize the computed solution $u_{S_\Delta^p}$, the mesh Δ, and the input data f, and g.

The estimators have the form (or similar form)

$$\mathcal{E}^U = \left(\sum_{\tau \in \Delta} \left(\eta^U(\tau)\right)^2 \right)^{\frac{1}{2}}, \tag{2.5a}$$

$$\mathcal{E}^L = \left(\sum_{\tau \in \Delta} \left(\eta^L(\tau)\right)^2 \right)^{\frac{1}{2}}. \tag{2.5b}$$

$\eta^U(\tau)$, and $\eta^L(\tau)$ are called the error indicators. We distinguish two kinds of estimators:

a) The guaranteed ones when $\mathcal{E}^L$, and $\mathcal{E}^U$ leads to estimate (2.4).

b) The correct estimator $\mathcal{E}$ when there exists constants C_L, and C_U such that

$$C_L \mathcal{E} \leq ||e_{S^p_\Delta}||_{\mathcal{U}} \leq C_U \mathcal{E}, \tag{2.6}$$

where C_L, and C_U are independent of the large class of meshes, of the input data (and solution u_{EX}) from a sufficiently large class.

The quality of an estimator is usually characterized by the effectivity index κ

$$\kappa = \frac{\mathcal{E}}{||e_{S^p_\Delta}||_{\mathcal{U}}}. \tag{2.7}$$

2.2 The basic estimators

Let us now list the basic a-posteriori estimators.

a) The subdomain residual estimator: Let $A(\Delta) = \{a_i\}$ be the set of the vertices of the mesh Δ. By σ_i we denote the patch associated to the vertex a_i. It is the union of the elements which have a vertex at a_i. Let $\hat{e}_{\sigma_i} \in \mathcal{U}_0(\sigma_i)$ be such that

$$B(\hat{e}_{\sigma_i}, v) = R(u_{S^p_\Delta}, v) = \int_D fv \; dx_1 dx_2 + \int_{\Gamma_N} gv \; ds - B(u_{S^p_\Delta}, v)$$
$$\forall v \in \mathcal{U}_0(\sigma_i), \tag{2.8a}$$

where $\mathcal{U}_0(\sigma_i) = \left\{ u \in H^1_0(\sigma_i), \; u = 0 \text{ on } \partial\sigma_i - \Gamma_N \right\}$. Then we can define

$$\mathcal{E}^2 = \sum_{\sigma_i} ||\hat{e}_{\sigma_i}||^2_{\mathcal{U}_0(\sigma_i)} \tag{2.8b}$$

and have

Theorem 2.1 There exists constants C_L, and C_U dependent only on the class of meshes (e.g. minimal angle condition, type of the elements, coefficients in the differential equation etc.) but are independent of u_{EX} and $u_{S^p_\Delta}$ such that (2.6) holds.
This estimator is one of the first proposed a-posteriori estiamtor, see[4-7]. Recently a modification of the definition of $\hat{e}_{\sigma_i}$ was suggested in [8-10]. The function $\hat{e}_{\sigma_i}$ is locally defined. Of course, we have to compute it numerically. It is possible to estimate the effects of the approximate solution of $\hat{e}_{\sigma_i}$ similarly as in [3]. The constants C_L, and C_U depends on the form of the patches and hence, if they are computed, they could lead to a conservative result. On the other hand, in practical computation we mostly have $\frac{1}{2} < \kappa < 2$ or better. The modifications in [9], [10] further improve the effectivity index in practical computation.

b) The explicit residual estimator: The residual $R(u_{S^p_\Delta}, v)$ defined in (2.8a) can be written in a more explicit form using integration by parts.

$$R(e_{S^p_\Delta}, v) = \sum_{\tau\in\Delta} \int_\tau r_\tau v \, dx_1 dx_2 + \sum_{\epsilon\in E} \int_\epsilon J_\epsilon v \, ds, \tag{2.9a}$$

where $\epsilon \in E$ are the sides of τ, i.e. E is the set of all edges in Δ.

$$r_\tau = f + \Delta u_{S^p_\Delta} \tag{2.9b}$$

is the residual on τ and

$$J_\epsilon = \left(\frac{\partial u_{S^p_\Delta}}{\partial n}\right)^+ - \left(\frac{\partial u_{S^p_\Delta}}{\partial n}\right)^- \tag{2.9c}$$

is the jump of the derivative of $u_{S^p_\Delta}$ in the edge ϵ. If $\epsilon \subset \Gamma_N$ then we use g instead of $\frac{\partial u_{S^p_\Delta}}{\partial n}$ and if $\epsilon \subset \Gamma_D$, we take $J(\epsilon) = 0$. It is easy to see that

$$||\hat{e}_{\sigma_i}||^2_{\mathcal{U}} \le C\Big(\sum_{\tau\in\sigma_i} |\tau| \int_\tau r^2 \, dx_1 dx_2 + \sum_{\epsilon\in\partial\tau-\partial\sigma_i,\ \tau\in\sigma_i} |\epsilon| \int_\epsilon \left(\frac{1}{2}J_\epsilon\right)^2 ds\Big), \tag{2.10}$$

where $|\tau|$ is the area of τ and $|\epsilon|$ is the length of ϵ. Hence we can define

$$\left(\eta^{EXPL}(\tau)\right)^2 = |\tau| \int_\tau r^2 \, dx_1 dx_2 + \sum_{\epsilon\in\partial\tau} |\epsilon| \int_\epsilon \left(\frac{1}{2}J_\epsilon\right)^2 ds \tag{2.11a}$$

and from Theorem 2.1 we have

Theorem 2.2a

$$||e_{S^p_\Delta}||^2_{\mathcal{U}} \le C \sum_{\tau\in\Delta} \left(\eta^{EXPL}(\tau)\right)^2 = \left(\mathcal{E}^{EXPL}\right)^2. \tag{2.11b}$$

Assuming that f is sufficiently smooth then we have

Theorem 2.2b

$$C\mathcal{E}^{EXPL} \le ||e_{S^p_\Delta}||_{\mathcal{U}}. \tag{2.11c}$$

The explicit estimator was proposed in [5], [11] and then used in many places. The correctness of this estimator can be proven by various means. The effectivity index of this estimator can be very poor specially when the triangles have small angles. For more see [3], [12], [14]. If we would have

$$\int_\tau r \, dx_1 dx_2 + \sum_{\epsilon\in\partial\tau} \int_\epsilon \frac{1}{2} J_\epsilon \, ds = 0 \tag{2.12}$$

then we can compute $\tilde{e}_\tau \in \mathcal{U}(\tau)$ so that

$$B_\tau(\tilde{e}_\tau, v) = \int_\tau rv\ dx_1dx_2 + \sum_{\epsilon\in\partial\tau} \int_\epsilon \frac{1}{2} J_\epsilon v\ ds \qquad \forall v \in \mathcal{U}(\tau), \tag{2.13}$$

where

$$B_\tau(u,v) = \int_\tau \nabla u \nabla v\ dx_1dx_2 \tag{2.14}$$

and $\mathcal{U}(\tau) = \left\{u \in H^1(\tau),\ B_\tau(u,v) = ||u||^2_{\mathcal{U}(\tau)} < \infty\right\}$ and

$$||\tilde{e}_\tau||^2_{\mathcal{U}(\tau)} \leq C\eta^2(\tau), \tag{2.15a}$$

where $\eta(\tau)$ is given in (2.10). We can then define the estimator

$$\mathcal{E} = \Big(\sum_{\tau\in\Delta} ||\tilde{e}_\tau||^2_{\mathcal{U}(\tau)}\Big)^{\frac{1}{2}}. \tag{2.15b}$$

It is necessary to compute $\tilde{e}_\tau$ numerically. We can utilize (2.13) using only the set of polynomial of degree $p+k$ instead of $\mathcal{U}(\tau)$. Further we can restrict this space so that the polynomials are zero at the vertices and on every side ϵ they are orthogonal to all polynomials of degree p with zero at all the ends of ϵ. We denote this set as $\mathrm{B}(p+k)$ (the buble set).
The condition (2.12) does not hold in general. It is possible to modify J_ϵ so that (2.12) holds. It will be discussed below. If (2.12) does not hold then $\tilde{e}_\tau$ satisfying (2.13) does not exist but $\tilde{e}_\tau^{\mathrm{B}(p+k)}$ still exists and we can define $\eta^{\mathrm{B}(p+k)} = ||\tilde{e}_\tau^{\mathrm{B}(p+k)}||^2_{\mathcal{U}_\tau}$ and $\mathcal{E}^{\mathrm{B}(p+k)} = \left(\sum_{\tau\in\Delta} ||\tilde{e}_\tau^{\mathrm{B}(p+k)}||^2_{\mathcal{U}_\tau}\right)^{\frac{1}{2}}$ and Theorems 2.2a, b hold for this estimator too. For details see [3]. Usually for $k = 2$ this estimator performs better that $\mathcal{E}^{EXPL}$.

c) The equilibrated estimator: Assume now that we can change the jump J_ϵ used earlier into J_ϵ^* so that

$$R(e_{S^p_\Delta}, v) = \sum_{\tau\in\Delta} \Big(\int_\tau r_\tau v\ dx_1dx_2 + \sum_{\epsilon\in\partial\tau} \int_\epsilon J_\epsilon^* v\ ds\Big) \tag{2.16}$$

and

$$\int_\tau r_\tau\ dx_1dx_2 + \sum_{\epsilon\in\partial\tau} \int_{\partial\tau} J_\epsilon^*\ ds = 0. \tag{2.17}$$

This is possible to do by local computations. For such construction see [3, 14-18]. Let $\hat{e}_\tau \in \mathcal{U}(\tau)$ be such that

$$B_\tau(\hat{e}_\tau, v) = \int_\tau rv\ dx_1dx_2 + \sum_{\epsilon\in\partial\tau} \int_\epsilon J_\epsilon^* v\ ds \qquad \forall v \in \mathcal{U}(\tau); \tag{2.18}$$

Because of (2.17) $\hat{e}_\tau$ exists. Now we have

$$||e_{S^p_\Delta}||_{\mathcal{U}} = \sup_v \frac{R(e_{S^p_\Delta}, v)}{||v||_{\mathcal{U}}} = \sup_v \frac{\sum_{\tau\in\Delta} B_\tau(\hat{e}_\tau, v)}{||v||_{\mathcal{U}}} \leq \Big(\sum_{\tau\in\Delta} ||\hat{e}_\tau||^2_{\mathcal{U}(\tau)}\Big)^2 . \tag{2.19}$$

With $\eta^{EQ}(\tau) = ||\hat{e}_\tau||_{\mathcal{U}(\tau)}$. We can define

$$\mathcal{E}^{EQ} = \Big(\sum_{\tau\in\Delta} ||\hat{e}_\tau||^2_{\mathcal{U}(\tau)}\Big)^{\frac{1}{2}} \tag{2.20}$$

and we have

Theorem 2.3a

$$||e_{S^p_\Delta}||_{\mathcal{U}(\tau)} \leq \mathcal{E}^{EQ} \tag{2.21}$$

and hence $\mathcal{E}^{EQ}$ is guaranteed upper estimate. For sufficiently smooth data we also have

Theorem 2.3b $\mathcal{E}^{EQ}$ is a correct lower estimator. For more details see [3].

Of course we have to compute $\hat{e}_\tau$ numerically in an approximate way. For example we can solve (2.18) by using polynomials of degree $p+k$. The error of this approximate solution can be estimated and a guaranteed upper estimate can be obtained. For more see [3]. It is possible to get a guaranteed lower estimate also. For details see [3].

d) The recovery estimator: This estimator is based on special averaging so that the gradient of u_{EX} is "recovered". By this we mean that we can (locally) compute q_i, $i = 1, 2$ on every τ so that

$$||\text{grad}u_{EX} - q||_{L^2(\tau)} < ||\text{grad}u_{EX} - \text{grad}u_{S^p_\Delta}||_{L^2(\tau)}, \tag{2.22}$$

q is then the recovered gradient. This allows then to define

$$\eta^{ZZ}(\tau) = ||\text{grad}u_{S^p_\Delta} - q||_{L^2(\tau)}. \tag{2.23}$$

This kind of estimator was proposed by Zienkiewicz, see [19-21], and is closely related to superconvergence. For more details see [3]. It is also possible to prove that this estimator is a correct one when the data are sufficiently smooth.

2.3 Rating of the estimators

Since the first a-posteriori error estimator proposed in 1978 (see [4], [5], [11]), there have been many estimators proposed and analyzed in the literature, see e.g. [3], [22-24] and references therein. The problem is how to assess their quality. This depends on the purpose of the computations. Sometimes it is sufficient only to have correct estimators with constants more or less known from experience so that in practice the effectivity index is $\frac{1}{2} < \kappa < 2$. Other times we would like to have a guaranteed error estimator. For more details see [3]. One powerful rating approach is the asymptotic one which is described in [3], [25-27]. Assume that the mesh is patchwise (cell) translation invariant on domain $\omega \subset D$ of diameter H with h being the diameter of the cell. Also, assume that

a) there is no pollution.
b) the exact solution is smooth.
c) the error of $u_{S^p_\Delta}$ is of order p for elements of degree p.

Then it is possible to find (asymptotically) the finite element solution by local cell computation only. It is possible to assume that u_{EX} is a polynomial of degree $p+1$ on ω. These results are closely related to the superconvergence. See [3], [28, 29] for more details. This relationship allows the effectivity index for every estimator to be computed asymptotically. The effectivity index κ depends on the mesh Δ (i.e. the cell), the exact solution (polynomial), and the differential equation. Given the class of meshes $\mathcal{M}$, the class of solutions $\mathcal{S}$ (for example only those satisfying homogenous differential equations, such as Laplace's equation), for every estimator $\mathcal{E}$ we can compute

$$\underline{\kappa}(\mathcal{M},\mathcal{S},\mathcal{E}) = \inf_{\Delta\in\mathcal{M},u_{EX}\in\mathcal{S}} \kappa(\Delta,u_{EX},\mathcal{E}) \tag{2.24}$$

and

$$\bar{\kappa}(\mathcal{M},\mathcal{S},\mathcal{E}) = \sup_{\Delta\in\mathcal{M},u_{EX}\in\mathcal{S}} \kappa(\Delta,u_{EX},\mathcal{E}) \tag{2.25}$$

and define the robustness index

$$\mathcal{R}(\mathcal{M},\mathcal{S},\mathcal{E}) = \max\left(\bar{\kappa}-1, \frac{1}{\underline{\kappa}}-1\right). \tag{2.26}$$

Comparing the estimators we prefer the one which has the smaller robustness index. In [3], [27] robustness indices for various estimators are given. Roughly the estimator based on recovery and the equilibrated estimator using the bubble shape functions have the smallest robustness indices.

Asymptotic robustness does not mean that the error estimator is either a guaranteed upper or lower one. As we have mentioned we can construct locally the guaranteed upper and lower estimates. For experience with these estimators and computation of the reliability interval see [3].

2.4 The estimates of the error in the computed data of interest

In the previous section we have discussed the error of the finite element solution measured in the energy norm. In practice often the aim of the computation is to obtain values of some functionals, such as stress intensity factors, flux at a point, and average values of the solution over an area. In general we are interested in the value $\Phi(u_{EX})$ but we are able to compute only $\Phi(u_{S^p_\Delta})$ and hence we are interested in the estimate of $\Phi(u_{EX}) - \Phi(u_{S^p_\Delta})$. Here we have to consider two classes of functionals Φ:

a) Linear bounded functionals on (entire) space $\mathcal{U}$, such as the average value of u_{EX} over a subdomain ω.
b) Functionals which are not bounded on the entire set $\mathcal{U}$ such as the point value of u, the point value of $\frac{\partial u}{\partial x}$ or the stress intensity factor.

Let us first address the error in the bounded functionals. Given a bounded functional Φ there exists $\phi_{EX} \in \mathcal{U}$ such that

$$B(\phi_{EX}, v) = \Phi(v) \qquad \forall v \in \mathcal{U} \tag{2.27}$$

and its finite element approximation. Then it is easy to prove (see [3])

Theorem 2.5 We have

$$\Phi(u_{EX}) - \Phi(u_{S^p_\Delta}) = B(u_{EX} - u_{S^p_\Delta}, \phi_{EX} - \phi_{S^p_\Delta}) = B(e_{S^p_\Delta}(u), e_{S^p_\Delta}(\phi)), \tag{2.28}$$

where $e_{S^p_\Delta}(u) = u_{EX} - u_{S^p_\Delta}$, and $e_{S^p_\Delta}(\phi) = \phi_{EX} - \phi_{S^p_\Delta}$. Using the estimators for $e_{S^p_\Delta}(u)$, and $e_{S^p_\Delta}(\phi)$ we get

$$\left|\Phi(u_{EX}) - \Phi(u_{S^p_\Delta})\right| \le \mathcal{E}(u)\ \mathcal{E}(\phi) \tag{2.29}$$

or

$$\left|\Phi(u_{EX}) - \Phi(u_{S^p_\Delta})\right| \approx \sum_\tau \left|B_\tau(\hat{e}_\tau(u), \hat{e}_\tau(\phi))\right|. \tag{2.30}$$

To get a guaranteed upper and lower bound we write

$$B(e_{S^p_\Delta}(u), e_{S^p_\Delta}(\phi)) = \frac{1}{4}\left(||e_{S^p_\Delta}(s\ u + \frac{1}{s}\phi)||^2_{\mathcal{U}} - ||e_{S^p_\Delta}(s\ u - \frac{1}{s}\phi)||^2_{\mathcal{U}}\right) \tag{2.31}$$

and hence

$$\begin{aligned}(\mathcal{E}^L(s\ u + \frac{1}{s}\phi))^2 - (\mathcal{E}^U(s\ u - \frac{1}{s}\phi))^2 &\le 4(\Phi(u_{EX}) - \Phi(u_{S^p_\Delta}))\\ &\le (\mathcal{E}^U(s\ u + \frac{1}{s}\phi))^2 - (\mathcal{E}^L(s\ u - \frac{1}{s}\phi))^2\end{aligned} \tag{2.32}$$

for any $s > 0$. The optimal value of s can be easily computed, see [3].

If the functional Φ is not bounded on the entire space $\mathcal{U}$, usually it is possible to write

$$\Phi(u_{EX}) = \Phi^*(u_{EX}) + \Psi(u_{EX}), \tag{2.33}$$

where $\Psi(u_{EX})$ can be computed directly from the input data, e.g. f and Φ^* is a bounded functional on the entire space $\mathcal{U}$. Then we can define

$$\Phi(u_{S^p_\Delta}) = \Phi^*(u_{S^p_\Delta}) + \Psi(u_{S^p_\Delta}) \tag{2.34}$$

and the error can be computed as before. For more details and numerical examples we refer to [3].

Computation of data of interest and its a-posteriori estimates was first introduced in [30]. Now there are many papers addressing the computation of data of interest. For more see [3], [31-33]. A-posteriori error estimation is typical of the verification problem as mentioned in the introduction. In this section we have outlined the basic state of the art when the finite element method is used as the numerical treatment of the elliptic partial differential equations.

3 The problem of stochastic differential equation

Let us now address the validation problem when the data are not exactly known. Let us consider the model problem with $x = (x_1, x_2)$,

$$-\left(\frac{\partial}{\partial x_1} a \frac{\partial u}{\partial x_1} + \frac{\partial}{\partial x_2} a \frac{\partial u}{\partial x_2}\right) = f \quad \text{on} \quad D, \tag{3.1a}$$

$$u = 0 \quad \text{on} \quad \partial D. \tag{3.1b}$$

We will assume that the coefficient a is a stochastic function, and f is a deterministic one. More precisely let $(\Omega, \mathcal{F}, P)$ be a probability space and $D \subset \mathbf{R}^2$, with Lipschitz boundary and $a : D \times \Omega \longrightarrow \mathbf{R}$. We will assume:

1. Assumption Q_1: The function $a(x, \Omega)$ is measurable. We will work with the natural σ-algebra, $\mathcal{F} = \sigma(a)$ which is the smallest one which makes a measurable function.
2. Assumption Q_2: There exists $0 < \alpha_0 < \alpha_1 < \infty$ such that

$$\alpha_0 \leq a(X, \Omega) \leq \alpha_1. \tag{3.2}$$

If X is a real valued random variable in $(\Omega, \mathcal{F}, P)$ with $X \in L^1(\Omega)$ we denote its expected value by

$$E(X) = \int_\Omega X(\omega)\, dP(\omega) = \int_{\mathbf{R}} X d\mu(x). \tag{3.3}$$

Here μ is the standard distribution measure for X defined on the Borelian sets on $\mathbf{R}$, with $B(\mathbf{R})$ given by

$$\mu(B) = P(X^{-1}(B)). \tag{3.4}$$

3. Assumption Q_3: There exists a density function for X, $\rho : \mathbf{R} \to \mathbf{R}^Y$ such that

$$E(X) = \int_{\mathbf{R}} X\rho(X)dX. \tag{3.5}$$

Let V be a Hilbert space which is the completion of the set $\left\{ v \in L^1_{loc}(D) \otimes L^1(\Omega) \middle| \, ||a^{\frac{1}{2}} \nabla v||^2_{L^2(D\times\Omega)} < \infty \right\}$ and

$$B(u,v) = E[\int_D a(\nabla u \cdot \nabla v)] = \int_D E[a\nabla u \cdot \nabla v] dx_1 dx_2 \tag{3.6}$$

is bounded coersive bilinear form on $V \times V$. Then the weak form of our model problem reads:
Given $f \in V'$, find $u \in V$ such that

$$B(u,v) = < f, v > \qquad \forall v \in V. \tag{3.7}$$

Because of our assumption Q_1, Q_2, and Q_3 there exists a unique solution $u(x,\omega) \in V$.

Let us make further assumptions:

4. Assumption Q_4: Let

$$a(x,\omega) = (E[a])(x) + \sum_{i=1}^{m} \sqrt{\lambda_i} a_i(x) Y_i(\omega) \tag{3.8}$$

where $Y_i(\omega)$ are real random variables which are mutually independent and $\Gamma_i = Y_i(\Omega)$, $i = 1, ..., m$ is a bounded interval. Moreover each Y_i has a smooth density function $\rho_i : \Gamma_i \to \mathbf{R}$ which are bounded away from zero. Further $a_i(x)$ are sufficiently smooth functions. Denote $\Gamma = \Pi_{i=1}^m \Gamma_i \subset \mathbf{R}^m$, and $\rho(y) = \Pi_{i=1}^m \rho_i(y_i)$, $y \in \Gamma$.

Now we have

Theorem 3.1 Under assumptions Q_1 and Q_4, the solution $u(x,\omega) = u(x, Y_1(\omega), ..., Y_m(\omega))$, $y = (Y_1(\omega), ..., Y_m(\omega))$ satisfies

$$\begin{aligned} B(u,v) &= \int_\Gamma \rho(y) \int_D a(x,y) \nabla u(x,y) \cdot \nabla v(x,y) dx\, dy \\ &= \int_\Gamma \int_D \rho(y) f(x) v(x,y) dx\, dy \end{aligned} \tag{3.9}$$

with $u(x,y) \in H^1_0(D) \times L^2(\Gamma)$ and for any $v \in H^1_0(D) \times L^2(\Gamma)$.

Theorem 3.1 shows that we can transform the stochastic equation problem into a deterministic one then can be solved by the finite element method. All theoretical finite element method results can be utilized. For more and numerical results refer to [35]. Using a specific form of the p-version of the finite element method leads to the use of Wiener chaos polynomials proposed in [36].

4 Conclusions

Trusting the computed data depends on

a) the mathematical model used - the validation;
b) the reliability of the approximate numerical solution characterized by a-posteriori error estimation;
c) dealing with uncertain input data.

References

[1] Roache, P.J.: *Verification and Validation in Computational Science and Engineering*, Hermosa Publishers, Albuquerque, 1998.
[2] Babuška, I., Chleboun, J.: Effects of uncertainties in the domain on the solution of Neumann boundary value problems in two spatial dimensions, Preprint, TICAM University of Texas at Austin, 2000.
[3] Babuška, I., Strouboulis T.: *Finite Element Method and its Reliablity*, Oxford University Press, 2001.
[4] Babuška, I., Rheinboldt, W.C.: A-posteriori bounds and adaptive procedures for the finite element method, Recent advances in engineering science, Proceedings of 15the annual meeting, Soc. of Engg. Science, University of Florida, Decision of Continuing Education, 1978, 413-423.
[5] Babuška, I., Rheinboldt, W.C.: Error estimates for adaptive finite element computations, SIAM J. Numer. Anal., 15, 1978, 736-754.
[6] Babuška, I., Miller, A.: A-posteriori error estimates and adaptive techniques for the finite element method, Technical Note BN-968, Institute for Physical Science and Technology, University of Maryland, 1981.
[7] Babuška, I., Miller, A.: A feedback finite element method with a posteriori error estimation: Part 1. The finite element method and some basic properties of the a posteriori error estimator, Comput. Methods Appl. Mech. Engrg., 61, 1987, 1-40.
[8] Verfürth, R.: A-posteriori error estimators for singularly perturbed reaction diffusion equations, Num. Math., 78, 1998, 479-493.
[9] Cartensen, C., Funken, S.A.: Fully reliable localized error control in FEM, Ber. de. Math., Seminar Kiel, 97-12, 1992.
[10] Morin, P., Nochetto, R.H., Siebert, G.: Preprint, 2000.
[11] Babuška, I., Rheinboldt, W. C.: A-posteriori error estimates for the finite element method, Int. J. Numer. Methods Engrg., 12, 1978, 1597-1615.

[12] Babuška, I., Duran, R., Rodriguez, R.: Analysis of the efficiency of an a-posteriori error estimator for linear triangular elements, SIAM J. Numer. Anal., 29, 1992, 947-964.
[13] Cartensen, C., Funken, S.A.: Constants in Clement interpolation error and residual based a-posteriori estimates in finite element method, Ber. de. Math., Seminar Kiel, 97-11, 1997.
[14] Ladeveze, P., Leguillon, D.: Error estimate procedure in the finite element method and applications, SIAM J. Numer. Anal., 20, 1983, 485-509.
[15] Bank, R.E., Weiser, A.: Some a posteriori error estimators for elliptic partial differential equations, Math. Comp., 44, 1985, 283-301.
[16] Ainsworth, M., Oden, J.T.: A-posteriori error estimation in finite element analysis, Comput. Methods Appl. Mech. Engrg.: Computaional Mechanics Advances, 142, 1997, 1-88.
[17] Ainsworth, M., Oden, J. T.: A unified approach to a-posteriori error estimation using element residual methods, Numer. Math., 65, 1993, 23-50.
[18] Oden, J. T., Demkowicz, L., Rachowicz, W., Westermann, T. A.: Toward a universal h-p adaptive finite element strategy: Part 2, A Posteriori Error Estimates, Comput. Methods Appl. Mech. Engrg., 77, 1989, 113-180.
[19] Zienkiewicz, O. C., Zhu, J. Z.: A simple error estimator and adaptive procedure for practical engineering analysis, Int. J. Numer. Methods Engrg., 24, 1987, 337-357.
[20] Zienkiewicz, O. C., Zhu, J. Z.: The superconvergence patch recovery and a posteriori error estimates. Part 1: The recovery technique, Int. J. Numer. Methods Engrg., 33, 1992, 1331-1364.
[21] Zienkiewicz, O. C., Zhu, J. Z.: The superconvergence patch recovery and a posteriori error estimates. Part 2: Error estimates and adaptivity, Int. J. Numer. Methods Engrg., 33, 1992, 1365-1382.
[22] Ainsworth, M., Oden, J.T.: *A-posteriori error estimation in Finite Element Analysis*, John Wiley, 2000.
[23] Verfürth, R.: *A review of a-posteriori error estimation and adaptive mesh-refinement techniques*, Wiley, Teubner, 1996.
[24] Eriksson, K., Estep, D., Hansbo, P., Johnson, C.: Introduction to adaptive methods for differential equations, Acta. Numer., Cambridge Univ. Press, 1995, 105-158.
[25] Babuška, I., Strouboulis, T., Upadhyay, C. S.: A model study of the quality of a posteriori error estimators for linear elliptic problems. Error estimation in the interior of patchwise uniform grids of triangles, Comput. Methods Appl. Mech. Engrg., 114, 1994, 307-378.
[26] Babuška, I., Strouboulis, T., Upadhyay, C. S.: A model study of the quality of a-posteriori error estimators for finite element solutions of linear elliptic problems with particular reference to the behaviour near the boundary, Int. J. Numer. Methods Engrg., 40, 1997, 2521-2577.
[27] Babuška, I., Strouboulis, T., Upadhyay, C. S., Gangaraj, S. K., Copps, K.: An objective criterion for assessing the reliability of a-posteriori error estimators in finite element computation, IACM bulletin, 9, 1994, 27-37.

[28] Babuška, I., Strouboulis, T., Upadhyay, C. S., Gangaraj, S. K.: Computer based proof of the existence of superconvergence points in the finite element method: Superconvergence of the derivatives in finite element solutions of Laplace's, Poisson and elasticity equation, Num. Meth. for PDE's, 12, 1996, 347-392.

[29] Wahlbin, L. B.: Superconvergence in Galerkin Finite Element Methods, *Lecture Notes in Mathematics*, Springer-Verlag, 1995.

[30] Babuška, I., Miller, A.: The post-processing approach in the finite element method - Part 3: A-posteriori error estimates and adaptive mesh selection, Int. J. Numer. Methods Engrg., 20, 1984, 2311-2324.

[31] Babuška, I., Strouboulis, T., Gangaraj, S. K.: Guaranteed computable bounds for the exact error in the finite element solution - Part 1 : One dimensional model problem, Comput. Methods Appl. Mech. Engrg., 176, 1999, 51-79.

[32] Prudhomme, S., Oden, J. T.: On goal oriented error estimation for elliptic problems: Applications to the control of pointwise errors, Comput. Methods Appl. Mech. Engrg., 176, 1999, 313-331.

[33] Paraschivoiu, M., Peraire, J., Patera, A. T.: A-posteriori finite element bounds for linear functional outputs of elliptic partial differential equations, Comput. Methods Appl. Mech. Engrg., 150, 1997, 289-312.

[34] Becker, R., Rannacher, R.: A feedback approach to error control in finite element methods: Basic analysis and examples, East-West J. Numer. Math., 4, 1996, 237-264.

[35] Deb, M.K., Babuška, I., Oden, J.T.: Solution of Stochastic Partial Differential equations using Galerkin method and finite element techniques, Preprint, 2000.

[36] Ghanem, R., Spanos, P.: *Stochastic Finite Elements: A Spectral Approach*, Springer, 1991.

Numerical Computations for Ill-conditioned Problems by Multiple-Precision Systems

Yuusuke Iso[1], Hiroshi Fujiwara[1] and Kimihiro Saito[2]

[1] Kyoto University, Yoshida-Honmachi, Sakyo-ku, Kyoto 606-8501 Japan
[2] IBM Japan,Ltd.

Abstract. We propose a use of some multiple-precision systems for numerical analysis of ill-conditioned problems, and we show efficiency of the systems through numerical examples. We also introduce the F-system which is a fast multiple-precision system designed by one of the authors.

1 Introduction

The aim of the paper is to show effective applications of multiple-precision systems to some ill-conditioned problems and to show a design and an implementation of a fast multiple-precision system (the F-system) proposed by one of the authors. The ill-conditioned problems mean the problems which contain numerically unstable processes in their numerical computations, and we focus especially on those arising from discretizations of ill-posed problems in the sense of Hadamard.

We should recall the influence of rounding errors in numerical computations on the floating point arithmetic. It is not assumed substantial in numerically stable processes, but it may become crucial for the ill-conditioned problems. Especially for discretizations of ill-posed problems, they are always crucial, because the problems are too sensitive for perturbations.

The ill-posedness is opposite concept to the well-posedness, of which the main concept consists in the stability against perturbations piercing to problems. We must fix a norm to estimate magnitude of the perturbations in order to discuss well-posedness of problems, and in most of cases of well-posed problems, we find such a nice discretization that we show stability of its numerical processes by the norm. But we remark that the discretization behave as an ill-conditioned problem even for the case, if the norm is not suitable to estimate the influence of the rounding errors. For the ill-posed problems, all their discretizations become ill-conditioned problems. We need some device to stabilize their numerical processes, because the main reason for instability of numerical solutions is rapid increase of the influence of the rounding errors in their numerical processes.

We propose effective use of multiple-precision systems for numerical treatments of ill-conditioned problems instead of stabilization techniques. In this paper, we mean multiple-precision systems as the systems where we can manipulate as many digits in the floating point arithmetic as we wish. They lead us to control the influence of the rounding errors to our request and are expected to give us new numerical tools to deal with ill-conditioned problems. We will show some numerical examples by multiple-precision systems in the present paper.

Multiple-precision systems have a fatal demerit of consumption of vast computing time, and we need some ideas to overcome the demerit in order to use them in practical computing. One of the authors has proposed a design to improve and has succeeded in an implementation as the Fast Multiple-precision system (F-system) in 2000. Refer to Fujiwara [1] for the details of the F-system.

In the following sections, we show some numerical results of an inverse scattering problems in §2 and those of a finite difference approach to the Cauchy-Riemann equation in §3. Finally we introduce our multiple-precision system.

The present research is partially supported by Sanwa Systems Development Co.,Ltd.

2 Application to an Acoustic Inverse Scattering

We deal with numerical analysis of an inverse scattering problem to determine an unknown obstacle in the 2D wave field governed by the Helmholtz equation. Our inverse problem is formulated by Kirsch-Kress [4], and we determine the obstacle from the measurement of the far-field pattern.

Prior to the inverse problem, we pose a direct problem connected with it. We denote $x \in \mathbb{R}^2$ by $x = (x_1, x_2)$. Consider an exterior boundary value problem of the Helmholtz equation in $\mathbb{R}^2$;

$$\Delta u^s + k^2 u^s = 0 \quad \text{in} \quad \mathbb{R}^2 \setminus \bar{D}, \tag{2.1}$$

$$u^i + u^s = 0 \quad \text{on} \quad \partial D, \tag{2.2}$$

$$\frac{\partial}{\partial |x|} u^s + \sqrt{-1}\, k u^s = o\left(|x|^{-1/2}\right) \quad |x| \to +\infty, \tag{2.3}$$

where D is a simply connected domain and k is the wave number. We denote the (given) incidental wave and the scattered wave by u^i and u^s respectively, and we pose the Sommerfeld radiation condition (2.3) to ensure the uniqueness in the exterior problem for u^s. We suppose smoothness of ∂D to be of C^2-class, and we call the open set D an obstacle. Since the fundamental solution to the 2D Helmholtz equation is

$$E(x) = \frac{i}{4} H_0^1(k|x|),$$

where H_0^1 is the Hankel function of order zero and of the first kind, the unique solution to (2.1) and (2.2) is written in

$$u^s(x) = \int_{\partial D} \left\{ u^s(y) \frac{\partial}{\partial n_y} E(x-y) - \frac{\partial u^s}{\partial n}(y) E(x-y) \right\} d\sigma_y \tag{2.4}$$

for $x \in \mathbb{R}^2 \setminus \bar{D}$, where n is the unit outward normal vector to ∂D. By an application of the asymptotic expansion of $E(x)$ to (2.4), there exists a function $u_\infty(\hat{x})$ on S^1 such that

$$u^s(x) = \frac{e^{ik|x|}}{\sqrt{|x|}} \left\{ u_\infty(\hat{x}) + O\left(\frac{1}{|x|}\right) \right\} \quad |x| \to +\infty, \tag{2.5}$$

where $\hat{x} = x/|x| \in S^1$. We call the function $u_\infty(\hat{x})$ the far-field pattern.

Our inverse problem is to determine an unknown obstacle D by the measurement of the far-field pattern $u_\infty(\hat{x})$. We follow a method by Kirsch-Kress [4] in order to reconstruct the obstacle using a priori information that the unknown obstacle D contains the unit circle S^1 and that the solution u^s of (2.1) and (2.2) is written by the single layer potential defined on S^1. The a priori information implies

$$u^s(x) = \int_{S^1} E(x - \hat{y})\mu(\hat{y})\, d\sigma_{\hat{y}}, \tag{2.6}$$

where $\mu(\hat{y})$ is a density function on S^1. From (2.6), we see

$$u^s(x) = \frac{e^{\frac{\pi}{4}i} e^{ik|x|}}{\sqrt{8\pi k|x|}} \left\{ \int_{S^1} e^{-ik(\hat{x}\cdot\hat{y})} \mu(\hat{y})\, d\sigma_{\hat{y}} + O\left(\frac{1}{|x|}\right) \right\} \quad |x| \to +\infty, \tag{2.7}$$

and its comparison with (2.5) leads us an integral equation for $\mu(\hat{y})$;

$$\frac{e^{\frac{\pi}{4}i}}{\sqrt{8\pi k}} \int_{S^1} e^{-ik(\hat{x}\cdot\hat{y})} \mu(\hat{y})\, d\sigma_{\hat{y}} = u_\infty(\hat{x}) \quad \hat{x} \in S^1. \tag{2.8}$$

We look for the density function $\mu(\hat{y})$ by the measured far-field pattern u_∞, and we give the solution $u^s(x)$ by (2.6). We seek the unknown boundary ∂D as the contour line of $u^s + u^i = 0$, and we are able to reconstruct ∂D numerically through discretization of this process. We call this idea the Kress method, and refer to Kress [5] for details.

We note that the kernel of the integral equation (2.8) is analytic, and the equation becomes ill-posed within the class of the Sobolev spaces. We need a discretized version of the Tikhonov regularization to stabilize discretizations for (2.8) in usual computations, but we apply a multiple-precision system instead of it in the present research. Since we restrict ourselves to show effectiveness of a multiple-precision system, we adopt the Fourier series method in discretization.

We explain our numerical computations. We firstly approach to the direct problem (2.1) and (2.2) by the boundary element technique to obtain the unknown Neumann data $\partial u^s/\partial n$ on ∂D. Since we have a closed formula to express the far-field pattern u_∞ by using the Dirichlet data u^s and the Neumann data $\partial u^s/\partial n$, we can calculate the far-field pattern for a known obstacle by the formula. We start reconstruction of ∂D by the Kress method from the calculated far-field pattern. We discretize this process and construct numerical examples shown later.

Let $x \in \mathbb{R}^2 \setminus D$ tend to $z \in \partial D$, we obtain a boundary integral equation

$$\int_{\partial D} E(z-y)q(y)\, d\sigma_y = \frac{1}{2}\, u^i(z) - \int_{\partial D} \frac{\partial}{\partial n_y} E(z-y)u^i(y)\, d\sigma_y \qquad (2.9)$$

for the (unknown) Neumann data $q := \partial u^s/\partial n$. The integral equation is uniquely solvable as far as k^2 belongs to the resolvent set $\rho(-\Delta)$. We will show numerical results for the case of $k = 10$ and $\partial D = \{(x_1, x_2) \mid x_1 = 2\cos\theta, x_2 = \frac{3}{2}\sin\theta,\ 0 \le \theta < 2\pi\}$, and the incidental wave u^i is a plane wave. We use the REAL*16 and the UBASIC with about 96 digits in radix-10 as our computational environments. The REAL*16 is one of the extended formats of IEEE754 [3] and available on a package of the FORTRAN, and it ensures about 32 digits in radix-10. The UBASIC is a Japanese software of a multiple-precision system working on personal computers and is widely used in the research of the number theory.

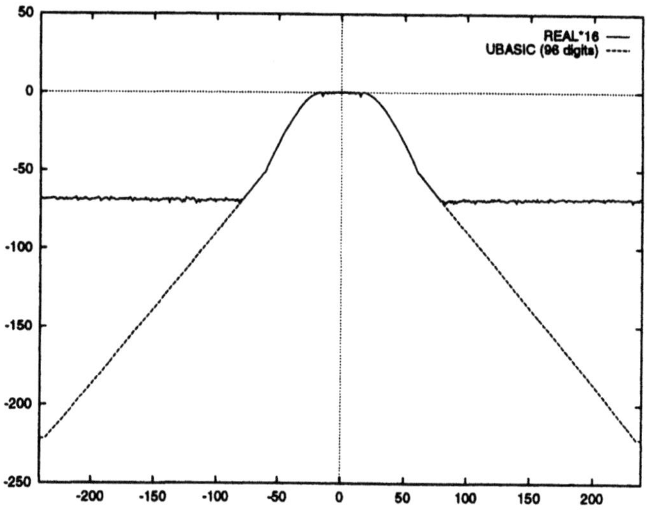

Fig. 1. the direct problem (horizontal axis : n, vertical axis : $\ln |a_n|$)

We expand the solution q to the equation (2.9) in the Fourier series and write our aimed solution q_M in

$$q_M(\theta) = \sum_{n=-M}^{M} a_n e^{in\theta} \quad (0 \le \theta < 2\pi), \qquad (2.10)$$

and we substitute it into (2.9) and apply the L^2-projection. Fig.1 shows the behaviors of $\{a_n\}_{n=-M}^{M}$ for $M = 256$. We notice that we can calculate $\{a_n\}$ up to $|n| \fallingdotseq 220$ in the UBASIC but that we obtain them only up to $|n| \fallingdotseq 90$ in REAL*16.

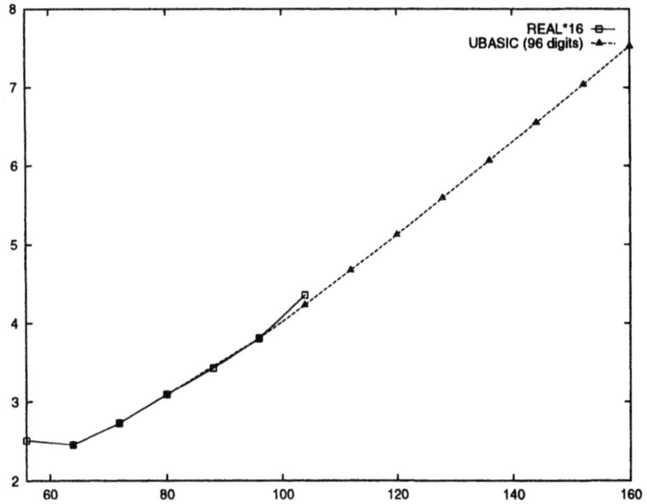

Fig. 2. the inverse problem (horizontal axis : N, vertical axis : $-\ln e_N$)

For the inverse problem, we calculate the unknown $\mu(\hat{y})$ in the equation (2.8) by the same manner from the computed far-field pattern $u_{\infty,M}$, which is obtained from u^i and q_M. We denote our numerical solution by μ_M.

We give the solution ∂D to the inverse problem in advance and try numerical reconstructions in the research, and we define the discretization error e_M by the maximum of $|u^i + u^s_M|$ on ∂D, where u^s_M is given by

$$u^s_M(x) = \int_{S^1} E(x - \hat{y})\mu_M(\hat{y})\, d\sigma_{\hat{y}} \quad x \in \partial D.$$

We observe in Fig.2 that we obtain accurate numerical solutions up to $n \fallingdotseq 90$ in REAL*16 and that we get them up to more than $n \fallingdotseq 160$ in the UBASIC. We notice from Fig.2 that numerical solutions by the Kress method seem converge exponentially to the exact solution. Some other examples are seen in Saito [8].

These simple numerical examples clearly explain the efficiency of applications of multiple-precision systems to ill-conditioned problems. But we must note that the UBASIC consumes very much time in each computation; it is aimed for the research of number theory and is not aimed for large scale problems arising from the discretizations of functional equations such as integral equations and PDE's etc. We should improve a multiple-precision system, if we wish to apply multiple-precision systems to numerical analysis connected with PDE's. We succeed in construction of a fast multiple-precision system (F-system), which is introduced in §4.

3 Cauchy-Riemann Equation

We deal with a finite difference approach to the initial value problem of the Cauchy-Riemann equation in this section and we give some numerical results given by the F-system. The problem is equivalent to the harmonic extension across a boundary of a domain, and it is one of the typical ill-posed problems.

We firstly refer to a theory of finite difference schemes in the class of analytic functions by Hayakawa [2]. Let A be a constant square matrix and consider the initial value problem

$$\frac{\partial}{\partial t}u(t,x) = A\frac{\partial}{\partial x}u(t,x), \tag{3.1}$$

$$u(0,x) = u_0(x), \tag{3.2}$$

where u and u_0 are vector valued functions. If $u_0(x)$ is analytic on an interval containing $x = 0$, there exists a unique analytic solution in a neighborhood $\mathcal{V}$ of the origin. We consider a finite difference scheme

$$u_h(t+\Delta t, x) = \lambda A u_h(t, x+\Delta x) + (I - \lambda A)u_h(t,x) \tag{3.3}$$

$$u_h(0,x) = u_0(x) \tag{3.4}$$

in order to approximate the solution in the neighborhood, where $\lambda = \Delta t/\Delta x > 0$ and I is the identity. According to the theory, we have the following theorem.

Theorem 1. *(Hayakawa [2]) Suppose $u_0(x)$ to be analytic near $x = 0$, and define $u_h(t,x)$ $(t \geq 0)$ by* (3.3) *and* (3.4) *for a fixed λ. Then there exists a domain D contained in the neighborhood $\mathcal{V}$ of the origin such that $u_h(t,x)$ uniformly converges to the exact solution to* (3.1) *and* (3.2) *on D as $\Delta t, \Delta x \to 0$.*

We remark that the theorem covers the case

$$A = \begin{pmatrix} 0 & 1 \\ -1 & 0 \end{pmatrix}, \tag{3.5}$$

and that the finite difference solution u_h converges to the exact solution without stability; the equation (3.1) becomes the Cauchy-Riemann equation for the case (3.5), and we know the scheme (3.3) to be unstable.

We will show some numerical examples by the scheme (3.3) for

$$\frac{\partial}{\partial t}\begin{pmatrix} u_1(t,x) \\ u_2(t,x) \end{pmatrix} = \begin{pmatrix} 0 & 1 \\ -1 & 0 \end{pmatrix}\frac{\partial}{\partial x}\begin{pmatrix} u_1(t,x) \\ u_2(t,x) \end{pmatrix},$$

$$\begin{pmatrix} u_1(0,x) \\ u_2(0,x) \end{pmatrix} = \begin{pmatrix} t^2 - x^2 \\ 2tx - 0.5 \end{pmatrix},$$

as follows. Dotted lines mean the exact solution and $\diamond$ means computed values by the F-system.

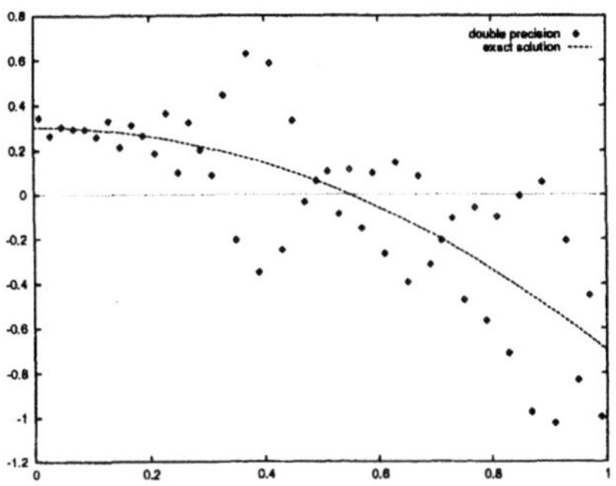

Fig.3. result by the double precision (horizontal axis : x, vertical axis : $u_1(t,x)(t = 0.55)$)

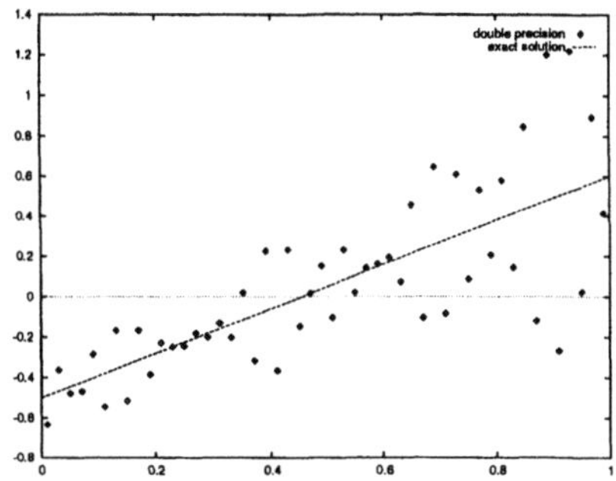

Fig.4. result by the double precision (horizontal axis : x, vertical axis : $u_2(t,x)(t = 0.55)$)

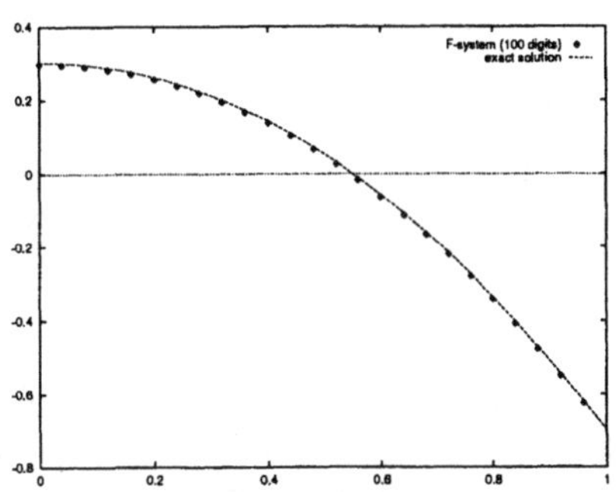

Fig.5. result by 100 digits in 10-radix (horizontal axis : x, vertical axis : $u_1(t,x)(t = 0.55)$)

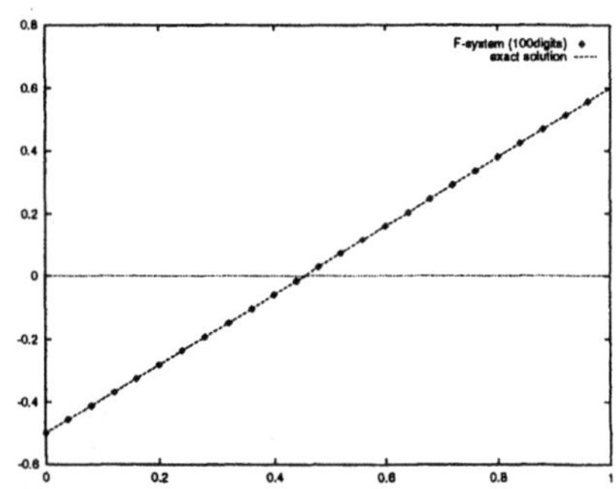

Fig.6. result by 100 digits in 10-radix (horizontal axis : x, vertical axis : $u_2(t,x)(t = 0.55)$)

We remark again that our problem is ill-posed and that the scheme is unstable. Fig.3 and Fig.4 show the instability, and meaningful calculations cannot be done in the double precision. We think that the fatal influence in unstable schemes should be the rounding errors, and we use a multiple-precision system to remove them virtually. Fig.5 and Fig.6 show good coincidence of computed values in 100 digits in radix-10 with the exact ones at $t = 0.55$. But we notice that 100 digits comparison is not enough for the case of $t = 0.65$ (Fig.7 and Fig.8), and we need 120 digits for the case (Fig.9 and Fig.10).

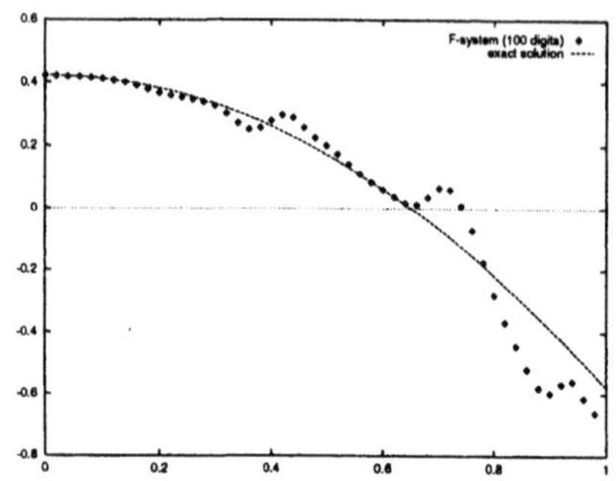

Fig.7. result by 100 digits in 10-radix (horizontal axis : x, vertical axis : $u_1(t,x)(t=0.65)$)

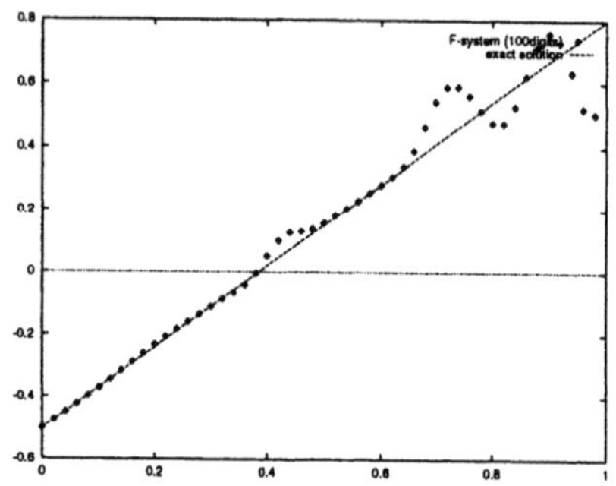

Fig.8. result by 100 digits in 10-radix (horizontal axis : x, vertical axis : $u_2(t,x)(t=0.65)$)

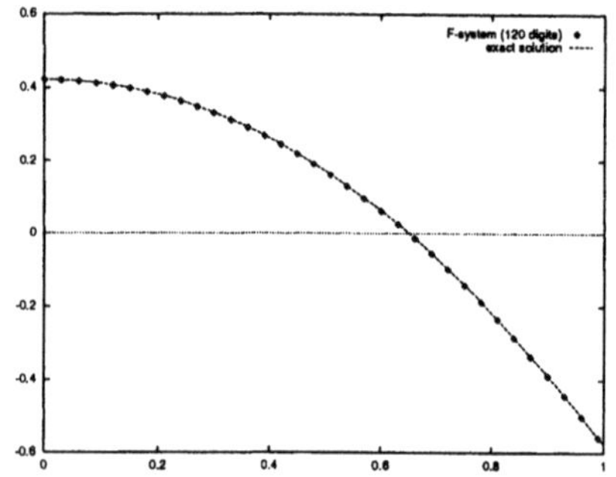

Fig.9. result by 120 digits in 10-radix (horizontal axis : x, vertical axis : $u_1(t,x)(t=0.65)$)

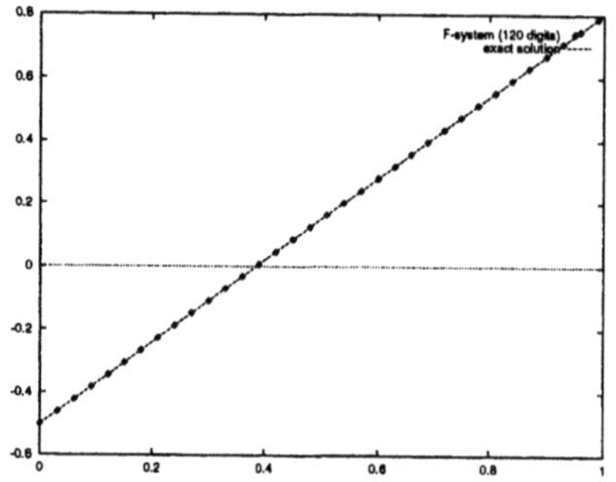

Fig.10. result by 120 digits in 10-radix (horizontal axis : x, vertical axis : $u_2(t,x)(t=0.65)$)

We conclude that a multiple-precision computation is a powerful tool for numerical computation for unstable problems, but we should know necessary digits in computations in advance. And the F-system is aimed for computation for large scale problems, and we can save much time in computation with the UBASIC. Refer to Fujiwara [1] for the detailed data of computing by the F-system.

4 Fast Multiple-precision System (F-system)

In this section, we give a brief introduction of the Fast Multiple-precision System (F-system) implemented by one of the authors. Refer to Fujiwara [1] for details.

Our aim is numerical treatment of ill-conditioned problems arising in engineering and physics, and we need a fast computer system in which we can represent and operate real numbers without the rounding errors. To this end, we choose a strategy that we can fix an arbitrary (finite) precision to represent and to calculate approximated real numbers within the prescribed precision. We call it a multiple-precision arithmetic system, which is considered one of the extended formats. Though we cannot remove the rounding errors even by the strategy, we *virtually* carry out exact numerical computations on the floating point arithmetic. The term 'virtual' means that we control significant digits to avoid the influence of the rounding errors as much as we wish, and we are able to obtain numerical results as precisely as we wish.

The principles of the design for the F-system are;

- fast computing,
- using less memory, which enables us to deal with large scale problems,
- fully designed in 64bit base; possibility of high performance with resources of the typical next generation computer environments,
- equipment of easy interfaces for accesses of low end users.

The most important features in the construction of a fast multiple-precision system are data structure and implementation of improved arithmetic algorithms.

We firstly explain a format of multiple-precision numbers in the F-system. We adopt an original extension of the floating point format designed in IEEE754 (see [3]). Each real number is normalized in the form $(-1)^s \times 2^e \times 1.F$, where the sign part s has one bit, the exponent part e has 63bit, and the fraction part $F = f_1 f_2 \cdots f_n$ has $64 \times n$ bit. (Each f_i has 64bit width.) This type of floating point multiple-precision numbers are stored in memory using the 64bit unsigned integer array, and accuracy of the numbers is $\log_{10} 2^{64n+1} (\sim 19.26 \times n)$ digits, and the above mentioned real number $(-1)^s \times 2^e \times 1.F$ is

$$(-1)^s \times 2^{e_b - b} \times \left(1 + \sum_{i=1}^{n} f_i \, 2^{-64i}\right),$$

where $b = 2^{62} - 1$.

One of the advantages of the design is cost performance in the use of memory. The ALU (Arithmetic Logical Unit) of 64bit CPU's is more advantageous in accuracy than the FPU (Floating Point Unit), and we can use all bits in the integer array which stores a fraction part of the multiple-precision format according to our design.

We are enough to use less memory than in the case of IEEE754 double format number array, and our system enables us to deal with larger scale problems. At the same time, since a number of necessary elements of the array decreases, and since the memory access and branch operators are less issued, we succeed in speed up in computation. We remark that our 64bit integer base representation and operation should be a remarkable feature of the design and have an advantage.

We secondly mention about improvement and implementation of arithmetic algorithms. In our system, the arithmetic for multiple-precision numbers is designed as external routines of the programming language C (LP 64 model). We choose the Alpha system as a 64bit computing environment in the present research. The four fundamental rules for multiple-precision arithmetic, multiplication by integer, and division by integer are implemented in the Alpha assembly language and are optimized for its architecture. The classical algorithms are used in addition and in subtraction, and we adopt the classical algorithm ($O(n^2)$) and Karatsuba-Offman's algorithm ($O(n^{\log_2 3})$) in multiplication (see [6]). In division, we implement an improved Ozawa's algorithm (see [7]).

Considering usability, we should design multiple-precision systems as modules of an existing programming language, and we implement our system as external routines of the programming language C. But we should note that the language C has a disadvantage in vulnerability of the argument types etc; users should be requested to parse formulas in their programming. These processes are burdens for the users, and we remark that we design the interface of the F-system by use of the polymorphism supplied by the programming language C++.

We are sure that a fast multiple-precision system, which is applicable to large scale problems, should become a powerful new tool in computational mathematics, and our research is one of the first steps to overcome difficulties in the computation of the ill-conditioned problems.

References

1. H. Fujiwara, *Design of fast multiple-precision arithmetic tool for the 64bit computer environments.* (preprint)
2. K. Hayakawa, *Convergence of Finite Difference Scheme and Analytic Data*, Publ.Res.Inst.Math.Sci., **24** (1988), 759–764
3. IEEE, *IEEE Standard 754-1985 for Binary Floating-Point Arithmetic.* (Reprinted in SIGPLAN **22(2)** (1987), 9–25)
4. A. Kirsch and R. Kress, *An optimization method in inverse acoustic scattering*, Boundary Elements IX Vol. 3, Springer-Verlag (1987), 3–18
5. R. Kress, Linear Integral Equation, Springer-Verlag (1989)
6. D. E. Knuth, The art of computer programming (3rd edition), Vol. 2, Addison Wesley (1998)
7. K. Ozawa, *A Fast $O(n^2)$ Division Algorithm for Multiple-Precision Floating-Point Numbers*, J. of Information Processing **14** (1991) 354–356
8. K. Saito, *Numerical analysis of an inverse scattering problem for the Helmholtz equation* (in Japanese) Master thesis (Dep. Math.,Kyoto Univ.) 1998 March

Numerical Verification Methods for Solutions of Free Boundary Problems

Mitsuhiro T. Nakao[1] and Cheon Seoung Ryoo[2]

[1] Faculty of Mathematics, Kyushu University 33
Fukuoka 812–8581,Japan
e-mail: mtnakao@math.kyushu-u.ac.jp
[2] Department of Mathematics, Kyungpook National University
Taegu 702-701, Korea
e-mail: ryoocs@knu.ac.kr

Abstract. In this paper, we consider numerical techniques which enable us to verify the existence of solutions for the free boundary problems governed by two kinds of elliptic variational inequalities(EVIs). Based upon the finite element approximations and the explicit a priori error estimates for some simple EVIs, we present effective verification procedures that, through numerical computation, generate a set which includes exact solutions. We describe a survey of the previous works as well as show some newly obtained results up to now.

1 Introduction

The authors have studied for years the numerical verification of solutions for elliptic partial differential equations([8][9][11]etc.) and elliptic variational inequalities(EVIs) using the finite element method and the constructive error estimates combining with Schauder's and Banach's fixed point theorem. Several results in our research are already published in [12],[17],[19]. In this paper, we briefly overview our recent research results including works not yet published.

In Section 2, so-called first kind problems of EVI are considered. Namely, in 2.1, we first give, a slightly detailed description of the basic principle and formulation of our numerical verification method for the solution of obstacle problems with a homogeneous condition. This should be an appropriate introduction to another applications of our idea. The basic approach of the method consists of the fixed point formulation of the problems and construction of the function set, in a computer, satisfying the validation condition of a certain infinite dimensional fixed point theorem. We also mention that it is possible to extend the method to more general problems with non-homogeneous obstacles. Moreover, in order to apply our method to the problem whose associated operator is not retractive in a neighborhood of the solution, a Newton-like method is introduced.

Next, in 2.2, we apply our method to another type of free boundary problem which appears in the elasto-plastic deformation theory. This problem causes some properties of non-smoothness in the associated finite dimensional equations. But, we can also overcome such a difficulty by applying the solution method for non-smooth problems developed by [3]. In the subsection 2.3, we briefly remark that our enclosure method can also be applied to the so-called simplified Sigorini problem which is a simplified version of a problem occurring in the elasticity theory.

Finally, in Section 3, we show the way to apply our approach to EVIs of the second kind appearing in the flow problems of a visco-plastic fluid in a pipe.

2 EVIs of the first kind

2.1 Obstacle problem

In this subsection, we consider the verification method for solutions of the obstacle problem which is known as a free boundary problem cahracterizing the contacted zone with the obstacle of an elastic membrane.

2.1.1 Homogeneous case

Here, 'homogeneous' stands for the case that obstacle $\psi \equiv 0$ in the whole domain.

Problem and basic formulation of verification Though the basic idea of verification is given in other places(e.g., [17],[19]), in order to keep the paper as self-contained as possible, we describe rather detailed formulation and verification procedure for the present case.

Let Ω be a bounded convex domain in R^n, $1 \le n \le 2$, with a piecewise smooth boundary $\partial\Omega$. We set $V \equiv H_0^1(\Omega) = \{v \in H^1(\Omega) : v|_{\partial\Omega} = 0\}$ and

$$a(u, v) = (\nabla u, \nabla v)$$

which is adopted as the inner product on V, where $(\cdot, \cdot)$ stands for the inner product on $L^2(\Omega)$. We define $K := \{v \in V : v \ge 0 \text{ a.e. on } \Omega\}$.

First, we note that, by the well-known result [7], for any $g \in L^2(\Omega)$, the problem:

$$a(u, v - u) \ge (g, v - u), \quad \forall v \in K, \quad u \in K, \tag{1}$$

has a unique solution $u \in V \cap H^2(\Omega)$, and the estimate

$$|u|_{H^2(\Omega)} \le \|g\|_{L^2(\Omega)} \tag{2}$$

holds (cf.[7]), where $|w|_{H^2}$ implies the semi-norm of w in $H^2(\Omega)$ defined by

$$|w|^2_{H^2(\Omega)} \equiv \sum_{i,j=1}^{n} \| \frac{\partial^2 w}{\partial x_i \partial x_j} \|^2_{L^2(\Omega)}.$$

Now consider the following EVI with a nonlinear right-hand side:

$$\begin{cases} \text{Find } w \in K \quad \text{such that} \\ a(w, v - w) \geq (f(w), v - w), \quad \forall v \in K. \end{cases} \tag{3}$$

Here, assume that f satisfies the hypotheses as follows:

A1. f is the continuous map from V into $L^2(\Omega)$.

A2. For each bounded subset $W \in V$, $f(W)$ is also bounded in $L^2(\Omega)$.

We take an appropriate finite dimensional subspace V_h of V for $0 < h < 1$. Usually, V_h is taken to be a finite element subspace with mesh size h. We then define K_h, an approximation of K, by

$$K_h = V_h \cap K = \{v_h | v_h \in V_h, \quad v_h \geq 0 \text{ on } \overline{\Omega}\}.$$

We also define the projection P_K from V onto K. That is, $v = P_K(w)$, the projection of $w \in V$ into K, is defined as the unique solution of the following problem:

$$v \in K: \quad a(v, \zeta - v) \geq a(w, \zeta - v), \quad \forall \zeta \in K. \tag{4}$$

And define the projection P_{K_h} from V onto K_h. That is, $v_h = P_{K_h}(w)$, the projection of w into K_h, is defined as follows:

$$v_h \in K_h: \quad a(v_h, \zeta - v_h) \geq a(w, \zeta - v_h), \quad \forall \zeta \in K_h. \tag{5}$$

Now, as one of the approximation properties of K_h, assume that

A3. For each $w \in K \cap H^2(\Omega)$, there exists a positive constant C_1, independent of h, such that

$$\|w - P_{K_h} w\|_V \leq C_1 h |w|_{H^2(\Omega)}. \tag{6}$$

Here, C_1 has to be numerically determined. For example, it is known that we may take $C_1 = \frac{\sqrt{5}}{\pi}$ for the linear element in the one dimensional case[17]. Furthermore, it will be readily seen that the same constant can be taken for the two dimensional bilinear element from the consideration on the proof of Theorem 5.1 in [17].

To verify the existence of a solution of (3) in a computer, we use the fixed point formulation.

First, note that, for each $w \in V$, there exists a unique $F(w) \in V$ such that

$$(\nabla F(w), \nabla v) = (f(w), v), \quad \forall v \in V, \tag{7}$$

which also implies that

$$\begin{cases} -\Delta F(w) = f(w) \text{ in } \Omega, \\ F(w) = 0 \text{ on } \partial\Omega. \end{cases} \tag{8}$$

Then the map $F : V \longrightarrow V$ is compact. By (7), problem (3) is equivalent to finding $w \in V$ such that

$$a(w, v - w) \geq a(F(w), v - w), \qquad \forall v \in K. \tag{9}$$

By using the definition (4) and (9), we now have the following fixed point problem for the compact operator $P_K F$.

$$\text{Find } w \in V \text{ such that } w = P_K F(w). \tag{10}$$

Verification condition We introduce two concepts, rounding and rounding error, which enable us to deal with the infinite dimensional problem by finite procedures, i.e., in a computer.

Now we define the dual cone of K_h by

$$K_h^* = \{w \in V : a(w, v) \leq 0, \quad \forall v \in K_h\},$$

and note that K_h^* is also a closed convex cone in V with vertex at the origin with is the only point common to K_h and K_h^*. From (5) it follows that K_h^* is the null set of the projection P_{K_h}. We have the following lemma which is from [13].

Lemma 1. *Any $w \in V$ can be uniquely decomposed into the sum of two orthogonal elements. That is,*

$$w = P_{K_h} w \bigoplus (I - P_{K_h}) w = P_{K_h} w \oplus P_{K_h^*} w.$$

Here, $\oplus$ denotes the sum of two orthogonal elements in the sense of V.

For any $w \in V$, we now define the rounding $R(P_K F(w)) \in K_h$ by the solution of the following problem:

$$a(R(P_K F(w)), v_h - R(P_K F(w))) \geq (f(w), v_h - R(P_K F(w))), \quad \forall v_h \in K_h.$$

Next, for any subset $W \subset V$, we define the rounding $R(P_K FW) \subset K_h$ by

$$R(P_K FW) = \{w_h \in K_h : w_h = R(P_K F(w)), \ w \in W\}.$$

Usually, $R(P_K FW)$ is enclosed and represented as a linear conbination of the base functions in V_h with interval coefficients.

Moreover, for $W \subset V$, we define $RE(P_K FW)$, the rounding error of $P_K FW$, as a subset of K_h^*, i.e.,

$$RE(P_K FW) = \{v \in K_h^* : \|v\|_V \leq C_0 h \|f(W)\|_{L^2}\}, \tag{11}$$

where

$$\|f(W)\|_{L^2} \equiv \sup_{w \in W} \|f(w)\|_{L^2}.$$

Here, $C_0 \equiv C_1C_2$, where C_1 is the same positive constant as in (6), and C_2 is determined by the following regularity estimate for the solution to (1) of the form

$$|u|_{H^2} \leq C_2||g||_{L^2}. \tag{12}$$

Thus we may take as $C_2 = 1$ for the present case from (2). Then, we have

$$P_K F(w) - R(P_K F(w)) \in RE(P_K F(w)), \quad \forall w \in W.$$

Therefore, the following verification condition is obtained by Schauder's fixed point theorem.

Lemma 2. *If there exists a nonempty, bounded, convex, and closed subset $W \subset K$ such that*

$$R(P_K FW) \oplus RE(P_K FW) \subset W, \tag{13}$$

then there exists a solution of $w = P_K F(w)$ in W.

We sometimes refer the above set W as *a candidate set*, which we generate in a computer so that it satisfies the condition (13).

Verification procedures We describe below the method to find a set W satisfying (13).
Consider the following approximate solution $w_h \in K_h$ of (1):

$$a(w_h, v_h - w_h) \geq (g, v_h - w_h), \quad \forall v_h \in K_h. \tag{14}$$

Since the bilinear form $a(\,\cdot\,,\,\cdot\,)$ is symmetric, (14) is reduced to the quadratic programming problem:

$$\min_{v \in K_h} \left[\frac{1}{2} a(v,v) - (g,v) \right]. \tag{15}$$

Let $\{\phi_j\}_{j=1\cdots M}$ be a basis of V_h with usual linear functions such that $\phi_j(x) \geq 0, \ \ \forall x \in \Omega$ and satisfying

$$\phi_j(x_i) = \begin{cases} 1, & i = j, \\ 0, & i \neq j, \end{cases}$$

where x_i is a node of the finite element mesh. Then (15) reduces to the following vector form:

$$\min_{w \geq 0} \left[\frac{1}{2} w' D w - P' w \right], \tag{16}$$

where $w \geq 0$ means the componentwise relation. Here, $D := (d_{ij})_{1\leq i,j\leq M}$ with $d_{ij} = (\nabla\phi_i, \nabla\phi_j)$, and w is the coefficient vector with $\{\phi_j\}$ of the function v in (15). Also, $P := \Big((g, \phi_j)\Big)_{1\leq j\leq M}$.
Furthermore, for any $\alpha \in R^+$, nonnegative real number, we set

$$[\alpha] \equiv \{\phi \in K_h^* : \quad \|\phi\|_V \leq \alpha\}.$$

Then, for a given candidate set $W = W_h \oplus [\alpha]$ with $W_h \subset K_h$, the computation of the rounding $R(P_K FW)$ reduces to enclose an interval vector $Z = (Z_j)$ and $Y = (Y_j)$ satisfying the following nonlinear system of equations(see [17] for details):

$$\begin{cases} Y - DZ = -(f(W), \phi_j), & 1 \leq j \leq M, \\ Y_j Z_j = 0, & 1 \leq j \leq M. \end{cases} \tag{17}$$

Here, $(f(W), \phi_j)$ is evaluated as an interval B_j such that $\{(f(w), \phi_j) | w \in W\} \subset B_j$. In order to solve (17) with guaranteed accuracy, we use some interval approaches for the nonlinear system of equations(e,g., [14]). Thus, using the solution of (17), we can enclose the set $R(P_K FW)$ in (13). Combining this with (11), we can successfully compute the left-hand side of (13) for any candidate set $W = W_h \oplus [\alpha]$.

Thus we can present a computational verification condition. In the actual computation, we use an iterative procedure with *δ-inflation* technique to find the set W satisfying (13)(*cf.*[10],[17]etc.). Several numerical examples for verification are presented in [17], for the one dimensional problem using the linear finite element.

2.1.2 Non-homogeneous case

For the non-homogeneous case, we define $K := \{v \in V : v \geq \psi \text{ a.e. on } \Omega\}$, where ψ is a given $H^2(\Omega)$ function such that $\psi \leq 0$ on $\partial\Omega$ and is not identically equal to 0. In this case, we take V_h as in 2.1.1 and define K_h by

$$K_h = \{v_h \in V_h : \quad v_h(p) \geq \psi(p), \forall p \in \mathcal{N}_h\},$$

where $\mathcal{N}_h$ denotes the set of all nodes associated with the subspace V_h. Note that, in general, $K_h \neq V_h \cap K$. Then, P_K and P_{K_h} are similarly defined as before, and we also have the constructive error estimates of the form, $\forall v_h \in K_h$ and $\forall v \in K$,

$$\begin{aligned} \|u_h - u\|^2_{H_0^1(\Omega)} \leq 2\|g + \Delta u\|_{L^2}(\|u - v_h\|_{L^2} + \|u_h - v\|_{L^2}) \\ + \|v_h - u\|^2_{H_0^1(\Omega)}. \end{aligned} \tag{18}$$

Based upon this inequality and the similar arguments to that in [6], by using the approximation property of K_h and the constructive regularity estimates of the form: $|u|_{H^2} \leq \Phi(g, \psi)$ for the solution u of the variational inequality of the same type as (1), we can obtain the desired error estimates. Here $\Phi(g, \psi)$ is a nonlinear and constructive function of g and ψ(see [6]). Detailed arguments and numerical examples will be given in forthcoming paper[18].

2.1.3 A Newton-type verification method

The idea of the enclosure method for solutions of obstacle problems in 2.1.1 is based upon simply sequential iterations for the original fixed point operator $P_K F$. Therefore, it is difficult to apply the method to the problem of which associated operator is not retractive in a neighborhood of the solution. In order to overcome such a difficulty, we introduce an another formulation using a Newton-like operator. The essential point is the way to devise the Newton-like operator for a kind of non-differentiable map which defines the original problem.

To formulate a Newton-type verification condition, we need a Fréchet derivative of the operator $P_K F$. However, $P_K F$ is not Fréchet differentiable at all. Therefore, we define the approximate Fréchet-like derivative $\widetilde{D}_K F(u_h)$ on V_h for some $u_h \in K_h$ instead of the Fréchet derivative. Assume that $\{\phi_j\}_{j=1\cdots M}$ is a basis of V_h, where $M = \dim V_h$, such that $\phi_j(x) \geq 0$ on Ω and satisfying

$$\phi_j(x_i) = \begin{cases} 1, & i = j, \\ 0, & i \neq j, \end{cases}$$

where x_i is a node of the finite element mesh.

And, for $v_h \in V_h$, we represent it such as

$$v_h = \sum_{j=1}^{M} v_{hj}\phi_j.$$

Here, $(v_{hj})_{j=1,\cdots,M}$ is called as the coefficient vector of v_h. Now we take a fixed subset $N_0 \subset \{1, 2, \cdots, M\}$, define V_{h,N_0}, the closed subspace of V_h, by

$$V_{h,N_0} = \{v_h | v_h \in V_h, v_{hj} = 0 \text{ for } j \notin N_0\}.$$

And let P_{h,N_0} be a H_0^1-projection from V onto V_{h,N_0} defined by

$$a(u - P_{h,N_0}u, v) = 0, \quad \forall v \in V_{h,N_0}, P_{h,N_0}u \in V_{h,N_0}.$$

In order to define $\widetilde{D}_K F(u_h) : V_h \longrightarrow V_{h,N_0}$, we differentiate the first equation of (17) in W at $W = u_h$ to get, for arbitrary $\delta \in V_h$,

$$\partial Y^* - D\partial Z^* = -\{(f'(u_h)\delta, \phi_j)\}_{1\leq j\leq M}. \tag{19}$$

Here, $\partial Y^* = (\widetilde{Y}_j^*)_{1\leq j\leq M}$ and $\partial Z^* = (\widetilde{Z}_j^*)_{1\leq j\leq M}$, where $\widetilde{Y}_j^* = 0$ for $j \in N_0$ and $\widetilde{Z}_j^* = 0$ for $j \notin N_0$, respectively.

Then we define the approximate Fréchet-like derivative of $P_K F(u)$ at $u = u_h$, as the linear map $\widetilde{D}_K F(u_h) : V_h \longrightarrow V_{h,N_0}$ such that, for each $\delta \in V_h$,

$$\widetilde{D}_K F(u_h)(\delta) := \sum_{j=1}^{M} \widetilde{Z_j}^* \phi_j.$$

We now assume that

A4. The restriction to V_{h,N_0} of the operator $P_{h,N_0}[I - \widetilde{D}_K F(u_h)] : V_h \longrightarrow V_{h,N_0}$ has the inverse operator

$$[P_{h,N_0} - \widetilde{D}_K F(u_h)]_h^{-1} : V_{h,N_0} \longrightarrow V_{h,N_0}.$$

Here, I means the identity map on V_h.

By using the above approximate Fréchet-like derivative, we define the Newton-like operator $N_h : V \longrightarrow V_h$ by

$$N_h(w) \equiv P_{K_h} w - [P_{h,N_0} - \widetilde{D}_K F(u_h)]_h^{-1} P_{h,N_0}(P_{K_h} - P_{K_h} P_K F)(w)).$$

Next we define the operator $T : V \longrightarrow V$ as follows:

$$T(w) \equiv N_h(w) + (I - P_{K_h}) P_K F(w).$$

Then T becomes a compact map on V and it follows the fixed point problem $w = P_K F w$ is equivalent to $w = T(w)$. Hence, we can formulate a Newton-type verification condition such as in [10]. By using this formulation, we succeed the enclosure for the solutions of problems for which the previously sequential iteration method could not work(see [19] for numerical examples).

2.2 Elasto-plastic torsion problems

In this subsection, we consider an enclosure metnod of solutions for elasto-plastic torsion problems governed by an EVI. The nonlinear elasto-plastic torsion problem is defined as the same type EVI as (3) with

$$K := \{v \in H_0^1(\Omega) : |\nabla v| \leq 1 \quad \text{a.e. on } \Omega\}. \tag{20}$$

As is well known(e.g., [5], [13]), two sub-domains Ω_p and Ω_e defined by

$$\Omega_p = \{x : x \in \Omega, |\nabla u| = 1\},$$

and

$$\Omega_e = \Omega \setminus \Omega_p = \{x : x \in \Omega, |\nabla u| < 1\}$$

correspond to the plastic and elastic regions, respectively. The elastic region Ω_e and the plastic region Ω_p are not known beforehand and should be determined, therefore $\partial\Omega_e \cap \partial\Omega_p$ is actually the free boundary of the problem (3). The problem (3) has been formulated as the problem of finding u satisfying

$$\begin{cases} -\Delta u = f(u) \ \text{ in } \Omega_e, \\ |\nabla u| = 1 \ \text{ in } \Omega_p, \\ u = 0 \ \text{ on } \partial\Omega. \end{cases} \tag{21}$$

The finite dimensional convex subset K_h is also defined similarly as before:

$$K_h := V_h \cap K = \{v_h | \ v_h \in V_h, \ \ |\nabla v_h| \leq 1 \text{ a.e. on } \Omega\}. \tag{22}$$

In order to formulate the verification procedure, we need a verified computational method for solving the finite dimensional part(rounding) and a constructive estimates for infinite dimensional part(rounding error) as in the previous subsection.

Following [7], we define the Lagrangian functional $\mathcal{L}$ associated with (1) by

$$\mathcal{L}(v,\mu) = \frac{1}{2}\int_\Omega |\nabla v|^2 dx - (g,v) + \frac{1}{2}\int_\Omega \mu(|\nabla v|^2 - 1)dx.$$

It follows, from [7], that if $\mathcal{L}$ has a saddle point $\{u,\lambda\} \in H_0^1(\Omega) \times L_+^\infty(\Omega)$, then u is a solution of (1), where $L_+^\infty(\Omega) = \{q \in L^\infty(\Omega) : q \geq 0 \text{ a.e. in } \Omega\}$. We use the Uzawa algorithm to solve (1)([7]).

Thus we can claculate the rounding $R(P_K F(W))$, for a candidate set W, by solving the following problem with guaranteed error bounds:

$$\begin{cases} \text{Find } \{u_h, \lambda_h\} \in K_h \times \Lambda_h \quad \text{such that} \\ \lambda_h = \max[\lambda_h + \rho(|\nabla u_h|^2 - 1), 0] \ \text{ with } \rho > 0. \\ \int_\Omega (1+\lambda_h)\nabla u_h \cdot \nabla v_h dx = (f(W), v_h), \forall v_h \in V_h, \ u_h \in V_h, \end{cases} \tag{23}$$

The problem (23) can be formulated as a system of nonlinear and nonsmooth (nondifferentiable) equations. A verification method for nonsmooth equations by a generalized Krawczyk operator is studied in [3]. We briefly describe the method presented by [3] in the below.

We consider the following equivalent system of nonlinear(and nondifferentiable) equations to (23) for a fixed $w \in W$

$$H(x) = 0. \tag{24}$$

Here, we assume that $H : R^n \longrightarrow R^n$ is locally Lipschitz continuous. The equivalence means that x^* solves (23) if and only if x^* solves (24). The method is based on the mean value theorem for local Lipschitz functions of the form

$$H(x) - H(y) \in co\partial H([x])(x-y), \ \text{ for all } x, y \in [x],$$

where $[x]$ stands for an interval vector, "co" denotes the convex hull, and ∂H the generalized Jacobian in Clarke's sense [4], which is also considered as a slope function, and

$$co\partial H([x]) := co\{V \in \partial H(x) : \ x \in [x]\}.$$

Let $[L_{[x]}]$ be an interval matrix such that $co\partial H([x]) \subseteq [L_{[x]}]$. Then for any $x, y \in [x] \subseteq R^n$ it holds that $H(x) - H(y) \in [L_{[x]}](x-y)$. Next an interval operator for nonsmooth equations is defined by

$$G(x, A, [x]) := x - A^{-1}H(x) + (I - A^{-1}[L_{[x]}])([x] - x). \tag{25}$$

The mapping $G(x, A, [x])$ is called a generalized Krawczyk operator. Therefore, the verification condition of solutions for (24) in $[x]$ is given by $G(x, A, [x]) \subseteq [x] \subset D$.

Thus, we can compute the solution of (23) with guaranteed accuracy. That is, we can enclose the rounding $R(P_K F(U))$.

On the other hand, for the calculation of the rounding error $RE(P_K F(U))$, the similar arguements in 2.1.1 can also be applied for the one dimensional problem. Actually, we can prove that the same constant $C_0 = \frac{\sqrt{5}}{\pi}$ is also valid for the present problem in the one dimensinal case, which implies that we can give a verification procedure besed on the same principle as before(see, [12]).

2.3 Signorini problem

A simplified Signorini problem is also given by the EVI of the form (3) with

$$K := \{v \in H_0^1(\Omega) : \mid v \geq 0 \quad \text{on } \partial\Omega\} \tag{26}$$

and

$$a(u, v) \equiv (\nabla u, \nabla v) + (u, v). \tag{27}$$

As well known, the solution u of this EVI can be characterized as a solution of the following free boundary problem finding u and two subsets Γ_0 and Γ_+ such that $\Gamma_0 \cup \Gamma_+ = \partial\Omega$ and $\Gamma_0 \cap \Gamma_+ = \emptyset$

$$\begin{cases} -\Delta u + u = f(u) \text{ in } \Omega, \\ u = 0 \text{ on } \Gamma_0, \dfrac{\partial u}{\partial n} \geq 0 \text{ on } \Gamma_0, \\ u > 0 \text{ on } \Gamma_+, \dfrac{\partial u}{\partial n} = 0 \text{ on } \Gamma_+, \end{cases} \tag{28}$$

where $\frac{\partial}{\partial n}$ the outer normal derivative on $\partial\Omega$. In the present case, the approximation subspace K_h is taken as

$$K_h := V_h \cap K = \{v_h | \ v_h \in V_h, \ \ v_h \geq 0 \text{ on } \partial\Omega\}. \tag{29}$$

For a candidate set W, the computation of rounding $R(P_K F(W))$ is also reduced to the quadratic programming problem as in 2.1.1([22], [7]).

Since the constant C_2 in (12) is easily estimated as $C_2 = 1$, the standard approximation property of the interpolation by K_h gives a constructive error estimates to compute the rounding error $RE(P_K F(W))$. The detailed computational procedures and numerical examples will be described in the forthcoming paper [20].

3 EVI of the second kind

In this section, we show that our idea of verification method can also be applied to the EVI of the second kind.

Now, we define the functional $j(v) = \int_\Omega |\nabla v| dx$. We consider the following problem of the flow of a viscous plastic fluid in a pipe:

$$\begin{cases} \text{Find } u \in H_0^1(\Omega) \quad \text{such that} \\ \\ a(u, v-u) + j(v) - j(u) \geq (f(u), v-u), \quad \forall v \in H_0^1(\Omega). \end{cases} \tag{30}$$

As in the previous section, we consider the following auxiliary problem associated with (30) for a given $g \in L^2(\Omega)$:

$$a(u, v-u) + j(v) - j(u) \geq (g, v-u), \forall v \in H_0^1(\Omega), u \in H_0^1(\Omega). \tag{31}$$

By the well known result [2], we have the following lemma.

Lemma 3. *([2]) There exists a unique solution $u \in H_0^1(\Omega) \cap H^2(\Omega)$ of (31) for any $g \in L^2$, such that*

$$\|u\|_{H^2(\Omega)} \leq \widehat{C} \|g\|_{L^2(\Omega)}.$$

This lemma follows by obvious modification in the proof of Theorem 15 in [2]. When we denote the solution u of (31) by $u = Ag$ and define the composite map F on $H_0^1(\Omega)$ by $F(u) \equiv Af(u)$, which is a little bit of different from the previously appeared symbol F in Section 2, we have

Theorem 4. *F is compact on $H_0^1(\Omega)$ and the problem (30) is equivalent to the fixed point problem*

$$u = F(u).$$

Proof. First, for a bounded subset $U \subset L^2(\Omega)$, we show that $AU \subset H_0^1(\Omega)$ is relatively compact. Secondly, prove that $A : L^2(\Omega) \to H_0^1(\Omega)$ is continuous. By Lemma 3, $AU \subset H^2(\Omega) \cap H_0^1(\Omega)$ and AU is bounded in $H^2(\Omega)$. Since U is bounded in $L^2(\Omega)$, by the Sobolev imbedding theorem, we have AU is relatively compact in $H_0^1(\Omega)$. Next, for arbitrary $f_1, f_2 \in L^2(\Omega)$, setting $u_1 = Af_1$ and $u_2 = Af_2$, by using (31), we obtain

$$a(u_1, u_2 - u_1) + j(u_2) - j(u_1) \geq (f_1, u_2 - u_1),$$

$$a(u_2, u_1 - u_2) + j(u_1) - j(u_2) \geq (f_2, u_1 - u_2).$$

With the above inequalities, we obtain $a(u_2 - u_1, u_2 - u_1) = -a(u_1, u_2 - u_1) + a(u_2, u_2 - u_1) \leq j(u_2) - j(u_1) - (f_1, u_2 - u_1) + j(u_1) - j(u_2) - (f_2, u_1 - u_2) = (f_2 - f_1, u_2 - u_1)$. Hence, by the Poincaré inequality, we have

$$\|u_2 - u_1\|^2_{H_0^1(\Omega)} \leq \|f_2 - f_1\|_{L^2(\Omega)} \|u_2 - u_1\|_{L^2(\Omega)} \leq \overline{C} \|f_2 - f_1\|_{L^2} \|u_2 - u_1\|_{H_0^1(\Omega)}.$$

Therefore, we obtain

$$\|u_2 - u_1\|_{H_0^1(\Omega)} \le \overline{C}\|f_2 - f_1\|_{L^2(\Omega)}.$$

That is, A is Lipschitz continuous as a map $L^2(\Omega) \to H_0^1(\Omega)$. Hence A is compact. The latter half in the theorem is straightforward from the definition of F.

We now define the approximate problem corresponding to (31) as

$$a(u_h, v_h - u_h) + j(v_h) - j(u_h) \ge (g, v_h - u_h), \forall v_h \in V_h, u_h \in V_h \quad (32)$$

In order to apply our verification method to enclose the solutions of (30), we need a guaranteed computation of the exact solution of the problem (32)(a rounding procedure), as well as the constructive error estimates between the solution of (31) and (32)(rounding error estimates).

A major difficulty in solving the problem (32) numerically is the processing of the nondifferentiable term $j(u) = \int_\Omega |\nabla u| dx$. One approach is the method of Lagrange multiplier on that term, whose continuous version is as follows [7]. Let us define $\Lambda = \{q \mid q \in L^2(\Omega) \times L^2(\Omega), |q(x)| \le 1 \text{ a.e. } x \in \Omega\}$ with $|q(x)| = \sqrt{q_1(x)^2 + q_2(x)^2}$. Then the solution u of (31) is equivalent to the existence of q satisfying

$$\begin{cases} a(u, v) + \displaystyle\int_\Omega q \cdot \nabla v = (g, v), \forall v \in H_0^1(\Omega), u \in H_0^1(\Omega), \\ q \cdot \nabla u = |\nabla u| \text{ a.e. }, q \in \Lambda, \end{cases} \quad (33)$$

which means the simultaneous equations for two unknown functions u and q.

Moreover, it is known that (33) is equivalent to the following problem:

$$\begin{cases} a(u, v) + \int_\Omega q \cdot \nabla v = (g, v), \forall v \in H_0^1(\Omega), u \in H_0^1(\Omega), \\ q = \dfrac{q + \rho \nabla u}{\sup(1, |q + \rho \nabla u|)}. \end{cases} \quad (34)$$

Here ρ is a positive constant. Let $\mathcal{T}_h$ be a triangulation of Ω, and let define L_h and Λ_h (approximation of $L^\infty(\Omega) \times L^\infty(\Omega)$ and Λ, respectively) by

$$L_h = \{q_h \mid q_h = \sum_{\tau \in \mathcal{T}_h} q_\tau \chi_\tau, \ q_\tau \in R^2\} \text{ and } \Lambda_h = \Lambda \cap L_h, \text{ respectively,}$$

where χ_τ is the characteristic function of τ.

Then our first purpose, computing the rounding $RF(U)$, is to enclose the solution of the following approximation problem of (34):

$$\begin{cases} a(u_h, v_h) + \int_\Omega q_h \cdot \nabla v_h = (g, v_h), \forall v_h \in V_h, u_h \in V_h, \\ q_h = \dfrac{q_h + \rho \nabla u_h}{\sup(1, |q_h + \rho \nabla u_h|)}. \end{cases} \quad (35)$$

The equation (35) leads to a kind of finite dimensional, nonlinear and non-differentiable problem. We use a slope function method proposed by Rump [15],[16] to enclose the solutions of (35) with $g = f(W)$ for a candidate set W.

On the other hand, the rounding error $RE(F(U))$ can be computed by using the following constructive error estimates:

Theorem 5. *Let u and u_h be solutions of (31) and (32), respectively. If $g \in L^2(\Omega)$, then there exists a constant $C(h)$ such that*

$$\|u_h - u\|_{H_0^1(\Omega)} \leq C(h)\|g\|_{L^2(\Omega)}. \tag{36}$$

Here, we may take $C(h) = \frac{\sqrt{5}}{\pi}h$ for the linear element in the one dimensional case, and C is also numerically estimated such that $C(h) \approx O(h^{\frac{1}{2}})$ for the two dimensional linear element.

The proof of this theorem would be described in the forthcoming paper[21].

Thus we can also implement the verification algorithm for the solution of (30) as in the previous section.

A NUMERICAL EXAMPLE.

Let $\Omega = (0,1)$. We considered the case $f(u) = Qu + 4$, where Q is a constant. We used the usual linear element with uniform mesh size $h = \frac{1}{M}$, where M denotes the total number of elememts.
The execution conditions are as follows:

$$\begin{cases} Q = 1. \\ \text{Numbers of elements(M)} = 41. \qquad \dim V_h = 40. \\ \text{Approximate solution}: u_h = \text{Galerkin approximation (32)}. \end{cases}$$

Approximate locations of free boundary : $x = 0.238095$ and $x = 0.761905$.

Verified results are as follows:

$$\begin{cases} \text{Iteration numbers}: 12. \\ H_0^1 - \text{error bound}: 0.027202. \\ \text{Maximum width of coefficient intervals of } \phi_j = 0.072109. \end{cases}$$

References

1. Bazaraa,M.S., Shetty,C.M.: Nonlinear Programming. John Wiley, New York, 1979
2. Brezis, H.: Monotonicity in Hilbert Spaces and Some Applications to Nonlinear Partial Differential Equations. in Contributions to Nonlinear Functional Analysis (ed. by E.Zarantonell), Academic-Pres, New York (1971) 101-116

3. Chen, X.: A verification method for solutions of nonsmooth equations. Computing **58** (1997) 281-294
4. Clarke, F.H.: Optimization and Nonsmooth Analysis. John Wiley, New York, 1983
5. Elliott, C.M., Ockendon, J.R.: Weak and variational methods for moving boundary problems. Pitman, Boston, 1980
6. Falk, R.S.: Error estimates for the approximation of a class of variational inequalities. Math. Comp. **28**(1974) 963-971
7. Glowinski,R.: Numerical Methods for Nonlinear Variational Problems. Springer, New York, 1984
8. Nakao,M.T.: A numerical approach to the proof of existence of solutions for elliptic problems. Japan Journal of Applied Math. **5** (1988) 313-332
9. Nakao,M.T.: A numerical verification method for the existence of weak solutions for nonlinear boundary value problems. Journal of Math. Analysis and Appl. **164** (1992) 489-507
10. Nakao, M.T.: Solving nonlinear elliptic problems with result verification using an H^{-1} residual iteration, Computing, Supplementum **9**(1993) 161-173
11. Nakao,M.T., Yamamoto,N.: Numerical verification of solutions for nonlinear elliptic problems using an L^∞ residual method. Journal of Mathematical Analysis and Applications **217**, (1998) 246-262
12. Nakao, M.T., Lee, S.H., Ryoo, C.S.:Numerical verification of solutions for elasto-plastic torsion problems. Computers & Mathematics with Applications **39** (2000) 195-204.
13. Rodrigues,J.F.: Obstacle problems in Mathematical Physics. Math. Stud. **134** North-Holland, Amsterdam, 1987
14. Rump,S.M.: Solving algebraic problems with high accuracy, A new approach to scientific computation. Academic Press, New York, 1983
15. Rump, S.M.: Inclusion of Zeros of Nowhere Differentiable n-Dimensional Functions. Reliable computing **3**(1997) 5-16
16. Rump, S.M.: Expansion and estimation of the range of nonlinear functions. Mathematics of Computation **65**(1996) 1503-1512
17. Ryoo,C.S. and Nakao,M.T.: Numerical verifications of solutions for variational inequalities. Numerische Mathematik **81** (1998) 305-320
18. Ryoo,C.S., Nakao,M.T.: Numerical verifications of solutions for variational inequalities II. (submitted)
19. Ryoo, C.S.: Numerical verification of solutions for obstacle problems using Newton-like method. Computers and Mathematics with Applications **39** (2000) 185-194
20. Ryoo,C.S.: Numerical verification of solutions for a simplified Signorini problem. (preprint)
21. Ryoo,C.S., Nakao,M.T.: Numerical verification of solutions for variational inequalities of the second kind. (preprint)
22. Scarpini, F., Vivaldi, M.A.: Error estimates for the approximation of some unilateral problems. RAIRO Numerical Analysis **11** (1977) 197-208
23. Yamamoto,N., Nakao, M.T.: Numerical verifications of solutions for elliptic equations in nonconvex polyponal domains. Numerische Mathematik, **65** (1993) 503-521
24. Yamamoto,N., Nakao,M.T.: Numerical verifications of solutions to elliptic equations using residual iterations with a higher order finite element. Journal of Comp. and Applied Math. **60** (1995) 271-279

Pattern Formation of Heat Convection Problems

Takaaki Nishida[1], Tsutomu Ikeda[2], and Hideaki Yoshihara[1]

[1] Kyoto University, Sakyo-ku, Kyoto 606-8502, Japan
[2] Ryukoku University, Seta, Ohtsu 520-2194, Japan

Abstract. We consider the Rayleigh-Bénard problem of the heat convection in the horizontal strip with the stress free boundary condition for the velocity and Dirichlet boundary condition for the temperature. We examine the pattern formation of the roll type solution, the rectangular type solution and the hexagonal type solution and see the stability of them and a better bifurcation diagram for the full system by using numerical computations.

1 Introduction

We consider the Rayleigh-Bénard problem for the heat convection using the Oberbeck-Boussinesq equations for the velocity, pressure and temperature in the dimensionless form :

$$\begin{aligned} \frac{1}{\mathcal{P}}\Big(\frac{\partial \boldsymbol{u}}{\partial t} + \boldsymbol{u}\cdot\nabla\boldsymbol{u}\Big) + \nabla p &= \Delta\boldsymbol{u} - \rho(T)\nabla z\ , \\ \frac{\partial T}{\partial t} + \boldsymbol{u}\cdot\nabla T &= \Delta T\ \ ,\quad \nabla\cdot\boldsymbol{u} = 0\ , \end{aligned}$$

in the horizontal domain $\{\, x \in \mathbf{R},\ y \in \mathbf{R},\ 0 < z < \pi \,\}$, where $\rho(T) = \mathrm{G} - \mathcal{R}T$ is assumed for the density of the fluid, $\mathcal{P}$ is the Prandtl number and $\mathcal{R}$ is the Rayleigh number.

When the temperature $T = \pi$ is given on the lower boundary and $T = 0$ on the upper boundary, the equilibrium state is the purely heat conduction solution, which exists for all parameter values $\mathcal{P} > 0, \mathcal{R} > 0$:

$$\boldsymbol{u} = 0\ ,\ \ T = \pi - z\ ,\ \ \rho = \mathrm{G} - \mathcal{R}(\pi - z)\ ,\ \ p = \mathrm{G}(\pi - z) - \mathcal{R}\Big(\frac{\pi^2}{2} - \pi z + \frac{z^2}{2}\Big) + p_a\ .$$

We assume the stress free boundary condition for the velocity on the both boundaries ($z = 0,\ \pi$), and Dirichlet boundary condition for the temperature as above.

We will consider the bifurcation problems from this equilibrium state under the assumption that all perturbations are periodic in the horizontal direction, especially with the periodicity $0 \le x \le 2\pi/a,\ 0 \le y \le 2\pi/b$. The system for the perturbation to the equilibrium state is given by

$$\begin{aligned} \frac{1}{\mathcal{P}}\Big(\frac{\partial \boldsymbol{u}}{\partial t} + \boldsymbol{u}\cdot\nabla\boldsymbol{u}\Big) + \nabla p &= \Delta\boldsymbol{u} + \mathcal{R}\theta\nabla z\ \ , \\ \frac{\partial \theta}{\partial t} + \boldsymbol{u}\cdot\nabla\theta &= \Delta\theta + w\ \ ,\quad \nabla\cdot\boldsymbol{u} = 0\ , \end{aligned}$$

where

$$0 \le x \le \frac{2\pi}{a}\ ,\ 0 \le y \le \frac{2\pi}{b}\ ,\ 0 \le z \le \pi\ .$$

Also we assume the usual even- or odd-ness conditions for the unknown functions :

$$\begin{aligned}
u(x,y,z) &= -u(-x,y,z) &&= u(x,-y,z)\ ,\\
v(x,y,z) &= v(-x,y,z) &&= -v(x,-y,z)\ ,\\
w(x,y,z) &= w(-x,y,z) &&= w(x,-y,z)\ ,\\
\theta(x,y,z) &= \theta(-x,y,z) &&= \theta(x,-y,z)\ ,\\
p(x,y,z) &= p(-x,y,z) &&= p(x,-y,z)\ .
\end{aligned}$$

Then the unknown functions have the expansions :

$$\begin{aligned}
u(t,x,y,z) &= \sum_{l,m,n} u_{lmn}(t)\ \sin alx\ \cos bmy\ \cos nz\ ,\\
v(t,x,y,z) &= \sum_{l,m,n} v_{lmn}(t)\ \cos alx\ \sin bmy\ \cos nz\ ,\\
w(t,x,y,z) &= \sum_{l,m,n} w_{lmn}(t)\ \cos alx\ \cos bmy\ \sin nz\ ,\\
\theta(t,x,y,z) &= \sum_{l,m,n} \theta_{lmn}(t)\ \cos alx\ \cos bmy\ \sin nz\ ,\\
p(t,x,y,z) &= \sum_{l,m,n} p_{lmn}(t)\ \cos alx\ \cos bmy\ \cos nz\ .
\end{aligned}$$

The incompressibility condition is given for each l, m, n by the following :

$$alu_{lmn} + bmv_{lmn} + nw_{lmn} = 0\ .$$

The function spaces for the solution are the following :

$$L^2_{a,b} = \{u,v,w,\theta,p \mid \sum_{l,m,n} \{u^2_{lmn} + v^2_{lmn} + w^2_{lmn} + \theta^2_{lmn} + p^2_{lmn}\} < \infty\}$$

$$\begin{aligned}
H^2_{a,b} = \{u,v,w,\theta,p \mid \sum_{l,m,n} \{((al)^2 + (bm)^2 + n^2)^2\ (u^2_{lmn} + v^2_{lmn} + w^2_{lmn} + \theta^2_{lmn})\\
+ ((al)^2 + (bm)^2 + n^2)\ p^2_{lmn}\} < \infty\}
\end{aligned}$$

We analyze the eigenvalue and eigenvector for the linearized system, where we rescale θ, p and use the parameter $R = \sqrt{\mathcal{PR}}$:

$$\begin{aligned}
\frac{\partial \boldsymbol{u}}{\partial t} + \nabla p &= \mathcal{P}\Delta \boldsymbol{u} + R\theta\nabla z\ ,\\
\frac{\partial \theta}{\partial t} &= \Delta\theta + Rw\ ,\\
\nabla\cdot\boldsymbol{u} &= 0\ .
\end{aligned}$$

The linearized system is selfadjoint and has real eigenvalues as follows. Substituting the above expressions for the solution such as :

$$\theta(t,x,y,z) = \theta_{lmn}(0)\, e^{\lambda t} \cos alx \cos bmy \sin nz ,$$

we have the eigenvalue problem for λ, which can be solved explicitly. The eigenvalues have the following forms for each l , m , n .

$$\lambda_3 = -\mathcal{P} A^2 ,$$

$$\lambda_{\pm} = \frac{1}{2}(-(1+\mathcal{P})A^2 \pm \sqrt{(1+\mathcal{P})^2 A^4 + 4\,\frac{(a^2l^2+b^2m^2)R^2 - \mathcal{P}A^6}{A^2}}) ,$$

where

$$A^2 = a^2l^2 + b^2m^2 + n^2 .$$

For each (l, m, n) - mode λ_3 and λ_- are always negative, but

$$\lambda_+ = 0 \quad \text{at} \quad R^2 = \mathcal{P}\mathcal{R} = \frac{\mathcal{P}A^6}{(a^2l^2+b^2m^2)} .$$

Therefore the critical Rayleigh number for fixed (a,b) is

$$\mathcal{R}_c = \inf_{l,m,n} \frac{(a^2l^2+b^2m^2+n^2)^3}{(a^2l^2+b^2m^2)} .$$

Thus we know ([9]) that
If $\mathcal{R} < \mathcal{R}_c$, then the heat conduction state is linearly stable.
If $\mathcal{R} > \mathcal{R}_c$, then the heat conduction state is linearly unstable.
Furthermore Joseph ([5]) proved by energy method that
if $\mathcal{R} < \mathcal{R}_c$, then the heat conduction state is globally nonlinearly stable.

Therefore when the biggest eigenvalue becomes $\lambda_+ = 0$ at the critical Rayleigh number $\mathcal{R}_c$ for fixed a and b and it is "simple", the usual stationary bifurcation theory can be applied and the stationary bifurcation occurs at the critical Rayleigh number. If we examine the function

$$f = \frac{(\alpha^2+1)^3}{\alpha^2}$$

we see that $\mathcal{R}_c$ attains the minimum value 6.75 at $\alpha = 1/\sqrt{2}$. ($\mathcal{R}_c = 6.75 \times \pi^4$ for the usual dimensionless system.) The minimum value 6.75 is attained at $a = 1/\sqrt{2}$, $(l,m,n) = (1,0,1)$, at $a = 1/2\sqrt{2}$, $(l,m,n) = (2,0,1)$, and so on. We have considered two dimensional problems, namely roll type solutions on the extended bifurcation curves for the case $a = 1/\sqrt{2}$ and proved the existence of the solutions not close to the first bifurcation point $\mathcal{R} = \mathcal{R}_c = 6.75$ by a computer assisted proof [12] . Here we consider the three dimensional problem to obtain not only roll type solutions but also rectangular type solutions and hexagonal type solutions.

2 Rectangular Type Solution and Hexagonal Type Solution

To see the pattern formation clearly we choose special aspect ratio $b/a = \sqrt{3}$ and $a = 1/2\sqrt{2}$, which is one of the most interesting case, where the critical Rayleigh number $\mathcal{R}_c = 6.75$ and the biggest eigenvalue $\lambda = 0$ for $\mathcal{R} = \mathcal{R}_c$ has a two-dimensional eigenspace. One eigen-function with $(l, m, n) = (2, 0, 1)$ corresponds to the roll type solution :

$$\theta = \theta_{201} \cos(2ax) \sin z \ .$$

Here we only express the temperature for the eigen-function, but the other unknowns have similar expressions. The other eigen-function with $(l, m, n) = (1, 1, 1)$ does to the rectangular type
solution :

$$\theta = \theta_{111} \cos(ax) \cos(\sqrt{3}ay) \sin z \ .$$

Furthermore a linear combination of them does to the hexagonal type solution:

$$\theta = \theta_{hex}\{\, 2\cos(ax)\cos(\sqrt{3}ay)\sin z + \cos(2ax)\sin z\,\} \ .$$

Thus the simple bifurcation theory does not apply directly to this case.

To obtain the roll type solution by the simple bifurcation theory we have to restrict the function space $H^{\,2}_{a,\sqrt{3}a}$ to the subspace H^2_{roll} as follows : For example the temperature has the following expansion.

$$\theta = \sum_{n=0}^{\infty} \sum_{l+n=even}^{\infty} \theta_{l,0,n} \cos(2alx) \sin(nz).$$

Then the eigenvalue $\lambda = 0$ becomes simple at $\mathcal{R} = \mathcal{R}_c = 6.75$, and the simple stationary bifurcation theory applies to this and we obtain the roll type solution which bifurcates from the point for $\mathcal{R} > \mathcal{R}_c$. (cf. [4])
To obtain the rectangular type solution we can restrict the function space $H^{\,2}_{a,\sqrt{3}a}$ to the subspace $H^{\,2}_{rect}$ as follows :

$$\begin{aligned}\theta \;=\; &\sum_{n=odd}^{\infty} \sum_{l+m=even,l,m=odd}^{\infty} \theta_{l,m,n} \cos(alx)\cos(\sqrt{3}amy)\sin(nz) \\ &+ \sum_{n=even}^{\infty} \sum_{l+m=even,l,m=even}^{\infty} \theta_{l,m,n} \cos(alx)\cos(\sqrt{3}amy)\sin(nz) \ .\end{aligned}$$

The other unknowns have similar expansions. In this subspace the simple bifurcation theory applies to the system and the stationary bifurcation of rectangular type solution occurs at the critical Rayleigh number.

The hexagonal cell solution has the invariance of $2\pi/3$ rotation in $x-y$ plane and we can restrict $H^{\,2}_{a,\sqrt{3}a}$ to the subspace H^2_{hexa} as follows :

$$
\begin{aligned}
u &= \sum_{l,m,n}^{\infty} \sum_{l+m=even}^{\infty} \{\, u_{l,m,n} \sin(alx) \cos(\sqrt{3}amy) \\
&\quad + (\frac{1}{2}u_{l,m,n} - \frac{\sqrt{3}}{2}v_{l,m,n}) \sin(a(\frac{l-3m}{2})x) \cos(\sqrt{3}a(\frac{l+3m}{2})y) \\
&\quad + (\frac{1}{2}u_{l,m,n} + \frac{\sqrt{3}}{2}v_{l,m,n}) \sin(a(\frac{l+3m}{2})x) \cos(\sqrt{3}a(\frac{l-3m}{2})y) \,\} \cos(nz) \,, \\
v &= \sum_{l,m,n}^{\infty} \sum_{l+m=even}^{\infty} \{v_{l,m,n} \cos(alx) \sin(\sqrt{3}amy) \\
&\quad + (\frac{\sqrt{3}}{2}u_{l,m,n} + \frac{1}{2}v_{l,m,n}) \cos(a(\frac{l-3m}{2})x) \sin(\sqrt{3}a(\frac{l+3m}{2})y) \\
&\quad + (\frac{\sqrt{3}}{2}u_{l,m,n} - \frac{1}{2}v_{l,m,n}) \cos(a(\frac{l+3m}{2})x) \sin(\sqrt{3}a(\frac{l-3m}{2})y) \,\} \cos(nz) \,, \\
\theta &= \sum_{l,m,n}^{\infty} \sum_{l+m=even}^{\infty} \theta_{l,m,n} \{\, \cos(alx) \cos(\sqrt{3}amy) \\
&\quad + \cos(a(\frac{l-3m}{2})x) \cos(\sqrt{3}a(\frac{l+3m}{2})y) \\
&\quad + \cos(a(\frac{l+3m}{2})x) \cos(\sqrt{3}a(\frac{l-3m}{2})y) \,\} \sin(nz) \,.
\end{aligned}
$$

The other unknowns have similar expansions to that of temperature. In this subspace we have the bifurcation of hexagonal type solution. Here we notice that by the global nonlinear stability theorem of Joseph all these bifurcation branches come out of the same bifurcation point to the direction of $\mathcal{R} > \mathcal{R}_c$

3 Bifurcation Diagram

Each solution given by the above bifurcation theory is stable in each restricted subspace of $H^{\,2}_{a,\sqrt{3}a}$ by the direction of the bifurcated branch mentioned above. However we do not know the stability of them in the original space $H^{\,2}_{a,\sqrt{3}a}$. Also we want to know those solutions on the extended bifurcation curves. To treat these problems we are forced to use the numerical computations. We restrict the function space $H^{\,2}_{a,\sqrt{3}a}$ to the finite dimensional space H^2_N by the condition $l+m+n \leq N$ and $l+m =$ even . The latter condition comes from that the rotation of $2\pi/3$ is invariant in the subspace. If we apply the Galerkin method in H^2_N to the stationary solutions of the original system, it is reduced to the finite dimensional system of bilinear algebraic equations.

Thus the stationary solutions of this finite dimensional system are obtained by Newton's method. We have taken Prandtl number $\mathcal{P} = 10$ and $N = 12$ in the computations. We have the following figures.

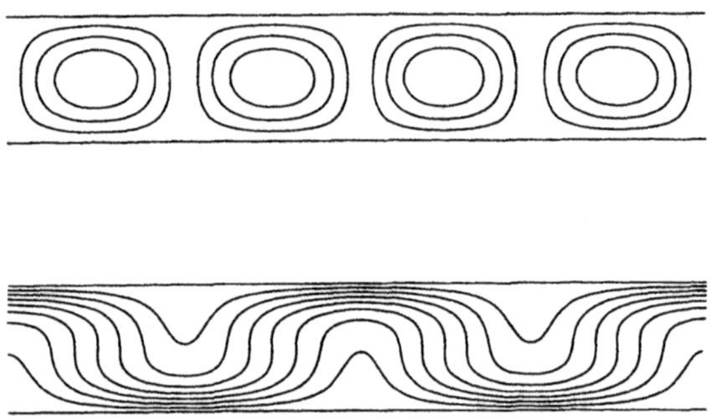

Fig. 1. Roll solution for $r = \mathcal{R}/\mathcal{R}_c = 2.0$. The upper showsthe stream line and the lower shows the isothermal line for $0 \leq x \leq 2\pi/a,\ 0 \leq z \leq \pi$ in $x - z$ plane.

Fig. 2. Roll solution for $r = \mathcal{R}/\mathcal{R}_c = 10.0$. The upper shows the stream line and the lower shows the isothermal line for $0 \leq x \leq 2\pi/a,\ 0 \leq z \leq \pi$ in $x - z$ plane.

Figure 3
Rectangular type solution for $r = \mathcal{R}/\mathcal{R}_c = 1.25$, which is stable. The upper shows the isothermal line for $0 \le x \le 4\pi/a$, $0 \le y \le 4\pi/\sqrt{3}a$ in $x - y$ plane at $z = \pi/2$ and the lower shows the contour line of $u^2 + v^2$ for $0 \le x \le 4\pi/a$, $0 \le y \le 4\pi/\sqrt{3}a$ in $x - y$ plane at $z = \pi$.

Figure 4
Rectangular type solution for $r = \mathcal{R}/\mathcal{R}_c = 1.5$, which is unstable. The upper shows the isothermal line for $0 \le x \le 4\pi/a$, $0 \le y \le 4\pi/\sqrt{3}a$ in $x - y$ plane at $z = \pi/2$ and the lower shows the contour line of $u^2 + v^2$ for $0 \le x \le 4\pi/a$, $0 \le y \le 4\pi/\sqrt{3}a$ in $x - y$ plane at $z = \pi$.

Figure 5
Hexagonal type solution for $r = \mathcal{R}/\mathcal{R}_c = 1.5$, which is unstable. The upper shows the isothermal line for $0 \leq x \leq 4\pi/a$, $0 \leq y \leq 4\pi/\sqrt{3}a$ in $x - y$ plane at $z = \pi/2$ and the lower shows the contour line of $u^2 + v^2$ for $0 \leq x \leq 4\pi/a$, $0 \leq y \leq 4\pi/\sqrt{3}a$ in $x - y$ plane at $z = \pi$.

Figure 6
Mixed (rectangular and hexagonal) type solution for $r = \mathcal{R}/\mathcal{R}_c = 1.5$, which is stable. The upper shows the isothermal line for $0 \leq x \leq 4\pi/a$, $0 \leq y \leq 4\pi/\sqrt{3}a$ in $x - y$ plane at $z = \pi/2$ and the lower shows the contour line of $u^2 + v^2$ for $0 \leq x \leq 4\pi/a$, $0 \leq y \leq 4\pi/\sqrt{3}a$ in $x - y$ plane at $z = \pi$.

Figure 7
Hexagonal type solution for $r = \mathcal{R}/\mathcal{R}_c = 2.0$, which is stable. The upper shows the isothermal line for $0 \le x \le 4\pi/a$, $0 \le y \le 4\pi/\sqrt{3}a$ in $x - y$ plane at $z = \pi/2$ and the lower shows the contour line of $u^2 + v^2$ for $0 \le x \le 4\pi/a$, $0 \le y \le 4\pi/\sqrt{3}a$ n $x - y$ plane at $z = \pi$.

Figure 8
Bifurcation diagram for $0.0 \le r = \mathcal{R}/\mathcal{R}_c \le 3.0$. The upper line represents the bifurcation curve of roll type solution, which is stable. The middle line does that of hexagonal type solution and the lower line does that of rectangular type solution. The curve which connects the rectangular branch with the hexagonal one does the mixed type solution, which is stable.

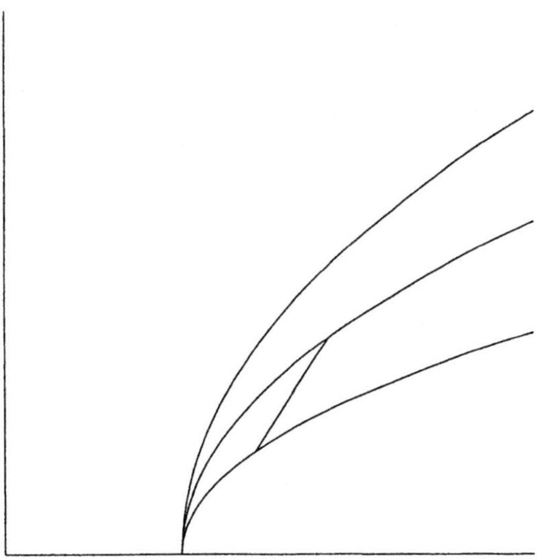

The stability of those solutions are examined by the linearization around those solutions and are reduced to the eigenvalue problem for the linearized time dependent system. These eigenvalue problems are solved by numerical computations. Now we summarize the stability and bifurcation diagram as follows. The roll type solution is always stable under our computations $1.0 < r \leq 3.0$, the rectangular type solution is stable for $1.0 \leq r \leq 1.41$ then it becomes unstable $r \approx 1.43$ and the mixed type solution bifurcates from it and is stable for $1.45 \leq r \leq 1.83$ and it goes in the hexagonal type solution at $r \approx 1.845$, then the hexagonal type solution becomes stable for $1.86 \leq r \leq 3.0$. The conceptional diagram is shown in Figure 8. We hope this bifurcation diagram will be justified by a computer assisted proof such as in [12].

References

1. Busse, F. H. : The oscillatory instability of convection rolls in a low Prandtl number fluid. J. Fluid Mech. **52** (1972) 97–112
2. Busse, F. H. and Bolton, E. W. : Instabilities of convection rolls with stress-free boundaries near threshold. J. Fluid Mech. **146** (1984) 115–125
3. Bolton, E. W. and Busse, F. H. : Stability of convection rolls in a layer with stress-free boundaries. J. Fluid Mech. **150** (1985) 487–498
4. Crandall, M. G. and Rabinowitz, P. H. : Bifurcation, perturbation of simple eigenvalues, and linearized stability. Arch. Rat. Mech. Anal. **52** (1973) 161–180
5. Joseph, D. D. : On the stability of the Boussinesq equations. Arch. Rat. Mech. Anal. **20** (1965) 59–71
6. Kagei, Y. and Wahl, W. v. : The Eckhaus criterion for convection roll solutions of the Oberbeck-Boussinesq equations. Intern. J. Nonlin. Mech. **32** (1997) 563–620
7. Kirchgässner, K. and Kielhöfer, H.: Stability and bifurcation in fluid dynamics. Rocky Mountain J. Math. **3** (1973) 275–318
8. Moore, D. R. and Weiss, N. O. : Two-dimensional Rayleigh-Bénard convection. J. Fluid Mech. **58** (1973) 289–312
9. Rayleigh, L. : On convection currents in a horizontal layer of fluid, when the higher temperature is on the under side. Phil. Mag. Ser. 6, **32** (1916) 529–546
10. Velte, W. : Stabilitätsverhalten und Verzweigung stationärer Lösungen der Navier-Stokesschen Gleichungen. Arch. Rat. Mech. Anal. **16** (1964) 97–125
11. Veronis, G.: Large amplitude Bénard convection. J. Fluid Mech. **26** (1966) 49–68
12. Watanabe, Y., Yamamoto, N., Nakao, M. and Nishida, T.: A numerical verification method for bifurcated solutions of Rayleigh-Bénard Problems . Preprint. (2000) 1–38
13. Yudovich, V. I. : On the onset of convection. P. M. M. (J. Appl. Math. Mech.) **30** (1966) 1193–1199

Mathematical Modeling and Numerical Simulation of Earth's Mantle Convection *

Masahisa Tabata and Atsushi Suzuki

Department of Mathematical Sciences, Kyushu University, Fukuoka, 812-8581 Japan

Abstract. Rheology and geometry are two important factors in the Earth's mantle convection phenomenon. That is, the viscosity is strongly dependent on the temperature and the phenomenon occurs in a spherical shell domain. Focusing our attention on these two factors, we describe a total approach of numerical simulation of the Earth's mantle convection, i.e., mathematical modeling, mathematical analysis, computational scheme, error analysis, and numerical result.

1 Introduction

The Earth's mantle convection is considered to have a close relation to the earthquake by the plate tectonics theory. It is, therefore, an important research subject for scientists, especially, in the countries where earthquakes often occur. Numerical simulations have been done by many authors to analyze the phenomenon. See for instance [1],[11] and the references therein. The main part of the mathematical model of this phenomenon consists of the Rayleigh-Bénard equations with infinite Prandtl number. While computations in the early stage had been done with the constant viscosity and/or in the box domain, Ratcliff et al. [8] pointed out the importance of rheology and geometry in this phenomenon. The former means that the viscosity of the mantle is strongly dependent on the temperature, and the latter means that the domain of the problem is a three-dimensional spherical shell. The corresponding mathematical model becomes a nonlinear system of the Stokes equations and the convection-diffusion equation in a spherical shell, coupled with the viscosity, the buoyancy and the convection. In this paper we review the mathematical model, prove the existence of the solution, present an efficient and mathematically justified finite element scheme, and show a numerical result using the scheme.

Experimental data on this phenomenon are very few compared to engineering problems. There exist temporal and spatial limitations to get them. It is, therefore, important to establish numerical methods mathematically justified. At the same time numerical schemes should be so efficient as to solve the three-dimensional problem in a reasonable computation time. In the isoviscosity case we developed a stabilized finite element scheme and proved

* Dedicated to Professors Masayasu Mimura and Takaaki Nishida on their 60th birthday

the convergence of the finite element solutions to the exact one [10]. We have made a computation code of the scheme and reported some three-dimensional numerical experiments [9]. This paper presents an extension of the scheme to the temperature dependent viscosity case.

The contents of this paper are as follows. In Section 2 we derive a mathematical model of the Earth's mantle convection, starting from the Rayleigh-Bénard problem. The existence and uniqueness of the solution of the nonlinear partial differential equations are discussed in Section 3. In Section 4 we present a finite element scheme for the equations. Considering the cost of three-dimensional problem, we use P1/P1/P1 element, i.e., velocity, pressure, and temperature are all approximated by the piecewise linear element, which implies that some stabilization method is required. In the isoviscosity case we have used the Galerkin least square (GLS) type stabilization [6,4] for the Stokes equations. In the temperature dependent viscosity case we use the penalty type stabilization [2], which reduces the computational cost but keeps the same convergence order as the GLS stabilization. In Section 5 we present a numerical result, which shows a clear effect of the temperature dependent viscosity. We give some concluding remarks in Section 6.

Throughout the paper we denote by c and $c(*)$ positive constants, where the latter is dependent on the argument.

2 Mathematical Model

In this section we review a mathematical model for the Earth's mantle convection problem. The mantle is considered to be an incompressible fluid in the spherical domain between the core and the surface of the Earth. The core is hot and the surface is cold. The direction of gravity is to the center of the Earth. Accordingly we suppose that the movement of the Earth's mantle convection is governed by the Rayleigh-Bénard equations

$$\begin{aligned}\rho\left\{\frac{\partial u}{\partial t} + (u\cdot\nabla)u\right\} + \nabla p - \nabla\cdot[2\mu D(u)] &= -\rho g e^{(r)},\\ \nabla\cdot u &= 0,\\ \frac{\partial\theta}{\partial t} + u\cdot\nabla\theta - \nabla\cdot(\kappa\nabla\theta) &= f,\end{aligned}$$

where $u = (u_1, u_2, u_3)^T$ is the velocity, p is the pressure, θ is the temperature, ρ is the density, μ is the viscosity, g is the gravity acceleration, $e^{(r)}$ is the unit radial vector, κ is the thermal diffusivity, f is the heat source, and $D(u)$ is the velocity rate tensor defined by

$$D(u) := \left(\nabla u + \nabla u^T\right)/2.$$

The density ρ is, in general, a function of the temperature and the pressure. We, however, introduce the Bussinesq approximation to ρ, i.e., the ρ of the buoyancy term is replaced by

$$\rho = \rho_0\{1 - \alpha(\theta - \theta_0)\}$$

and the ρ of the inertia term is replaced by ρ_0, where α is the thermal expansion coefficient, and ρ_0 and θ_0 are representative density and temperature, respectively. The viscosity μ of mantle is strongly dependent on the temperature. We treat μ as a function of the temperature θ and the position x. Thus we consider the rheology of mantle. As for the importance to introduce this complex rheology we refer to Ratcliff et al.[8]. We assume that the others, g, κ, and α are positive constants.

Using the scales d, d^2/κ, κ/d, $\mu_0\kappa/\rho_0 d^2$, $\Delta\theta$, κ/d^2, and μ_0 for x, t, u, p, θ, f, and μ, and translating θ, we obtain non-dimensional equations

$$\frac{1}{Pr}\left\{\frac{\partial u}{\partial t} + (u\cdot\nabla)u\right\} + \nabla p - \nabla\cdot[2\mu(\theta)D(u)] = Ra\theta e^{(r)},$$

$$\nabla\cdot u = 0,$$

$$\frac{\partial\theta}{\partial t} + u\cdot\nabla\theta - \nabla^2\theta = f,$$

where d is the depth of mantle, $\Delta\theta$ is the difference of temperature, μ_0 is the representative viscosity, and Pr and Ra are Prandtl and Rayleigh numbers defined by

$$Pr := \frac{\mu_0}{\kappa\rho_0}, \quad Ra := \frac{\rho_0 g\alpha\Delta\theta d^3}{\kappa\mu_0}.$$

Since Pr is of order $10^{23} \sim 10^{24}$ in the mantle convection, we omit the inertia term. Scaling again t by $1/Ra$ and u, p, and f by Ra, we obtain

$$-\nabla\cdot[2\mu(\theta)D(u)] + \nabla p = \theta e^{(r)}, \tag{1}$$

$$\nabla\cdot u = 0, \tag{2}$$

$$\frac{\partial\theta}{\partial t} + u\cdot\nabla\theta - \frac{1}{Ra}\nabla^2\theta = f\ . \tag{3}$$

We recall again that the viscosity $\mu = \mu(x,\theta)$ is a positive function of x and θ.

Equations (1)–(3) hold in a spherical domain

$$\Omega := \{x \in \mathbb{R}; R_1 < |x| < R_2\},$$

where $|x|$ is the Euclidian norm of $x = (x_1, x_2, x_3)^T$, and R_1 and R_2 are positive constants. In the case of the Earth $R_1 = 11/9$ and $R_2 = 20/9$. Let Γ_1 and Γ_2 be inner and outer boundaries and Γ be the whole boundary. The slip boundary conditions for u and Dirichlet boundary conditions for θ

$$u\cdot n = 0, \tag{4}$$

$$D(u)n \times n = 0, \tag{5}$$

$$\theta = \theta_\Gamma \tag{6}$$

are imposed on Γ, where n is the exterior unit normal and θ_Γ is a given temperature field on the boundary. Initial condition for θ at $t = 0$,

$$\theta = \theta^0 \tag{7}$$

completes a mathematical model of the Earth's mantle convection, where θ^0 is a given temperature field.

3 Existence and Uniqueness of the Solution

In this section we discuss the existence and uniqueness of the solution of (1)–(7). After dividing the equations into a Stokes problem and a convection-diffusion problem, we investigate the whole problem.

If the temperature is known, (1) and (2) are Stokes equations with a variable viscosity ν and an external force g,

$$\nu = \mu(\cdot, \theta), \quad g = \theta e^{(r)} .$$

We consider a Stokes problem in the spherical domain Ω

$$-\nabla \cdot [2\nu D(u)] + \nabla p = g, \tag{8}$$

$$\nabla \cdot u = 0, \tag{9}$$

subject to the slip boundary conditions (4) and (5). Since (4) is an essential boundary condition, it is natural to introduce the function space

$$W := \{v \in H^1(\Omega)^3;\ v \cdot n = 0 \text{ on } \Gamma\}.$$

However, as was discussed in [10], the velocity is not determined uniquely in W. There are three freedoms of rigid body movements

$$v^{(i)} := e^{(i)} \times x \quad \text{for} \quad i = 1, 2, 3,$$

where $e^{(i)}$ is the unit vector to the x_i-direction. Eliminating the freedoms, we seek the velocity in

$$V := \{v \in W;\ (v, v^{(i)}) = 0 \ \ (i = 1, 2, 3)\},$$

and the pressure in

$$Q := \{q \in L^2(\Omega);\ (q, 1) = 0\},$$

where $(\cdot, \cdot)$ means the $L^2(\Omega)^3$- or $L^2(\Omega)$-inner product.

Lemma 1. *Suppose that*

$$g \in H^{-1}(\Omega)^3, \quad \langle g, v^{(i)} \rangle = 0 \ (i = 1, 2, 3),$$
$$\nu \in L^\infty(\Omega), \quad \nu \geq \nu_0 \quad \text{in } \Omega,$$

where ν_0 is a positive constant and $\langle \cdot, \cdot \rangle$ denotes the dual product. Then, there exists a unique solution $(u, p) \in V \times Q$ of (8), (9), (4) *and* (5), *and the estimate*

$$||u||_{H^1(\Omega)^3} + ||p||_{L^2(\Omega)} \leq c(\nu_0)||g||_{H^{-1}(\Omega)^3}$$

holds.

Lemma 1 can be proved in a similar way to Lemma 2 of [10].

Next we consider the convection-diffusion equation (3) in temperature, supposing that the velocity u is known. Let T be a positive constant. We consider (3) on the time interval $(0, T)$.

Lemma 2. *Suppose that*

$$\begin{aligned} &u \in L^2(0,T; L^3(\Omega)^3), \; \nabla \cdot u \in L^2(0,T; L^3(\Omega)), \\ &f \in L^2(0,T; H^{-1}(\Omega)), \; \theta_\Gamma \in H^1(0,T; H^{1/2}(\Gamma)), \; \theta^0 \in L^2(\Omega) \; . \end{aligned}$$

Then, there exists a unique solution of (3), (6) *and* (7)

$$\theta \in L^2(0,T; H^1(\Omega)) \cap L^\infty(0,T; L^2(\Omega)),$$

and the estimate

$$\begin{aligned} &||\theta||_{L^2(0,T;H^1(\Omega)) \cap L^\infty(0,T;L^2(\Omega))} \leq c(||\nabla \cdot u||_{L^2(0,T;L^3(\Omega))}) \\ &\times \{||\theta^0||_{L^2(\Omega)} + ||f||_{L^2(0,T;H^{-1}(\Omega))} + ||\theta_\Gamma||_{H^1(0,T;H^{1/2}(\Gamma))}\} \end{aligned}$$

holds. Furthermore, if

$$f \in L^\infty(0,T; L^\infty(\Omega)), \; \theta_\Gamma \in L^\infty(0,T; L^\infty(\Gamma)), \; \theta^0 \in L^\infty(\Omega),$$

we have for $t \in [0, T]$

$$\begin{aligned} &||\theta(t)||_{L^\infty(\Omega)} \\ &\leq t\,||f||_{L^\infty(0,t;L^\infty(\Omega))} + \max\{||\theta_\Gamma||_{L^\infty(0,t;L^\infty(\Gamma))} + ||\theta^0||_{L^\infty(\Omega)}\} \; . \end{aligned} \tag{10}$$

Outline of the proof. Evaluating the nonlinear term $u \cdot \nabla\theta$ by the assumption on u, we get the first part. The second part is obtained from the maximum principle of the convection-diffusion equation. □

We now consider the whole problem (1)–(7) on the time interval $(0, T)$. We suppose that a positive function $\mu = \mu(x, \theta)$

$$\mu : \; \bar{\Omega} \times \mathbb{R} \to (0, +\infty), \tag{11}$$

is continuous in x and continuously differentiable in θ.

Theorem 3. *Suppose* (11) *and*

$$f \in L^\infty(0,T;L^\infty(\Omega)),\ \theta^0 \in L^\infty(\Omega),$$
$$\theta_\Gamma \in H^1(0,T;H^{1/2}(\Gamma)) \cap L^\infty(0,T;L^\infty(\Gamma))\ .$$

Then, there exist a solution (u,p,θ) *of* (1)–(7),

$$u \in L^\infty(0,T;H^1(\Omega)^3),\ p \in L^\infty(0,T;L^2(\Omega)),$$
$$\theta \in L^2(0,T;H^1(\Omega)) \cap L^\infty(0,T;L^\infty(\Omega))\ .$$

Furthermore, if

$$u \in L^\infty(0,T;W^{1,\infty}(\Omega)^3),$$

the solution is unique.

Outline of the proof. For the existence of the solution we make a sequence of approximate solutions $\{(u_{\Delta t},p_{\Delta t},\theta_{\Delta t})\}$, where $\Delta t \downarrow 0$. Let Δt be a positive constant. The approximate solution is constructed step by step on the interval $I_k := (k\Delta t,(k+1)\Delta t)$ for $k=0,\cdots,\lfloor T/\Delta t\rfloor - 1$. Let $\theta^k_{\Delta t}$, the value of $\theta_{\Delta t}$ at time $k\Delta t$, be known ($\theta^0_{\Delta t} := \theta^0$). Putting $\nu = \mu(\cdot,\theta^k_{\Delta t})$, we solve the Stokes equations (8) and (9) subject to (4) and (5). We define $(u_{\Delta t},p_{\Delta t})$ on I_k by this solution. We solve the convection-diffusion equation (3) on I_k with $u = u^k_{\Delta t}$ and the initial value $\theta_{\Delta t}(k\Delta t) = \theta^k_{\Delta t}$ to get $\theta_{\Delta t}$ on the interval. Thus, we obtain a step function $(u_{\Delta t},p_{\Delta t})$ from $(0,T)$ to $H^1(\Omega)^3 \times L^2(\Omega)$ and a continuous function $\theta_{\Delta t}$ from $[0,T]$ to $H^{-1}(\Omega)$. From Lemmas 1 and 2 the sequence $\{(u_{\Delta t},p_{\Delta t},\theta_{\Delta t})\}$ is uniformly bounded in

$$L^\infty(0,T;H^1(\Omega)^3) \times L^\infty(0,T;L^2(\Omega))$$
$$\times L^2(0,T;H^1(\Omega)) \cap L^\infty(0,T;L^\infty(\Omega)),$$

which shows that a common positive constant ν_0 can be chosen in solving the Stokes problems in all I_k. Choosing an appropriate subsequence, we get a limit function (u,p,θ), which is a solution of (1)–(7). Here we use the compactness argument (see [7], [12], for example) to treat the nonlinear term $u \cdot \nabla\theta$.

The uniqueness is proved by the standard Gronwall inequality on the difference of supposed two temperatures, where the assumption on u is used to treat the nonlinear term $\mu(\theta)$. □

4 Finite Element Scheme

Here we present a finite element approximation to the problem (1)–(7). Considering the cost of computation in three-dimensional problems, we employ a cheap element combination P1/P1/P1, that is, velocity, pressure, and temperature are all approximated by the piecewise linear element. As is well-known, the combination of P1/P1 element does not work for the Stokes problem. We, therefore, use a stabilization method. Considering again the cost of computation, we employ the stabilization of penalty type [2]. Since Ra is high in our problem, (3) is convection-dominant. To solve the equation stably, we use the stream upwind Petrov/Galerkin method [5], [4].

Let Ω_h be a polyhedral approximation to Ω and $\mathcal{T}_h$ be a partition of $\bar{\Omega}_h$ by tetrahedra, where h is the maximum diameter of tetrahedral elements. The boundary of Ω_h is denoted by Γ_h. We consider a regular family of subdivisions $\{\mathcal{T}_h\}$, $h \downarrow 0$, satisfying the inverse assumption [3]. Let S_h $(\subset H^1(\Omega_h) \cap C^0(\bar{\Omega}_h))$ be the P1 finite element space whose degrees of freedom are on the vertices of tetrahedra. We introduce finite element spaces W_h, V_h, Q_h, and Ψ_h corresponding to W, V, Q, and $\Psi := H_0^1(\Omega)$, respectively,

$$
\begin{aligned}
W_h &:= \left\{ v_h \in S_h^3 \; ; \; (v_h \cdot n_\Omega)(P) = 0 \; (\forall P) \right\}, \\
V_h &:= \left\{ v_h \in W_h \; ; \; (v_h, v^{(i)})_h = 0 \; (i = 1, 2, 3) \right\}, \\
Q_h &:= \{ q_h \in S_h \; ; \; (q_h, 1)_h = 0 \}, \\
\Psi_h &:= \{ \psi_h \in S_h \; ; \; \psi_h(P) = 0 \; (\forall P) \},
\end{aligned}
$$

where P stands for nodal point on Γ_h, n_Ω is the unit outer normal to Γ. Since we use the P1 element, every nodal point P on Γ_h is on Γ. We employ $H^1(\Omega_h)^3$-norm for W_h and V_h, $L^2(\Omega_h)$-norm for Q_h, and $H^1(\Omega_h)$-norm for Ψ_h, respectively. We define an affine space $\Psi_h(\theta_\Gamma)$ by

$$
\Psi_h(\theta_\Gamma) := \{ \psi_h \in S_h \; ; \; \psi_h(P) = \theta_\Gamma(P) \; (\forall P) \},
$$

where P stands again for the nodal point on Γ_h and $\theta_\Gamma = \theta_\Gamma(x)$ is supposed to be continuous.

We prepare the following bilinear and trilinear forms for u, $v \in H^1(\Omega)^3$, $q \in L^2(\Omega)$, and θ, $\psi \in H^1(\Omega)$,

$$
\begin{aligned}
a(\mu, u, v) &:= 2 \int_\Omega \mu D(u) : D(v) \, dx, \\
b(v, q) &:= - \int_\Omega \nabla \cdot v \, q \, dx, \\
c_0(\theta, \psi) &:= \frac{1}{Ra} \int_\Omega \nabla\theta \cdot \nabla\psi \, dx, \\
c_1(u, \theta, \psi) &:= \frac{1}{2} \left\{ \int_\Omega (u \cdot \nabla)\theta \, \psi \, dx - \int_\Omega (u \cdot \nabla)\psi \, \theta \, dx \right\} .
\end{aligned}
$$

Remark 4.
(i) In the finite element method every integral over Ω is replaced by that over Ω_h. We express such integrals by adding the subscript h. For example, $a_h(\cdot, \cdot, \cdot)$ is the trilinear form corresponding to $a(\cdot, \cdot, \cdot)$, whose integration is done over Ω_h, and $(\cdot, \cdot)_h$ is the inner product in $L^2(\Omega_h)$.
(ii) In S_h^3, the rigid body rotation $v^{(i)}$, $i = 1, 2, 3$, can be reproduced. Especially, $v^{(i)}$ belongs to W_h.

Let Δt be a time increment and set the total time step number $N_T := [T/\Delta t]$. We denote by v_h^n the value of v_h at $t = n\Delta t$ for an integer $n \in [0, N_T]$. Let X be a Banach space. We define $\ell^q(X)$ norm for a sequence $v_h \equiv \{v_h^n\}_{n=0}^{N_T} \subset X$ by

$$||v_h||_{\ell^q(X)} := \{\Delta t \sum_{n=0}^{N_T} ||v_h^n||_X^q\}^{1/q},$$

where q (≥ 1) is a real number and extended naturally to ∞.

We approximate the time derivative $\partial\theta/\partial t$ at $t = (n+1)\Delta t$ by the difference $D_{\Delta t}\theta^n := (\theta^{n+1} - \theta^n)/\Delta t$. A stabilized finite element approximation to (1)–(7) is to find $\{(u_h^n, p_h^n, \theta_h^n)\}_{n=0}^{N_T} \subset V_h \times Q_h \times \Psi_h(\theta_\Gamma)$ satisfying

$$a_h(\mu(\theta_h^n), u_h^n, v_h) + b_h(v_h, p_h^n) = (\theta_h^n e_h^{(r)}, v_h)_h, \tag{12}$$

$$b_h(u_h^n, q_h) - \delta \sum_{K\in\mathcal{T}_h} h_K^2 (\nabla p_h^n, \nabla q_h)_K = 0, \tag{13}$$

$$\begin{aligned} &(D_{\Delta t}\theta_h^n, \psi_h)_h + c_{0\,h}(\theta_h^{n+1}, \psi_h) + c_{1\,h}(u_h^n, \theta_h^{n+1}, \psi_h) \\ &+ \sum_{K\in\mathcal{T}_h} \tau_K (D_{\Delta t}\theta_h^n + u_h^n \cdot \nabla\theta_h^{n+1}, u_h^n \cdot \nabla\psi_h)_K \\ &= (f_h^{n+1}, \psi_h)_h + \sum_{K\in\mathcal{T}_h} \tau_K (f_h^{n+1}, u_h^n \cdot \nabla\psi_h)_K \end{aligned} \tag{14}$$

for any $(v_h, q_h, \psi_h) \in V_h \times Q_h \times \Psi_h$ and $n \in [0, N_T]$ with an initial condition θ_h^0. Here $(\cdot,\cdot)_K$ represents the L^2-inner product on element K, and θ_h^0 is an approximation to θ^0. A positive constant δ is a stability parameter for the Stokes equations and τ_K is also a stability parameter for the convection-diffusion equation defined by

$$\tau_K := \min\left\{\frac{\Delta t}{2}, Ra\frac{h_K^2}{12}, \frac{h_K}{2U_K}\right\},$$

where h_K is the diameter of element K, $U_K = |u_h(G_K)|$ and G_K is the barycenter of K.

Remark 5. The stabilization method of GLS type [6] includes the term $\nabla \cdot [\mu(\theta_h^n)\nabla u_h^n]$. It does not vanish element-wise for the P1 element unless μ is constant, which increases considerably the computation cost in the three-dimensional problem. That is the reason why we use the penalty type stabilization. The convergence rates are same for these two methods as we use the P1 element.

We can show that (12) and (13) is uniquely solvable in (u_h^n, p_h^n) and that (14) is uniquely solvable in θ_h^{n+1}. When θ_h^n is given, (u_h^n, p_h^n) is obtained from (12) and (13). Substituting the u_h^n to (14), θ_h^{n+1} is solved. Hence, starting from the initial value θ_h^0, we can obtain the finite element solution (u_h, p_h, θ_h).

Suppose that f, θ_Γ, and θ^0 satisfy the conditions in Theorem 3. Then, we know a priori bound of θ from (10). Modifying μ outside the bound, we can take a positive constant μ_0 such that

$$\mu(x,\xi) \geq \mu_0 \quad \text{for } (x,\xi) \in \bar{\Omega} \times \mathbb{R} .$$

Under such a modification we can show that the scheme (12)–(14) is unconditionally stable and that the finite element solutions converge to the exact one.

Theorem 6. *Let* (u,p,θ) *be a solution of* (1)–(7) *such that*

$$\begin{aligned} u \in& C([0,T]\,; H^2(\Omega)^3) \cap H^1(0,T\,; H^1(\Omega)^3), \\ p \in& C([0,T]\,; H^1(\Omega)), \\ \theta \in& H^1(0,T\,; H^2(\Omega)) \cap H^2(0,T\,; L^2(\Omega)). \end{aligned}$$

Suppose that the initial value θ_h^0 *satisfies*

$$\|\theta_h^0 - \theta^0\|_{L^2(\Omega)} \leq c\,h \|\theta^0\|_{H^1(\Omega)} .$$

Then, there exist positive constants $c_* = c_*(T; u, p, \theta)$ *and* h_0 *such that for any* $\Delta t > 0$ *and* $h \in (0, h_0]$,

$$\begin{aligned} &\|\theta_h - \theta\|_{\ell^\infty(L^2)}, \quad \|\frac{1}{\sqrt{Ra}}(\theta_h - \theta)\|_{\ell^2(H^1)}, \|\sqrt{\tau}\, u_h \cdot \nabla(\theta_h - \theta)\|_{\ell^2(L^2)}, \\ &\qquad \|\sqrt{\tau}\,(D_{\Delta t} + u_h \cdot \nabla)(\theta_h - \theta)\|_{\ell^2(L^2)} \leq c_*\,(\Delta t + h), \\ &\qquad \|u_h - u\|_{\ell^\infty(H^1)}, \quad \|p_h - p\|_{\ell^\infty(L^2)} \leq c_*\,(\Delta t + h), \end{aligned}$$

where $\phi_h = (D_{\Delta t} + u_h \cdot \nabla)\psi_h$ *means that* $\phi_h^{n+1} = D_{\Delta t}\psi_h^n + u_h^n \cdot \nabla \psi_h^{n+1}$ *and*

$$\|\sqrt{\tau}\,\phi_h\|_{L^2(\Omega)} = \left\{ \sum_K \tau_K \|\phi_h\|_{L^2(K)}^2 \right\}^{1/2} .$$

This theorem extends the result obtained in the isoviscosity case [10] to the temperature dependent viscosity case. The complete proof will be given in a forthcoming paper.

5 Numerical Result

We present a numerical result of (1)–(7). Since the temperature is normalized, θ_Γ is simply given by

$$\theta_\Gamma = 1 \text{ on } \Gamma_1, \quad \theta_\Gamma = 0 \text{ on } \Gamma_2.$$

We use a linearized Arrhenius law for the viscosity [8],

$$\mu(\theta) = \exp[-(\theta - 1/2)\log b],$$

where b is a positive number describing the contrast of viscosity, that is, μ is independent of x, normalized at $\theta = 1/2$, and the ratio of the maximum and minimum viscosity is equal to b. Heat source is $f = 0$ and the initial temperature θ^0 is

$$\theta^0(r, \varphi, \psi) = \theta^*(r) + \epsilon \sin \pi(\frac{R_2 - r}{R_2 - R_1}) Y_3^2(\varphi, \psi), \tag{15}$$

where (r, φ, ψ) is the spherical coordinates, $\theta^*(r)$ is the conductive solution defined by

$$\theta^*(r) = \frac{R_1}{R_2 - R_1}(\frac{R_2}{r} - 1),$$

Y_n^m is the normalized spherical harmonic function of degree n and order m, and $\epsilon = 0.1$. This initial condition was used in [8]. We set $Ra = 7,000$.

We performed a numerical simulation for this problem by the stabilized finite element scheme (12)–(14). Figure 1 shows the mesh and Table 1 shows the data for the computation. We set $b = 1,000$. Starting from the initial temperature (15), we got a numerically stationary solution (u_h, p_h, θ_h). In the left column of Fig.2 we show the isothermal surfaces of θ_h=0.2, 0.5, 0.8, and in the right column we show the corresponding results of the isoviscosity case $\mu \equiv 1$. In the variable viscosity case the viscosity increases and the velocity decreases at the place where the temperature is low. The plume heads, therefore, flatten much more than in the isoviscosity case. Such phenomenon can be observed clearly in Fig. 2. More detailed observation including results of other viscosity contrasts and Rayleigh numbers will be reported in the forthcoming paper.

6 Concluding Remarks

We have solved a mathematical model with temperature dependent viscosity for the Earth's mantle convection in the whole three-dimensional domain. Although we use the cheap element P1/P1/P1, the computational cost of the full three-dimensional problem is pretty large. In order to reduce the cost we have exploited symmetries of the domain and made a parallel computation code, where no symmetries of the solution are supposed. We will discuss on this subject in a forthcoming paper. For the computation of much higher Rayleigh numbers such effort of reducing the computation cost is necessary.

Acknowledgments

The authors thank to Professor S. Honda at Hiroshima University for stimulating discussions. This study was supported by the Japan Society for the Promotion of Science under Grants-in-Aid for Scientific Researches (A), No.10304007

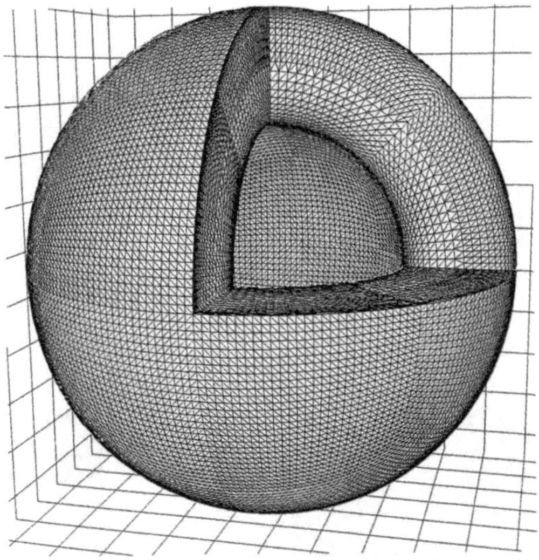

Fig. 1. Mesh

Table 1. Discretization parameters

# of nodes	# of elements	h	Δt
324,532	1,868,544	0.145	2.5

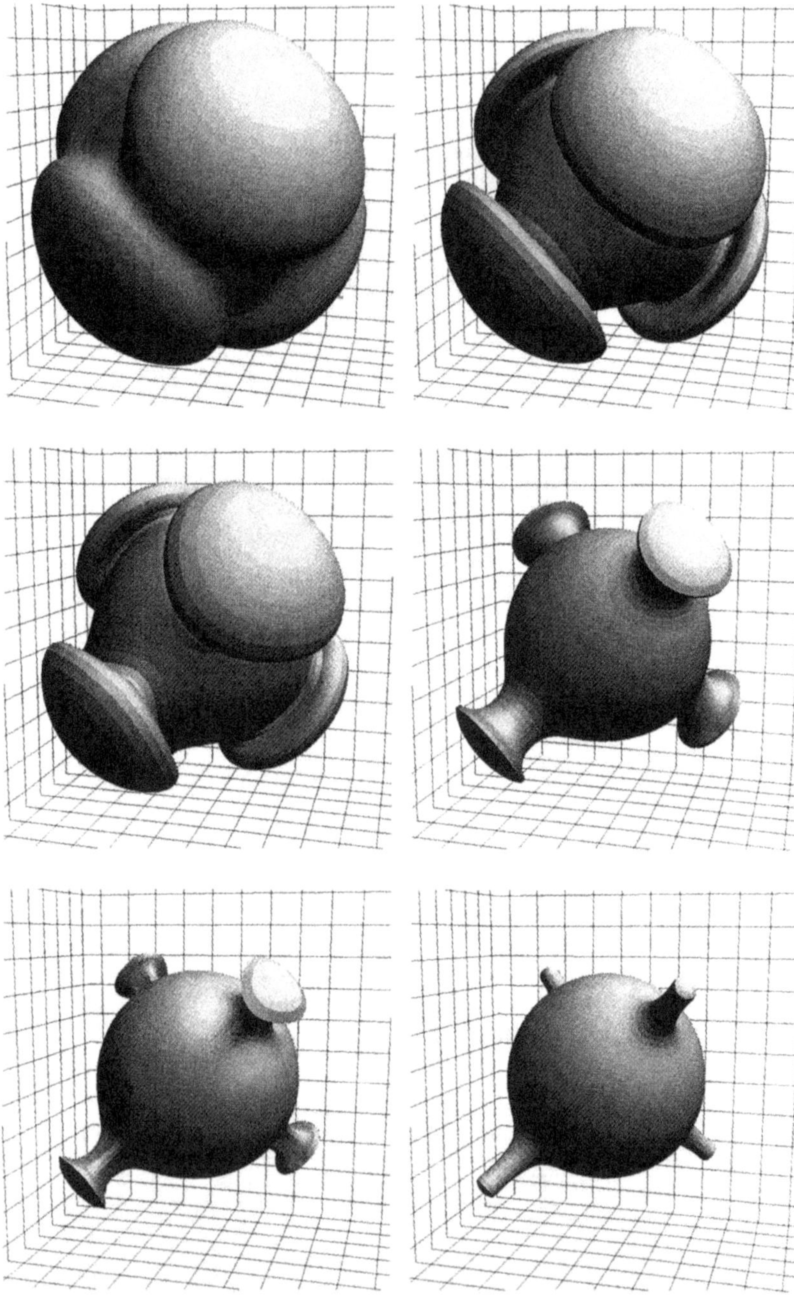

Fig. 2. Isothermal surfaces of θ_h=0.2(top), 0.5(middle), 0.8(bottom) in the cases b=1,000(left) and 1(right)

and (B), No.11554003, and by the Ministry of Education, Culture, Sports, Science and Technology of Japan under a special coordination fund for promoting researches on 'Superplume project – International research project on whole Earth dynamics'.

References

1. J. R. Baumgardner. Three-dimensional treatment of convective flow in the earth's mantle. *Journal of Statistical Physics*, 39:501–511, 1985.
2. F. Brezzi and J. Pitkäranta. On the stabilization of finite element approximations of the Stokes equations. In W. Hachbush, editor, *Efficient Solutions of Elliptic Systems*, volume 10 of *Notes on Numerical Fluid Mechanics*, pages 11–19. Braunschweig, 1984.
3. P. G. Ciarlet. *The Finite Element Method for Elliptic Problems*. North-Holland, 1978.
4. L. P. Franca, S. L. Frey, and T. J. R. Hughes. Stabilized finite element methods: I. Application to the advective-diffusive model. *Computer Methods in Applied Mechanics and Engineering*, 95:253–276, 1992.
5. T. J. R. Hughes and A. Brooks. A theoretical framework for Petrov-Galerkin methods with discontinuous weighting functions: Application to the streamline-upwind procedure. In R. H. Gallagher et al., editors, *Finite Elements in Fluids*, volume 4, pages 47–65. John Wiley & Sons, 1982.
6. T. J. R. Hughes and L. P. Franca. A new finite element formulation for computational fluid dynamics: VII. the Stokes problem with various well-posed boundary conditions: symmetric formulations that converge for all velocity/pressure spaces. *Computer Methods in Applied Mechanics and Engineering*, 65:85–96, 1987.
7. J. L. Lions. *Quelques méthodes de résolution des problèmes aux limites non linéaires*. Dunod, Paris, 1969.
8. J. T. Ratcliff, G. Schubert, and A. Zebib. Three-dimensional variable viscosity convection of an infinite Prandtl number Boussinesq fluid in a spherical shell. *Geophysical Research Letters*, 22:2227–2230, 1995.
9. A. Suzuki, M. Tabata, and S. Honda. Numerical solution of an unsteady Earth's mantle convection problem by a stabilized finite element method. *Theoretical and Applied Mechanics*, 48:371–378, 1999.
10. M. Tabata and A. Suzuki. A stabilized finite element method for the Rayleigh-Bénard equations with infinite Prandtl number in a spherical shell. *Computer Methods in Applied Mechanics and Engineering*, 190:387–402, 2000.
11. P. J. Tackley. Self-consistent generation of tectonic plates in time-dependent, three-dimensional mantle convection simulations 1. pseudoplastic yielding. *Geochemistry Geophysics Geosystems, An Electronic Journal of the Earth Sciences*, 1(2000GC000036):1–45, 2000.
12. R. Temam. *Navier-Stokes Equations - Theory and Numerical Analysis*. North-Holland, 1977.

Theoretical and Numerical Analysis on 3-Dimensional Brittle Fracture

Kohji Ohtsuka

Hiroshima Kokusai Gakuin University, 6-20-1, Aki-ku, Hiroshima 739-0321, Japan

Abstract. In this paper, we propose a mathematical formulation of 3-dimensional quasi-static brittle fracture under varying loads. We give a precise formulation of internal cracking and surface cracking in 3D elastic bodies. In Sections 2 and 3, we provide geometrical results and results on Sobolev spaces. Most criteria in fracture mechanics are based on Griffith's energy balance theory, which is explained briefly in Section 5. GJ-integral is proposed by the author (1981), which is a generalization of J-integral widely used in 2D fracture problems. GJ-integral expresses the variation of energies with respect to crack extensions and relates to Griffith's energy balance theory. Under varying loads, we cannot use Griffith's energy balance theory directly, however GJ-integral is applicable to such cases too. For practical use, we must study the combination with numerical calculation. In the last section, we give an error estimate for finite element approximation of GJ-integral.

1 Introduction

In this paper, we adopt Einstein's convention, that is, $x_i y_i$ means the sum $\sum_i x_i y_i$. By $H^m(Q)$ we denote Sobolev space of order m defined on (a domain or a surface) Q. For a function f, we denote by $f|_Q$ the restriction of f on Q. For a function space $\mathcal{H}$, we denote by $\mathcal{H}^m$ the product space $\mathcal{H} \times \cdots \times \mathcal{H}$ and by $\mathcal{H}'$ the dual space of $\mathcal{H}$.

Consider an elastic body whose undeformed shape is $\Omega_\Sigma = \Omega \setminus \Sigma$. Here Ω is a bounded domain in $\mathbb{R}^3$ with a Lipschitz boundary Γ and Σ is the undeformed shape of a crack, which is C^∞-smooth surface with C^∞-smooth boundary $\partial\Sigma$. We assume that Σ lies on a C^∞-smooth oriented open manifold S (see Fig. 1). The surface Σ is called a *crack surface.*

We consider the case where the crack extension is considered to occur in a quasi-static manner. Therefore, when we refer to time t we use it as a parameter that delineates the history of the sequence of events such as loading or crack extension, which are observed in the interval $[0, T]$ of time t. In the crack extension process, the part Γ_D of Γ is fixed (Dirichlet condition) and the surface force $\boldsymbol{g}(t)$ is given on the remainder part $\Gamma_N = \Gamma \setminus \overline{\Gamma_D}$. The body force $\boldsymbol{f}(t)$ also is given inside Ω as shown in Fig. 2.

In this paper, we consider only *smooth virtual crack extensions* $\Sigma(t) \subset \Omega$, but of two types. One is *internal cracking* satisfying $\overline{\Sigma(t)} \cap \Gamma = \emptyset$ for all $0 \le t \le T$ as shown in Fig. 2. We write the set of all smooth virtual internal crack extensions by $\mathrm{SC}_{in}(\Sigma|S)$. Another is *surface cracking* satisfying $\overline{\Sigma(t)} \cap \Gamma = L(t)$ for all $0 \le t \le T$ as shown in Fig. 3, where $L(t)$ denotes the smooth curve on Γ. We denote by $\mathrm{SC}_{su}(\Sigma|S)$ the set of all smooth virtual surface crack extensions. In surface crack extension, we assume that Γ is C^∞-class, so we can divide $\partial\Sigma(t)$ to two

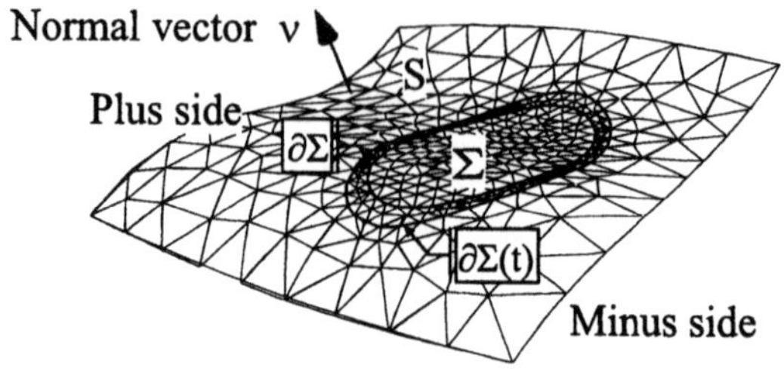

Fig. 1. 2-dimensional oriented C^∞-manifold S and the normal vector $\boldsymbol{\nu}$ directed from the minus side to the plus side. The crack Σ extends along S.

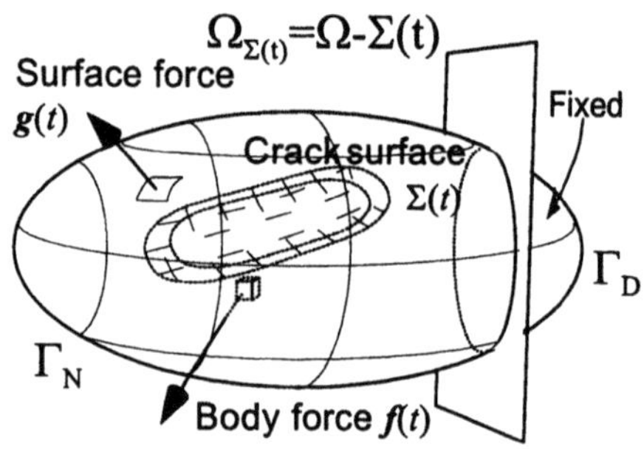

Fig. 2. Elastic body with crack $\Sigma(t)$ subjected to arbitrary loadings.

smooth curves $\partial_1\Sigma(t)$ and $\partial_2\Sigma(t) = L(t)$ as shown in Fig. 3. The surface $\Sigma(t)$ is called C^∞-manifold with corner. By considering the smooth extension of mapping in the atlas of $\Sigma(t)$ for each t, we can extend $\Sigma(t)$ to a C^∞-manifold $\Lambda(t)$ with the smooth boundary $\partial\Lambda(t)$. Then we assume the form $\Sigma(t) = \Lambda(t) \cap \Omega$ in what follows. Moreover, under some engineering environment (e.g. pressure vessel), we must consider the surface force $\boldsymbol{p}(t)$ on $\Sigma(t)$. If there is no need to distinguish $\mathrm{SC}_{in}(\Sigma|S)$ and $\mathrm{SC}_{su}(\Sigma|S)$, then we write them $\mathrm{SC}(\Sigma|S)$.

The plan of this paper is as follows: In Section 2, we explain $\mathrm{SC}(\Sigma|S)$ precisely. In Section 3, the density and trace theorem in Sobolev spaces defined on Ω_Σ (or $\Omega_{\Sigma(t)}$) are given. In Section 4, we construct a weak solution of quasi-static fracture in virtual crack extension. The main of this paper is Section 5 as below.

The *real crack extension is selected from virtual crack extensions by a criterion. Least-energy principle* is widely used, and applied to *fracture mechanics* in the famous studies by Griffith [5,6]. The developments of his studies are called *Griffith's energy balance theory* today, in which the *crack extension force* is called *energy release rate* (= variation of energies with respect to crack extensions). J-integral is widely used in fracture mechanics in 2D cases. In [26] and [2], it is shown independently that J-integral equals energy release rate (for a mathematical proof, refer

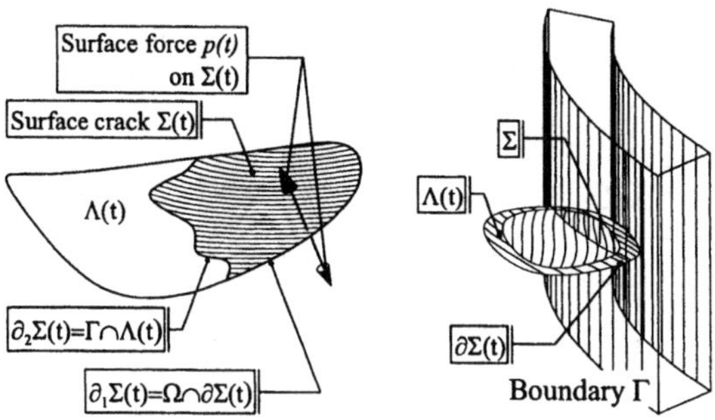

Fig. 3. Surface crack has the form $\Sigma(t) = \Lambda(t) \cap \Omega$ whose boundary is divided to $\partial_1\Sigma(t)$ and $\partial_2\Sigma(t)$.

to [13]). In Section 5, a generalization of J-integral is given in 3D cases, named GJ-integral by the author [14] which is the functional

$$(\boldsymbol{u}, \omega, \boldsymbol{X}) \mapsto J_\omega(\boldsymbol{u}, \boldsymbol{X})$$

with parameters; the displacement vector $\boldsymbol{u} = \boldsymbol{u}(0)$, the domain ω and the vector field $\boldsymbol{X}$. The important property of GJ-integral is the following: If there is no singularity inside ω, that is $\boldsymbol{u}|_{\omega\cap\Omega_\Sigma} \in H^2(\omega \cap \Omega_\Sigma)^3$, then $J_\omega(\boldsymbol{u}, \boldsymbol{X}) = 0$ for all vector field $\boldsymbol{X}$. However, $J_\omega(\boldsymbol{u}, \boldsymbol{X}) \neq 0$, if $\boldsymbol{u}$ has some kind of singularities inside ω, where $\boldsymbol{X}$ is derived from the movement of singularities. The theory on GJ-integral has been constructed in the papers [16,22] and is applied to various sensitivity analysis. The equivalence between *energy release rate* and GJ-integral has been proved in [14] for 3D internal crack extension and in [18] for 3D surface crack extension. The results of theoretical research in crack problems suggest that a weak solution (displacement vector) $\boldsymbol{u}$ has the following singularies (see e.g. [3])

$$u_i = k_j \rho_{\partial\Sigma}^{1/2} \varphi_{i,j} + u_{R,i} \quad \text{near } \partial\Sigma; \quad \rho_{\partial\Sigma}(x) = \min_{y\in\partial\Sigma} |x - y|, \tag{1}$$

for $i = 1, 2, 3$, where u_i are components of $\boldsymbol{u}$ and $u_{R,i} \in H^2(\text{near } \partial\Sigma)$ are regular terms. In the singular terms, the functions $\varphi_{i,j}$ are smooth functions defined near $\partial\Sigma$, and k_j are functions defined on $\partial\Sigma$. Without using (1), we will prove that *GJ-integral expresses a crack extension force* and if (1) is true,

$$J_\omega(\boldsymbol{u}, \boldsymbol{X}) = J_\omega(\boldsymbol{u}_S, \boldsymbol{X}), \quad \boldsymbol{u}_S = (k_j \rho_{\partial\Sigma}^{1/2} \varphi_{1,j}, k_j \rho_{\partial\Sigma}^{1/2} \varphi_{2,j}, k_j \rho_{\partial\Sigma}^{1/2} \varphi_{3,j}) \tag{2}$$

with $\boldsymbol{X}$ obtained from the crack extension (see Theorem 11). In 2D cases, (2) is proved in [13,19]. In Section 5, we will prove that $J_\omega(\boldsymbol{u}, \boldsymbol{X})$ has the expression

$$J_\omega(\boldsymbol{u}, \boldsymbol{X}) = \langle \delta\phi_0(\gamma), J_\gamma(\boldsymbol{u}, \boldsymbol{e}_1) \rangle_{\partial\Sigma} \tag{3}$$

where $\delta\phi_0(\gamma)$ is the *speed of crack extension* and $J_\gamma(\boldsymbol{u}, \boldsymbol{e}_1)$ is a distribution ($\gamma \in \partial\Sigma$). Studies in fracture mechanics (see e.g. [25,11]) indicate that

$$\langle \delta\phi_0(\gamma), J_\gamma(\boldsymbol{u}, \boldsymbol{e}_1) \rangle_{\partial\Sigma} = \int_{\partial\Sigma} \left\{ \frac{1}{E} \left(K_1^2(\gamma) + K_2^2(\gamma) \right) + \frac{1}{2\mu} K_3^2(\gamma) \right\} \delta\phi_0(\gamma) d\gamma \tag{4}$$

in the case of isotropic elasticity, where E is Young's modulus and μ Lamé's elastic coefficient. K_j are called *stress intensity factors* and using stress tensor σ_{ij}, they are defined by

$$K_1(\gamma) = \sqrt{2\pi}(\lim_{x\to\gamma} \rho_{\partial\Sigma}(x)^{1/2}\sigma_{ij}(x))\nu_i(\gamma)\nu_j(\gamma) \qquad \text{(opening mode)}, \tag{5}$$

$$K_2(\gamma) = \sqrt{2\pi}(\lim_{x\to\gamma} \rho_{\partial\Sigma}(x)^{1/2}\sigma_{ij}(x))\nu_i(\gamma)e_{1,j}(\gamma) \qquad \text{(shearing mode)},$$

$$K_3(\gamma) = \sqrt{2\pi}(\lim_{x\to\gamma} (\rho_{\partial\Sigma}(x)^{1/2}\sigma_{ij}(x))\nu_i(\gamma)e_{2,j}(\gamma) \qquad \text{(tearing mode)},$$

where $e_{i,j}$ are the components of $\boldsymbol{e}_i$ (see Fig. 4).

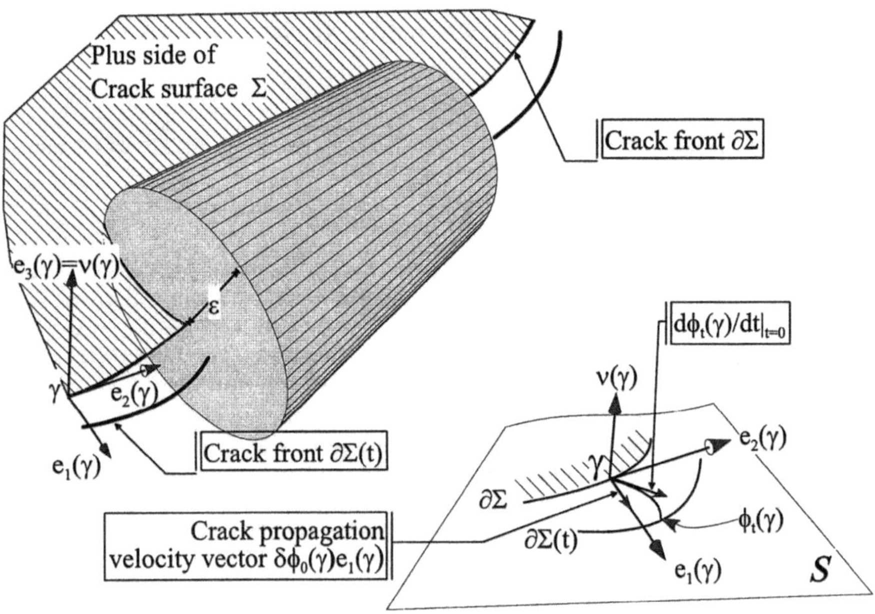

Fig. 4. Smooth crack extension and the velocity vector $\delta\phi_0(\gamma)\boldsymbol{e}_1(\gamma)$ at $\gamma \in \partial\Sigma$.

If $\gamma \mapsto J_\gamma(\boldsymbol{u}, \boldsymbol{e}_1)$ is continuous, we have the follwoing criterion of crack extension from Griffith's energy balance theory;

$$\text{if } \max_{\gamma\in\partial\Sigma} J_\gamma(\boldsymbol{u}, \boldsymbol{e}_1) \geq \alpha_c, \quad \text{then the crack will extend}, \tag{6}$$

where $\alpha_c > 0$ is an experimental value (see also [15,20]).

Although Griffith's energy balance theory is only applicable in the case of constant loading, (3) and (6) hold under varying loading, and (2) is also valid if (1) is true.

In the last section, we will explain how to calculate GJ-integral by the finite element method and give an error estimate of the approximate value.

This research has been partially supported by Grant-in-Aid (B)(1)#10440035 for Scientific Research in Japan.

2 Geometry on Crack Extension

2.1 Internal crack extension

We define the class of smooth virtual crack extensions as follows (see Fig.4).

(SC$_{in}$1) $\Sigma = \Sigma(0) \subset \Sigma(t) \subset \Sigma(t') \subset S$ if $0 \le t < t' \le T$.
(SC$_{in}$2) The crack extends along the closed C^∞-manifold S so that $\Sigma(t)$ may be C^∞-submanifolds of S with boundary $\partial\Sigma(t)$.
(SC$_{in}$3) For each $t \in [0,T]$, there is a C^∞-diffeomorphism $\phi_t : \partial\Sigma \to \partial\Sigma(t)$, and the map $\phi_\cdot : \partial\Sigma \times [0,T] \to S$ is of class C^∞.

Let us denote by $\mathrm{SC}_{in}(\Sigma|S)$ the set of all virtual crack extensions $\{\Sigma(t)\}$ satisfying (SC$_{in}$1)–(SC$_{in}$3).

Now we introduce the local coordinate system near the edge $\partial\Sigma$ of the initial crack Σ as shown in Fig. 4. Let $\{\mathbf{e}_1(\gamma), \mathbf{e}_2(\gamma), \mathbf{e}_3(\gamma)\}$ be the vectors at $\gamma \in \partial\Sigma$ such that $\mathbf{e}_3(\gamma)$ is unit normal to $\Sigma \subset S$ directed from minus side to plus side. We denote also $\mathbf{e}_3(\gamma)$ by $\boldsymbol{\nu}(\gamma)$. The vector $\mathbf{e}_1(\gamma)$ is the outward unit normal to $\partial\Sigma$ along S and $\mathbf{e}_2(\gamma)$ unit tangent to $\partial\Sigma$.

There is a number $\epsilon > 0$ such that, for all point $x \in U_\epsilon(\partial\Sigma) = \{x \in \mathbb{R}^3 : \rho_{\partial\Sigma}(x) < \epsilon\}$, we get only one point $y \in \partial\Sigma$ satisfying $|x - y| = \rho_{\partial\Sigma}(x)$. For any point $P \in S \cap U_\epsilon(\partial\Sigma)$, we get uniquely the point $\gamma_P \in \partial\Sigma$ and geodesic line connecting P and γ_P. The geodesic lines are defined as the locally shortest curves connecting two points on S (see Fig. 5). For each point $Q \in U_\epsilon(\partial\Sigma)$, we get the point $P_Q \in S$ uniquely which lies on the normal straight line through the point Q, and we get also the geodesic line connecting P_Q and $\gamma_Q \in \partial\Sigma$. So we can introduce the curvilinear coordinate system $(F, U_\epsilon(\partial\Sigma))$ called *tubular neighborhood.*

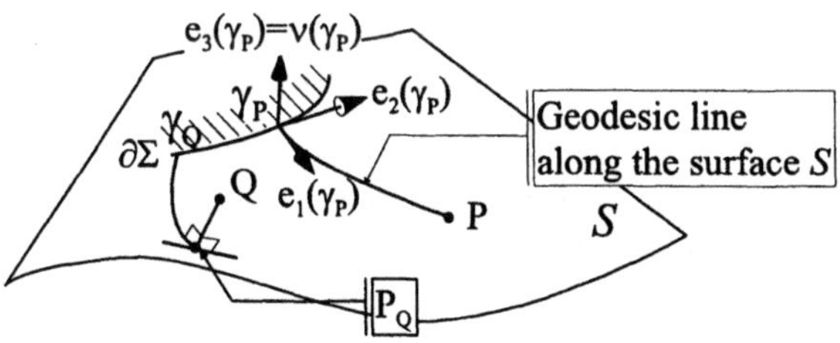

Fig. 5. For $P \in S \cap U_\epsilon(\partial\Sigma)$, $\gamma_P \in \partial\Sigma$ is uniquely determined, also for $Q \in U_\epsilon(\partial\Sigma)$, we get uniquely $\gamma_Q \in \partial\Sigma$ as shown above.

Lemma 1. *There is a constant $\epsilon > 0$ and an open neighborhood $U_\epsilon(\partial\Sigma)$ of $\partial\Sigma$ in $\mathbb{R}^3$ such that the map*

$$F: (\gamma, \xi_1, \xi_2) \to (x_1, x_2, x_3) \in U_\epsilon(\partial\Sigma),\ x_i = F_i(\gamma, \xi_1, \xi_2) \quad \textit{for } i = 1, 2, 3$$

is a C^∞-diffeomorphism of $\partial\Sigma \times D_\epsilon$ onto $U_\epsilon(\partial\Sigma)$, where D_ϵ denotes the disc with the radius ϵ. Here the path $\xi_2 \mapsto F(\gamma, \xi_2, 0)$, $|\xi_2| < \epsilon$ is the geodesic line along S starting from $\gamma \in \partial\Sigma$ and $F(\gamma, \xi_1, \xi_2) = \xi_2 \boldsymbol{\nu}(F(\gamma, \xi_1, 0))$ (see Fig. 5). Moreover, we have

$$\begin{aligned} F(\gamma, 0, 0) &= \gamma \qquad \text{for all } \gamma \in \partial\Sigma \\ S \cap U_\epsilon(\partial\Sigma) &= \{F(\gamma, \xi_1, 0);\ \gamma \in \partial\Sigma,\ -\epsilon < \xi_1 < \epsilon\} \\ \Sigma \cap U_\epsilon(\partial\Sigma) &= \{F(\gamma, \xi_1, 0);\ \gamma \in \partial\Sigma,\ -\epsilon < \xi_1 < 0\}. \end{aligned} \tag{7}$$

In smooth crack extension, the movement of crack is given by the disjoint path $t \mapsto \phi_t(\gamma)$ connecting γ and the point on $\partial\Sigma(t)$ for each $0 \le t \le T$ and $\gamma \in \partial\Sigma$, which make the vector field $d\phi_t/dt|_{t=0}$ on $\partial\Sigma$. In the paper [14], it is proved that the crack extension depend only on the $\mathbf{e}_1$- component $\delta\phi_0(\gamma)\mathbf{e}_1(\gamma)$ of $d\phi_t/dt|_{t=0}$, that is, $\delta\phi_0(\gamma) = \left(\delta\phi_t(\gamma)/dt|_{t=0}\right) \cdot \mathbf{e}_1(\gamma)$ by the inner product,

In (3), we called $\delta\phi_0(\gamma)$ the *speed of the virtual crack extension* at the initial crack (or $t = 0$). Also we call $\delta\phi_0(\gamma)\mathbf{e}_1(\gamma)$ the *velocity vector of crack extension.*

2.2 Surface crack extension

As stated in Introduction, we assume the undeformed shape of surface cracks has the form $\Sigma(t) = \Lambda(t) \cap \Omega$.

(SC$_{su}$1) $\Lambda = \Lambda(0) \subset \Lambda(t) \subset \Lambda(t') \subset S$ if $0 \le t < t' \le T$.

(SC$_{su}$2) The crack extends along the C^∞-manifold S so that $\Lambda(t)$ may be C^∞-submanifolds of S with boundary $\partial\Sigma(t)$.

(SC$_{su}$3) For each $t \in [0, T]$, there is a C^∞-diffeomorphism $\phi_t : \partial\Lambda \to \partial\Lambda(t)$, and the map $\phi_\cdot : \partial\Lambda \times [0, T] \to S$ is of class C^∞.

(SC$_{su}$4) The closure $\overline{\Lambda(t)}$ is a 2-dimensional C^∞-submanifold of S with C^∞-boundary $\partial\Lambda(t)$ *intersecting transversally* with Γ at two points $\lambda_1(t)$ and $\lambda_2(t)$.

Let us denote by SC$_{su}(\Sigma|S)$ the set of all virtual crack extensions $\{\Sigma(t)\}$ satisfying (SC$_{su}$1)–(SC$_{su}$4). We can construct the tubular neighborhood of $\partial\Lambda$ as given in Lemma 1. But we must notice that the geodesic line starting $\lambda_i(0)$, $i = 1, 2$ along S goes away from Γ, in general. So we use the spray (see e.g. [10, IV, 5]) which fits into the shape of Γ and $\partial\Lambda$. The usual argument on the existence of tubular neighborhood leads to the following Lemma (see, e.g. [10, IV, 5]).

Lemma 2. *There is a constant $\epsilon > 0$ and an open neighborhood $U_\epsilon(\partial\Lambda)$ of $\partial\Lambda$ in $\mathbb{R}^3$ such that the map*

$$F: (\gamma, \xi_1, \xi_2) \to (x_1, x_2, x_3) \in U_\epsilon(\partial\Sigma),\ x_i = F_i(\gamma, \xi_1, \xi_2) \quad \textit{for } i = 1, 2, 3$$

is a C^∞*-diffeomorphism of* $\partial\Sigma \times D_\epsilon$ *onto* $U_\epsilon(\partial\Sigma)$*. Moreover writing* $D_\epsilon^\pm = D_\epsilon \cap \{x_2 \in \mathbb{R}^2;\ x_2 \gtrless 0\}$*, we have*

$$F(\gamma,0,0) = \gamma \qquad \text{for all } \gamma \in \partial\Lambda \tag{8}$$

$$S \cap U_\epsilon(\partial\Lambda) = \{F(\gamma,\xi_1,0);\ \gamma \in \partial\Lambda,\ -\epsilon < \xi_1 < \epsilon\},$$

$$\Lambda \cap U_\epsilon(\partial\Lambda) = \{F(\gamma,\xi_1,0);\ \gamma \in \partial\Lambda,\ -\epsilon < \xi_1 < 0\},$$

$$\partial_1\Sigma \cap U_\epsilon(\partial\Lambda) = \{F(\gamma,\xi_1,0);\ \gamma \in \partial\Sigma,\ -\epsilon < \xi_1 < 0\},$$

$$\Omega_\Sigma^\pm \cap U_\epsilon(\partial\Lambda) = \cup_{i=1}^2 \left\{F(\lambda_i(0),\xi_1,\xi_2);\ (\xi_1,\xi_2) \in D_\epsilon^\pm\right\}.$$

3 Density and Trace Theorem

To formulate the boundary condition, we must define the value $v_{\Sigma(t)}^\pm$ on the plus side or the minus side of the surface $\Sigma(t)$. To construct the density and trace theorems in 3D case, we will approximate Ω_Σ by domains with local Lipschitz property. However Euclidean distance in $\mathbb{R}^3$ is not fit for this approximation. Now we introduce the new distance ℓ_{Ω_Σ}: For a domain $Q \subset \mathbb{R}^3$, we denote the distance $\ell_Q(x,y)$ of two points $x,y \in Q$ by the infimum of lengths of all piecewise linear lines connecting x and y inside Q. If Q has local Lipschitz property, then $\ell_Q(x,y)$ is equivalent to the Euclidean distance $|x-y|$.

We prepare some lemmas to show the density theorem.

Lemma 3. *(see [18, Lemma 2.1]) Let* Q *and* $\mathcal{O}$ *be bounded domain in* $\mathbb{R}^3$ *and* Φ *a mapping from* Q *into* $\mathcal{O}$*. Assume that* Q *is covered by a family* $\{Q_i\}_{i=1,\cdots,m}$ *of domains with local Lipschitz property in* $\mathbb{R}^3$ *satisfying the following:*

There is a constant C *such that*

$$|\Phi(x) - \Phi(y)| \le C|x-y| \text{ for any } x,y \in Q_i,\ i = 1,\cdots,m. \tag{9}$$

The image

$$Q_i^* = \Phi(Q_i) \text{ is a domain with local Lipschitz property} \tag{10}$$

for each i*. Then there is a constant* C_0 *such that*

$$\ell_{\mathcal{O}}(\Phi(x),\Phi(y)) \le C_0\ell_Q(x,y) \qquad \text{for all } x,y \in Q. \tag{11}$$

We now construct the mapping Φ_α which maps the neighborhood of $\partial\Sigma$ onto domains with local Lipschitz property (see Fig. 6)

$$\widetilde{U}_\epsilon(\Sigma) = \{F(\gamma,\xi_1,\xi_2);\ \gamma \in \partial\Sigma,\ |\xi_i| < \sqrt{\epsilon/2}, i = 1,2\} \subset U_\epsilon(\partial\Sigma),$$

$$\Xi(\alpha) = \left\{F(\gamma,\xi_1,\xi_2);\ \lambda \in \partial\Sigma,\ 0 < \xi_1 < \sqrt{\epsilon/2},\ 0 < \xi_2 < \alpha\xi_1\right\},$$

for $0 \le \alpha < \sqrt{\epsilon/8}$. For $\alpha > 0$, $\Xi(\alpha)$ become the domain with local Lipschitz property. We consider the family of mappings $\{\Phi_\alpha\}_{0\le\alpha<\sqrt{\epsilon/2}}$ defined by

$$\Phi_\alpha(x) = \begin{cases} F(\gamma(x),\xi_1(x),[\xi_2(x) + \operatorname{sign}(\xi_2(x))\,\alpha\xi_1(x)]/2) & x \in \Xi(\alpha) \setminus \Sigma \\ x & x \in \widetilde{U}_\epsilon(\partial\Sigma)) \setminus \Xi(\alpha) \end{cases} \tag{12}$$

where $(\gamma(x), \xi_1(x), \xi_2(x)) = F^{-1}(x)$ and $\text{sign}(\xi) = \xi/|\xi|$. Here we notice that $\Phi_\alpha(\Omega_\Sigma) = \Omega_{\alpha,\Sigma}$; $\Omega_{\alpha,\Sigma} = (\Xi_0(\alpha,\alpha) \setminus \overline{\Xi_0(\alpha/2,\alpha)}) \cup (\Omega_\Sigma \setminus \overline{U_{\alpha^2/2}(\Sigma)})$ has local Lipschitz property.

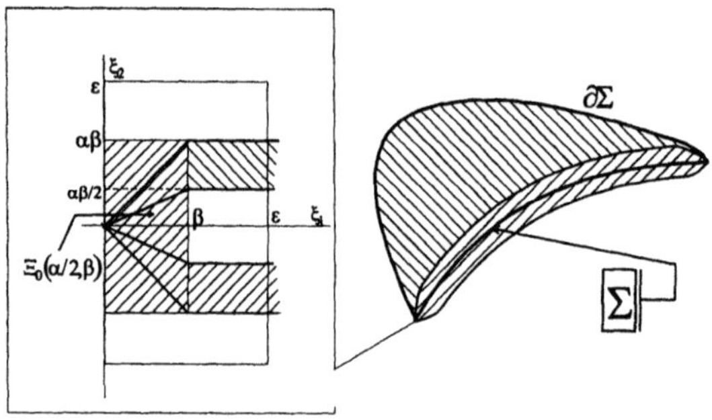

Fig. 6. $\Xi_0(\alpha, \beta)$ and the cross section of $\Omega \setminus \overline{\Omega_{\alpha,\Sigma}}$

Lemma 4. *(see [18, Lemma 2.3]) There is a constant $\alpha_0 > 0$ such that the map Φ_α defined in (12) become bijection with Lipschitz constant $C_\alpha > 0$*

$$C_\alpha^{-1} \ell_{\Omega_\Sigma}(x,y) \le |\Phi_\alpha(x) - \Phi_\alpha(y)| \le C_\alpha \ell_{\Omega_\Sigma}(x,y) \tag{13}$$

for all $x, y \in \Omega_\Sigma$ independent of x, y.

Theorem 5. *(see [18, Theorem 2.5]) Let Σ be a smooth crack. Let $C_\ell^{0,1}(\Omega_\Sigma)$ be the set of all Lipschitz continuous functions with respect to the distance $\ell_{\Omega_\Sigma}(x,y)$, that is, for $\varphi \in C_\ell^{0,1}(\Omega_\Sigma)$, there is a constant C_φ independent of x, y such that*

$$|\varphi(x) - \varphi(y)| \le C_\varphi \ell_{\Omega_\Sigma}(x,y) \qquad \text{for all } x, y \in \Omega_\Sigma.$$

Then $C_\ell^{0,1}(\Omega_\Sigma)$ is dense in $H^1(\Omega_\Sigma)$.

For a function $v \in C_\ell^{0,1}(\Omega_\Sigma)$, we can define the value of v on Σ by

$$v^\pm(x) = \lim_{\epsilon \to 0} v(x \pm \epsilon \boldsymbol{\nu}(x))$$

for $x \in \Sigma$ which are Lipschitz continuous with respect to Euclidean metric on Σ.

Theorem 6 ([17]). *Let $C_0^\infty(\Omega_\Sigma)$ be the space of all functions $\varphi \in C^\infty(\mathbb{R}^3)$ and*

$$\text{supp}\,\varphi = \overline{\{x;\ \varphi(x) \ne 0\}} \subset \Omega_\Sigma.$$

If $v \in H^1(\Omega_\Sigma)$ and $v|_\Gamma = 0$, $v|_\Sigma^\pm = 0$, then there is a sequence v_j of functions in $C_0^\infty(\Omega_\Sigma)$ such that $v_j \to v$ in $H^1(\Omega_\Sigma)$.

For two points $x,\,y\in\Sigma(t)$, we define $\ell_{\Sigma(t)}(x^+,y^-)$ by the limit

$$\lim_{\epsilon\downarrow 0}\ell_{\Omega_{\Sigma(t)}}(x+\epsilon\nu(x),y-\epsilon\nu(y)).$$

Also, for two points $x,\,y\in\Gamma_{\Sigma(t)}=\Gamma\setminus\overline{\Sigma(t)}$, taking the sequence of points

$$x_j,\,y_j\in\Omega_{\Sigma(t)}\,(j=1,\cdots,\infty);\qquad x_j\to x,\,y_j\to y\quad\text{as } j\to\infty,$$

we can define the distance $\ell_{\Gamma_{\Sigma(t)}}(x,y)$ by $\lim_{j\to\infty}\ell_{\Omega_{\Sigma(t)}}(x_j,y_j)$. Let us define Hilbert space $H^\alpha(\Sigma(t);\ell_{\Sigma(t)})$ for $\alpha>0$ by the subspace of $H^\alpha(\Sigma(t))\times H^\alpha(\Sigma(t))$ whose element (v_1,v_2) satisfy the following

$$\int_{\Gamma_{\Sigma(t)}}|v_1(x)-v_2(y)|^2\ell_{\Sigma(t)}(x^+,y^-)^{-2(\alpha+1)}ds_xds_y<\infty,$$

and has the norm $\|(v_1,v_2)\|_{\alpha,\ell_{\Sigma(t)}}$ defined by

$$\left\{\sum_{i=1}^{2}\|v_i\|_{0,\Sigma(t)}^2+\int_{\Gamma_{\Sigma(t)}}\int_{\Gamma_{\Sigma(t)}}|v_1(x)-v_2(y)|^2\ell_{\Sigma(t)}(x^+,y^-)^{-2(\alpha+1)}ds_xds_y\right\}^{1/2}.$$

Hilbert space $H^\alpha(\Gamma_{\Sigma(t)};\ell_{\Gamma_{\Sigma(t)}})$ is also defined by the norm

$$\|v\|_{\alpha,\ell(\Gamma_{\Sigma(t)})}=\left\{\|v\|_{0,\Gamma_{\Sigma(t)}}^2+\int_{\Gamma_{\Sigma(t)}}\int_{\Gamma_{\Sigma(t)}}|v(x)-v(y)|^2\ell_{\Gamma_{\Sigma(t)}}(x,y)^{-2(\alpha+1)}ds_xds_y\right\}^{1/2}.$$

In the case of *internal crack extension*, we have $H^\alpha(\Gamma)=H^\alpha(\Gamma_{\Sigma(t)};\ell_{\Gamma_{\Sigma(t)}})$ and

$$H^\alpha(\Sigma(t);\ell_{\Sigma(t)})=\left\{(v_1,v_2)\in H^1(\Omega_{\Sigma(t)})^2:\ v_1-v_2\in H_{00}^{1/2}(\Sigma(t))\right\}$$

where $H_{00}^{1/2}(\Sigma(t))$ is given by the norm

$$\|v\|_{1/2,00,\Sigma(t)}=\left\{\|v\|_{1/2,\Sigma(t)}^2+\int_{\Sigma(t)}\rho_{\partial\Sigma(t)}(x)^{-1}|v(x)|^2ds_x\right\}^{1/2}.$$

Theorem 7. *([18, pp.336–339]) For v in $C_\ell^{0,1}(\Omega_{\Sigma(t)})$, we have the estimation*

$$\|v|_{\Sigma(t)}\|_{1/2,\ell_{\Sigma(t)}}^2+\|v|_{\Gamma_{\Sigma(t)}}\|_{1/2,\ell(\Gamma_{\Sigma(t)})}^2\le C_1\|v\|_{1,\Omega_{\Sigma(t)}}^2,\tag{14}$$

with a constant $C_1>0$ independent of v.

Conversely, for each function $(\varphi^+,\varphi^-)\in H^{1/2}(\Sigma(t);\ell_{\Sigma(t)})$, $\varphi_\Gamma\in H^{1/2}(\Gamma_{\Sigma(t)};\ell_{\Gamma_{\Sigma(t)}})$ there is a function $v_\varphi\in H^1(\Omega(t))$ such that $v_\varphi|_{\Sigma(t)}{}^\pm=\varphi^\pm$, $v_\varphi|_{\Gamma_{\Sigma(t)}}=\varphi_\Gamma$. Moreover the following estimation is valid.

$$\|v_\varphi\|_{1,\Omega(t)}\le C_2\left\{\|(\varphi^+,\varphi^-)\|_{1/2,\ell_{\Sigma(t)}}^2+\|\varphi_\Gamma\|_{1/2,\ell(\Gamma_{\Sigma(t)})}^2\right\}^{1/2},\tag{15}$$

with a constant $C_2>0$ independent of $\varphi^\pm$, $\varphi_{\Gamma_{\Sigma(t)}}$.

Theorem 8 (generalized Stokes formula). *Let $\boldsymbol{\sigma} \in H^1(\Omega_{\Sigma(t)})^9$ and $\operatorname{div}\boldsymbol{\sigma} \in L^2(\Omega_{\Sigma(t)})^3$. Then the following generalized Stokes formula is true for all $\boldsymbol{w} \in H^1(\Omega_{\Sigma(t)})^3$.*

$$\int_{\Omega_\Sigma} ((\operatorname{div}\boldsymbol{\sigma})\cdot\boldsymbol{w} + \boldsymbol{\sigma}\cdot\operatorname{grad}\boldsymbol{w})\,dx = -\left\langle \boldsymbol{\sigma}^+\cdot\boldsymbol{\nu}, \boldsymbol{w}^+\right\rangle_\Sigma + \left\langle \boldsymbol{\sigma}^-\cdot\boldsymbol{\nu}, \boldsymbol{w}^-\right\rangle_\Sigma + \langle \boldsymbol{\sigma}_\Gamma\cdot\boldsymbol{n}, \boldsymbol{w}_\Gamma\rangle_\Gamma . \tag{16}$$

where $\langle\cdot,\cdot\rangle_\Sigma$ denotes the duality between $(H^{1/2}(\Sigma(t))^3)'$ and $H^{1/2}(\Sigma(t))^3$ and $\langle\cdot,\cdot\rangle_\Gamma$ the duality between $(H^{1/2}(\Gamma_{\Sigma(t)};\ell_{\Gamma_{\Sigma(t)}})^3)'$ and $H^{1/2}(\Gamma_{\Sigma(t)};\ell_{\Gamma_{\Sigma(t)}})^3$.

Proof. The proof is similar to [24] or refer to [23].

4 Fracture Problems in 3D elasticity

Under the virtual crack extension $\{\Sigma(t)\}_{0\le t\le T}$, let us consider the brittle fracture phenomenon which means the fracture process is described by a sequence of linear elastic solids defined on $\Omega_{\Sigma(t)}$. Let $\boldsymbol{u}(t)$ be the displacement vector and the constitution law is expressed by Hooke's tensor $c_{ijkl}(x)$, $i,j,k,l = 1,2,3$ with the following properties

$$c_{ijkl} = c_{jikl}, \qquad c_{ijkl} = c_{klij},$$
$$c_{ijkl}\xi_{ij}\xi_{kl} \ge \alpha\xi_{ij}\xi_{ij} \qquad \text{with } \alpha > 0, \qquad \text{for all } \xi_{ij} \in \mathbb{R};\ i,j,k,l = 1,2,3.$$

By Hooke's tensor, the stress tensors $\sigma_{ij}(\boldsymbol{u}(t))$ are expressed by the stain tensors $\varepsilon_{ij}(\boldsymbol{u}(t)) = (\partial_i u_j(t) + \partial_j u_i(t))$ as $\sigma_{ij}(\boldsymbol{u}(t)) = c_{ijkl}\varepsilon_{kl}(\boldsymbol{u}(t))$.

In the case of *surface crack extension*, we assume that $\Gamma_D \subset \Gamma_{\Sigma(t)}$ for all $0 \le t \le T$. For each t, the displacement vector $\boldsymbol{u}(t)$ is given by the solution of the boundary value problems $\mathrm{P}_{\mathcal{L}(t),\Sigma(t)}$: For a given $\mathcal{L}(t) = (\boldsymbol{f}(t), \boldsymbol{g}(t), \boldsymbol{p}) \in L^2(\mathbb{R}^3)^3 \times L^2(\Gamma_N)^3 \times L^2(S)^3$, find $\boldsymbol{u}(t)$ such that

$$\begin{cases} -\partial_j\sigma_{ij}(\boldsymbol{u}(t)) = f_i(t) & \text{in } \Omega_{\Sigma(t)} \\ \sigma_{ij}(\boldsymbol{u}(t))n_j = g_i(t) & \text{on } \Gamma_N \\ \sigma_{ij}(\boldsymbol{u}(t))^+\nu_j = \sigma_{ij}(\boldsymbol{u}(t))^-\nu_j = p_i(t) & \text{on } \Sigma(t) \\ \boldsymbol{u}(t)^+ = \boldsymbol{u}(t)^- & \text{on } \partial\Sigma(t) \\ \boldsymbol{u}(t) = 0 & \text{on } \Gamma_D \end{cases} \tag{17}$$

where $f_i(t)$, $g_i(t)$, $p_i(t)$ are the components of $\boldsymbol{f}(t)$, $\boldsymbol{g}(t)$, $\boldsymbol{p}|_{\Sigma(t)}$ respectively, $n_j(j = 1,2,3)$ the components of outward unit normal on Γ, $\varphi^\pm$ the value of function φ on the plus side or the minus side of $\Sigma(t)$, respectively

Now we will reformulate (17) to the variational problem.

4.1 Variational formula

For each t, the variational formula of $\mathrm{P}_{\mathcal{L}(t),\Sigma(t)}$ is formulated as follows:

For a given load $t \mapsto \mathcal{L}(t) = (\boldsymbol{f}(t), \boldsymbol{g}(t), \boldsymbol{p}) \in C^2([0,T]; L^2(\mathbb{R}^3)^3) \times C^2([0,T]; L^2(\Gamma_N)^3) \times L^2(S)^3$, find $\boldsymbol{u}(t) \in V(\Omega_{\Sigma(t)})$ which minimise the potential energy functional

$$\mathcal{E}(\boldsymbol{v}; \mathcal{L}(t), \Omega_{\Sigma(t)}) = \int_{\Omega_{\Sigma(t)}} \{E(x, \boldsymbol{v}) - \boldsymbol{f}(t) \cdot \boldsymbol{v}\}\, dx - \int_{\Gamma_N} \boldsymbol{g}(t) \cdot \boldsymbol{v}\, ds - \int_{\Sigma(t)} \boldsymbol{p} \cdot \boldsymbol{v}\, ds \tag{18}$$

over the space

$$V(\Omega_{\Sigma(t)}) = \left\{ \boldsymbol{v} \in H^1(\Omega_{\Sigma(t)})^3;\ \boldsymbol{v} = 0 \quad \text{on } \Gamma_D, \right\}$$

where $E(x, \boldsymbol{v}) = c_{ijkl}(x)\varepsilon_{kl}(\boldsymbol{v})\varepsilon_{ij}(\boldsymbol{v})/2$.

Theorem 9 (Existence). *Let $\{\Sigma(t)\} \in \mathrm{SC}_{su}(\Sigma|S)$. For each t, there is the unique solution $\boldsymbol{u}(t)$ satisfying*

$$\int_{\Omega(t)} \sigma_{ij}(\boldsymbol{u}(t))\varepsilon_{ij}(\boldsymbol{v})\, dx = \int_{\Omega_{\Sigma(t)}} \boldsymbol{f}(t) \cdot \boldsymbol{v}\, dx + \int_{\Gamma_N} \boldsymbol{g}(t) \cdot \boldsymbol{v}\, ds + \int_{\Sigma(t)} \boldsymbol{p} \cdot \boldsymbol{v}\, ds \tag{19}$$

for all $\boldsymbol{v} \in V(\Omega_{\Sigma(t)})$. Moreover, $\boldsymbol{u}(t)$ satisfy weakly the conditions in (17).

5 The crack extension force in brittle fracture

In the case of *surface crack extension*, that complicates matters, so we only write on *internal crack extension* in what follows. However by [18], we can get similar results stated below for surface crack extension.

5.1 Griffith's energy balance theory

Under the constant loading $\mathcal{L}(t) = \mathcal{L} = (\boldsymbol{f}, \boldsymbol{g}, 0)$ and a virtual crack extension $\{\Sigma(t)\}$, the difference of energies is written by

$$\mathcal{E}(\boldsymbol{u}(0); \mathcal{L}, \Omega_\Sigma) - \mathcal{E}(\boldsymbol{u}(t); \mathcal{L}, \Omega_{\Sigma(t)}) = \frac{1}{2} \int_{\Omega_{\Sigma(t)}} E(x, \boldsymbol{u} - \boldsymbol{u}(t))dx \geq 0,$$

so the difference of energies become the *crack extension force*. On the other hand, there is a *resisting force*, such as bonding strength, intermolecular force, etc. Griffith[5,6] used the surface force of glass as the resisting force. In the fracture mechanics, the simplest assumption is to consider the constant resisting force α_c per unit surface.

If there is a real crack extension $\{\Sigma^R(t)\}$, then the following inequality will satisfy,

$$\mathcal{E}(\boldsymbol{u}(0); \mathcal{L}, \Omega_\Sigma) - \mathcal{E}(\boldsymbol{u}(t); \mathcal{L}, \Omega_{\Sigma^R(t)}) \geq \alpha_c |\Sigma^R(t) \setminus \Sigma|, \tag{20}$$

where $|\Sigma^R(t) \setminus \Sigma|$ is the area of crack extension $\Sigma^R(t) \setminus \Sigma$. Dividing the both side of (20) by $|\Sigma^R(t) \setminus \Sigma|$ and taking the limit, we have the inequality

$$\mathcal{G}(\mathcal{L}; \Omega_{\Sigma^R(\cdot)}) \geq \alpha_c. \tag{21}$$

Here for an arbitrary virtual crack extension $\{\Sigma(t)\}$,

$$\mathcal{G}(\mathcal{L};\Omega_{\Sigma(\cdot)}) = \lim_{t\downarrow 0} |\Sigma(t)\setminus\Sigma|^{-1}\left[\mathcal{E}(\boldsymbol{u}(0);\mathcal{L},\Omega_{\Sigma(0)}) - \mathcal{E}(\boldsymbol{u}(t);\mathcal{L},\Omega_{\Sigma(t)})\right],$$

is called *energy release rate*. Moreover, the real crack extension $\{\Sigma^R(t)\}$ will take the maximum value of energy release rate over all virtual crack extensions, that is,

$$\mathcal{G}(\mathcal{L};\Omega_{\Sigma^R(\cdot)}) = \max_{\{\Sigma(t)\}} \mathcal{G}(\mathcal{L};\Omega_{\Sigma(\cdot)}).$$

If we find a virtual crack extension $\{\Sigma^*(t)\} \in \mathrm{SC}(\Sigma|S)$ such that $\mathcal{G}(\mathcal{L};\Omega_{\Sigma^*(\cdot)}) \geq \alpha_c$, then from the inequality

$$\mathcal{G}(\mathcal{L};\Omega_{\Sigma^R(\cdot)}) \geq \mathcal{G}(\mathcal{L};\Omega_{\Sigma^*(\cdot)}) \geq \alpha_c,$$

the crack will extend. We now find the *criterion of crack extensions* as follows:

Find the supremum over internal smooth crack extensions,

$$\mathcal{G}^*(\mathcal{L};\mathrm{SC}_{in}(\Sigma|S)) = \sup_{\{\Sigma(t)\}\in\mathrm{SC}_{in}(\Sigma|S)} \mathcal{G}(\mathcal{L};\Omega_{\Sigma(\cdot)}). \tag{22}$$

Because $\mathcal{G}^*(\mathcal{L};\mathrm{SC}(\Sigma|S)) \geq \alpha_c$ implies that $\mathcal{G}(\mathcal{L};\Omega_{\Sigma^R(\cdot)}) \geq \alpha_c$.

5.2 Generalized J-integral for internal crack extension

Let ω be a bounded domain in $\mathbb{R}^3$. We call the domain ω "regular relative to Ω_Σ" if the identity

$$\begin{aligned}\int_{\omega\cap\Omega_\Sigma} v\partial_i w dx = &-\int_{\omega\cap\Omega_\Sigma} \partial_i v w dx + \int_{\partial(\omega\cap\Omega)} vw\, n_i ds \\ &-\int_{\omega\cap\Sigma}\left(v^+w^+\,\nu_i - v^-w^-\,\nu_i\right) ds\end{aligned} \tag{23}$$

holds for all v, $w \in H^1(\Omega_\Sigma)$ and each $i = 1, 2, 3$. Here we notice the direction of $\boldsymbol{\nu}$ on S.

For each solution $\boldsymbol{u}$ of $\mathrm{P}_{\mathcal{L}(0),\Sigma(0)}$, we define the GJ-integral by

$$J_\omega(\boldsymbol{u},\mathcal{X}) = P_\omega(\boldsymbol{u},\mathcal{X}) + R_\omega(\boldsymbol{u},\mathcal{X})$$

as a functional depending on the domain ω and a vector field $\mathcal{X} \in C^\infty(\mathbb{R}^3)^3$, where

$$P_\omega(\boldsymbol{u},\mathcal{X}) = \int_{\partial(\omega\cap\Omega)} \{E(x,\boldsymbol{u})(\mathcal{X}\cdot\boldsymbol{n}) - \sigma_{ij}(\boldsymbol{u})n_j(\mathcal{X}\cdot\nabla u_i)\}\, ds, \tag{24}$$

$$R_\omega(\boldsymbol{u},\mathcal{X}) = \int_{\omega\cap\Omega_\Sigma} \{\sigma_{ij}(\boldsymbol{u})\partial_j\mathcal{X}_k\partial_k u_i - \boldsymbol{X}\nabla_x E(x,\boldsymbol{u}) - \boldsymbol{f}(\mathcal{X}\cdot\nabla u) - E(x,\boldsymbol{u})\mathrm{div}\mathcal{X}\}\, dx.$$

Here $\boldsymbol{n} = (n_1, n_2, n_3)$ is the outward unit normal $\boldsymbol{n}$ on $\partial(\omega\cap\Omega)$ and ds the surface element of $\partial(\omega\cap\Omega)$. The integral $R_\omega(\boldsymbol{u},\mathcal{X})$ is well-defined for all solutions $\boldsymbol{u}$ of $\mathrm{P}_{\mathcal{L}(0),\Sigma(0)}$, however $P_\omega(\boldsymbol{u},\mathcal{X})$ needs the regularity of $\boldsymbol{u}$. We notice that $P_\omega(\boldsymbol{u},\mathcal{X})$ contains no integral over $\omega\cap\Sigma$.

Theorem 10 (refer to [13,14,19]). *For a* $\{\Sigma(t)\} \in \mathrm{SC}_{in}(\Sigma|S)$,

$$\mathcal{G}(\mathcal{L}, \Omega_{\Sigma(\cdot)}) = J_\omega(\boldsymbol{u}; \boldsymbol{X}) \left(\int_{\partial\Sigma} \delta\phi_0(\gamma) d\gamma \right)^{-1}, \tag{25}$$

where ω *stands for an arbitrary small domain containing the crack front* $\partial\Sigma$, $\boldsymbol{X}$ *the vector field obtained from parallel extension of* $\delta\phi_0(\gamma)\boldsymbol{e}_1(\gamma)$, $\gamma \in \partial\Sigma$ *over* $\overline{U(\partial\Sigma)}$ *and* $d\gamma$ *the line element of* $\partial\Sigma$ *(see Fig. 4). Moreover, we have*

$$J_\omega(\boldsymbol{u}; \boldsymbol{X}) = \langle \delta\phi_0(\gamma), J_\gamma(\boldsymbol{u}, \boldsymbol{e}_1) \rangle_{\partial\Sigma} \tag{26}$$

where $J_\gamma(\boldsymbol{u}, \boldsymbol{e}_1)$ *is the distribution on* $\partial\Sigma$ *(see e.g. [7] for the distribution on the manifold).*

Proof. The mathematical proof of (25) is given in [14]. Here we notice that $J_\omega(\boldsymbol{u}, \boldsymbol{X})$ are independent of ω. For h_1, $h_2 \in C^\infty(\partial\Sigma)$ and α, $\beta \in \mathbb{R}$, we can construct the vector fields $\boldsymbol{X}_{\alpha h_1 + \beta h_2}$ by the parallel displacement of $(\alpha h_1 + \beta h_2)\boldsymbol{e}_1$ along geodesic line on S and the normal surfaces in the tubular neighborhood $(F, U_\epsilon(\partial\Sigma))$ (called Levi-Civita connection of $\delta\phi_0(\gamma)\boldsymbol{e}_1(\gamma)$ in $(F, U_\epsilon(\partial\Sigma))$, see e.g. [9]). Then $\boldsymbol{X}_{\alpha h_1 + \beta h_2} = \alpha \boldsymbol{X}_{h_1} + \beta \boldsymbol{X}_{h_2}$ (same to the relation $\nabla_{\alpha X + \beta Y} = \alpha\nabla_X + \beta\nabla_Y$ for covariant derivative in Riemannian manifold), so we have

$$J_\omega(\boldsymbol{u}, \boldsymbol{X}_{\alpha h_1 + \beta h_2}) = \alpha J_\omega(\boldsymbol{u}, \boldsymbol{X}_{h_1}) + \beta J_\omega(\boldsymbol{u}, \boldsymbol{X}_{h_2}).$$

Let $\eta_{\partial\Sigma} \in C_0^\infty(\Omega)$ be the cut-off function near $\partial\Sigma$, that is, $\eta_{\partial\Sigma} \equiv 1$ near $\partial\Sigma$ and $\eta_{\partial\Sigma} \equiv 0$ outside some neighborhood of $\partial\Sigma$. By the independence of ω, we can derive

$$J_\omega(\boldsymbol{u}, \boldsymbol{X}_h) = R_\omega(\boldsymbol{u}, \eta_{\partial\Sigma}\boldsymbol{X}_h)$$

for all $h \in C^\infty(\partial\Sigma)$. Using Schwarz inequality, we obtain the estimation

$$J_\omega(\boldsymbol{u}, \boldsymbol{X}_h) \le C \left(\|\boldsymbol{u}\|_{1,\Omega_\Sigma}^2 + \|\boldsymbol{f}\|_{0,\mathbb{R}^3}^2 \right) \|h\|_{C^1(\partial\Sigma)}$$

with some constant $C > 0$ independent of h, $\boldsymbol{u}$, $\boldsymbol{f}$. Therefore, $h \mapsto J_\omega(\boldsymbol{u}, \boldsymbol{X}_h)$ is the linear continuous functional defined on $C^1(\partial\Sigma)$. Since $C^\infty(\partial\Sigma) \subset C^1(\partial\Sigma)$ topologically, we then complete the proof of Theorem.

Theorem 11. *If the solution* $\boldsymbol{u}$ *of* $P_{\mathcal{L}(0),\Sigma(0)}$ *is in* $H^2(\text{near } \partial\Sigma)^3$, *then*

$$J_\omega(\boldsymbol{u}, \boldsymbol{X}) = 0, \tag{27}$$

for all vector field $\boldsymbol{X}$ *obtained from smooth crack extension. Moreover, we assume that* $\boldsymbol{u}$ *is decomposed to* $\boldsymbol{u}_S + \boldsymbol{u}_R$ *as follows; There is a constant* $M > 0$ *such that* $\left|\sqrt{\rho_{\partial\Sigma}}(x)\nabla\boldsymbol{u}_S(x)\right| < M$ *for all* $x \in \overline{U_\epsilon(\partial\Sigma)})$ *and* $\boldsymbol{u}_R \in H^2(U_\epsilon(\partial\Sigma) \cap \Omega_\Sigma)^3$. *Then*

$$J_\omega(\boldsymbol{u}, \boldsymbol{X}) = \lim_{|\omega| \to 0} P_\omega(\boldsymbol{u}_S, \boldsymbol{X}). \tag{28}$$

Remark 12. The assumption $\boldsymbol{u} = \boldsymbol{u}_S + \boldsymbol{u}_R$ will be true by [3], however obtained results in [3] are regularity and asymptotic representations on pseudodifferential equations defined on Σ. Then there is a small gap between their retults and the assumption.

Proof. The identity (27) is proved in [14].

Using the independence of ω and $\lim_{|\omega|\to 0} R_\omega(\boldsymbol{u}, \boldsymbol{X}) = 0$, we only check that $\lim_{\delta\to 0} P_{U_\delta(\partial\Sigma)}(\boldsymbol{u}, \boldsymbol{X}) = \lim_{\delta\to 0} P_{U_\delta(\partial\Sigma)}(\boldsymbol{u}_S, \boldsymbol{X})$,

$$P_{U_\delta(\partial\Sigma)}(\boldsymbol{u}, \boldsymbol{X}) = P_{U_\delta(\partial\Sigma)}(\boldsymbol{u}_S, \boldsymbol{X}) + P_{U_\delta(\partial\Sigma)}(\boldsymbol{u}_R, \boldsymbol{X}) + P_{U_\delta(\partial\Sigma)}(\boldsymbol{u}_S; \boldsymbol{u}_R, \boldsymbol{X}),$$

$$\begin{aligned} P_{U_\delta(\partial\Sigma)}(\boldsymbol{u}_S; \boldsymbol{u}_R, \boldsymbol{X}) &= \int_{U_\delta(\partial\Sigma)} \{\sigma_{ij}(\boldsymbol{u}_S)\varepsilon_{ij}(\boldsymbol{u}_R)(\boldsymbol{X}\cdot\boldsymbol{n}) \\ &\quad -\sigma_{ij}(\boldsymbol{u}_S)\nu_j(\boldsymbol{X}\cdot\nabla u_{i,R}) - \sigma_{ij}(\boldsymbol{u}_R)\nu_j(\boldsymbol{X}\cdot\nabla u_{i,S})\}\, ds, \end{aligned}$$

where $\boldsymbol{u}_R = (u_{i,R})$. By the assumption, $\sqrt{\rho_{\partial\Sigma}}\nabla\boldsymbol{u}_S$ is uniformly bounded on $U_\epsilon(\partial\Sigma)$, we have the estimation

$$\left|P_{U_\delta(\partial\Sigma)}(\boldsymbol{u}_S; \boldsymbol{u}_R, \boldsymbol{X})\right| \le C\delta^{-1/2} \int_{\partial U_\delta(\partial\Sigma)} |\nabla\boldsymbol{u}_R| ds. \tag{29}$$

with a constant $C > 0$ independent of $\boldsymbol{u}_R$ and δ. By Schwartz inequality, we obtain

$$\begin{aligned} \int_{\partial U_\delta(\partial\Sigma)} |\nabla\boldsymbol{u}_R| ds &\le \int_{\partial U_\delta(\partial\Sigma)} 1\, ds \left(\int_{\partial U_\delta(\partial\Sigma)} |\nabla\boldsymbol{u}_R|^2 ds\right)^{1/2} \\ &\le C_1\delta \|\nabla\boldsymbol{u}_R\|_{0,\partial U_\delta(\partial\Sigma)}. \end{aligned} \tag{30}$$

By the partition of unity and the change of variables, $\|\nabla\boldsymbol{u}_R\|^2_{0,\partial U_\delta(\partial\Sigma)}$ become the finite sum of the following integrals

$$\int_{-L}^{L} d\ell \int_{-\pi}^{\pi} |\partial_j U(\ell, \delta, \theta)|^2\, d\theta \qquad \text{for some } L > 0, \quad j = 1, 2, 3,$$

where $\partial_j U$ stands for some element of $\partial_j u_{i,R}$ after using partition of unity and the change of variables. We can assume that $\partial_j U(\ell, r, \theta) \equiv 0$ for $r > R_0$ with some $R_0 > 0$. By the expression

$$\partial_j U(\ell, \delta, \theta) = -\int_\delta^{R_0} \frac{\partial}{\partial r}\partial_j U(\ell, r, \theta) dr,$$

we have the estimation

$$|\partial_j U(\ell, \delta, \theta)| \le R_0 \left(\int_\delta^{R_0} \left|\frac{\partial}{\partial r}\partial_j U(\ell, r, \theta)\right|^2 dr\right)^{1/2}.$$

Then we can derive

$$\begin{aligned} \int_{-L}^{L} d\ell \int_{-\pi}^{\pi} |\partial_j U(\ell, \delta, \theta)|^2\, d\theta &\le R_0 \int_{-L}^{L} d\ell \int_{-\pi}^{\pi} d\theta \int_\delta^{R_0} \left|\frac{\partial}{\partial r}\partial_j U(\ell, r, \theta)\right|^2 dr \\ &\le R_0 C_2 \|\boldsymbol{u}_R\|^2_{2,\Omega_\Sigma} \end{aligned}$$

with $C_2 > 0$ depending only on the change of variables. We arrive the estimation

$$\|\nabla \boldsymbol{u}_R\|_{0,\partial U_\epsilon(\partial\Sigma)} \leq C_3 \|\boldsymbol{u}_R\|_{2,\Omega_\Sigma} \tag{31}$$

with a constant $C_3 > 0$ independent of $\boldsymbol{u}_R$ and δ. Combining (29) – (31), we obtain the following

$$|J(\boldsymbol{u}_S; \boldsymbol{u}_R, \boldsymbol{X})| \leq C_4 \delta^{1/2} \|\boldsymbol{u}_R\|_{2,\Omega_\Sigma},$$

which leads (28).

Fracture criterion (internal crack extension): We assume that $\gamma \mapsto J_\gamma(\boldsymbol{u}, \boldsymbol{e}_1)$ (see (26)) is in $C^0(\partial\Sigma)$, then

$$\mathcal{G}^*(\mathcal{L}; \mathrm{SC}_{in}(\Sigma|S)) = \sup_{h \in \mathcal{S}_{in}(\partial\Sigma)} \langle h, J_\gamma(\boldsymbol{u}, \boldsymbol{e}_1)\rangle_{\partial\Sigma} \tag{32}$$

$$\langle h, J_\gamma(\boldsymbol{u}, \boldsymbol{e}_1)\rangle_{\partial\Sigma} = \int_{\partial\Sigma} h(\gamma) J_\gamma(\boldsymbol{u}, \boldsymbol{e}_1) d\gamma,$$

where

$$\mathcal{S}_{in}(\partial\Sigma) = \left\{ h \in C^1(\partial\Sigma) : \ h > 0, \int_{\partial\Sigma} h(\gamma) d\gamma = 1 \right\}.$$

From the Riesz-Markov-Kakutani theorem, we can extend the domain of operator

$$h \to \int_{\partial\Sigma} h(\gamma) J_\gamma(\boldsymbol{u}, \boldsymbol{e}_1) d\gamma$$

to $rca(\partial\Sigma)$, where $rca(\partial\Sigma)$ stands for the space of all regular countable additive scalar valued set functions on the σ-field of all Borel sets in $\partial\Sigma$ (see e.g. [4]). Let us denote the duality between $C(\partial\Sigma)$ and $rca(\partial\Sigma)$ by $\langle \cdot, \cdot \rangle_{\partial\Sigma}$.

Thus the fracture criterion is reformulated as follows;

Find $h_{\max} \in rca(\partial\Sigma)$ such that

$$\langle h_{\max}, J_\gamma(\boldsymbol{u}, \boldsymbol{e}_1)\rangle_{\partial\Sigma} \geq \langle h, J_\gamma(\boldsymbol{u}, \boldsymbol{e}_1)\rangle_{\partial\Sigma} \qquad \text{for all } h \in \widetilde{\mathcal{S}}_{in}(\partial\Sigma),$$

$$\widetilde{\mathcal{S}}_{in}(\partial\Sigma) = \{ h \in rca(\partial\Sigma) \mid h \geq 0, |h| = 1 \}.$$

Since $rca(\partial\Sigma)$ is the dual space of $C^0(\partial\Sigma)$, we have

$$|\langle h, J_\gamma(\boldsymbol{u}, \boldsymbol{e}_1)\rangle_{\partial\Sigma}| \leq \max_{\gamma \in \partial\Sigma} J_\gamma(\boldsymbol{u}, \boldsymbol{e}_1)|h|.$$

The compactness of $\partial\Sigma$ imply the existence of the maximum value of $J_\gamma(\boldsymbol{u}, \boldsymbol{e}_1)$ at $\gamma_{\max} \in \partial\Sigma$. Let us denote by $\delta_{\gamma_{\max}}$, Dirac's distribution at $\gamma_{\max}$, i.e., $\delta_{\gamma_{\max}}$ satisfy the equality

$$\langle \psi, \delta_{\gamma_{\max}} \rangle_{\partial\Sigma} = \psi(\gamma_{\max}) \qquad \text{for all } \psi \in C^0(\partial\Sigma),$$

and furthermore $\delta_{\gamma_{\max}} \geq 0$ and $|\delta_{\gamma_{\max}}| = 1$. Then $\delta_{\gamma_{\max}} \in \widetilde{\mathcal{S}}_{in}(\partial\Sigma)$. From this,

$$\mathcal{G}^*(\mathcal{L}; \mathrm{SC}_{in}(\Sigma|S)) = J_{\gamma_{\max}}(\boldsymbol{u}, \boldsymbol{e}_1). \tag{33}$$

In [1], they adopt Irwin criterion (see e.g. [8]) when $S = \{(x_1, x_2, x_3) : x_3 = 0\}$, that is, if the virtual crack extension is not real, then $K_1(t;\gamma) \le K_{1c}$, and real crack extension implies that $K_1(t;\gamma) = K_{1c}$. Here $K_1(t;\gamma)$ are stress intensity factors (opening mode) near $\Sigma(t)$ with time-like parameter t, and K_{1c} the critical value determined by experiment. Using asymptotic expansion of $K_1(t;\gamma)$, they also rewrite Irwin's criterion the variational inequality and solve it. The uniqueness, smoothness and numerical realization (example) are given. The asymptotic expansion technique (refer to [1]) is very powerful, however it's very difficult even if we study them under simple situation.

5.3 Internal crack extension under varying load

Griffith's energy balance theory is not applicable under varying load. However, we get same result on *fracture criterion* from the following results.

Theorem 13 ([23]). *Under the constant loading, we have*

$$-\frac{d\mathcal{E}}{dt}(\boldsymbol{u}(t);\mathcal{L},\Omega_{\Sigma(t)})\Big|_{t=+0} = \lim_{t\downarrow 0}\frac{1}{2t}\langle\sigma_{ij}(\boldsymbol{u})\nu_j, [\![u_i(t)-u_i]\!]\rangle_{\Sigma(t)}, \tag{34}$$

and under the varying loading, we obtain

$$\begin{aligned}-\frac{d\mathcal{E}}{dt}(\boldsymbol{u}(t);\mathcal{L}(t),\Omega_{\Sigma(t)})\Big|_{t=+0} &= \lim_{t\downarrow 0}\frac{1}{2t}\langle\sigma_{ij}(\boldsymbol{u})\nu_j, [\![u_i(t)-u_i]\!]\rangle_{\Sigma(t)} \\ &\quad + \lim_{t\downarrow 0} t^{-1}\left\{\int_{\Omega_{\Sigma(t)}}(\boldsymbol{f}(t)-\boldsymbol{f})\boldsymbol{u}\,dx + \int_{\Gamma_N}(\boldsymbol{g}(t)-\boldsymbol{g})\boldsymbol{u}\,ds\right\}.\end{aligned} \tag{35}$$

Under the varying load $\mathcal{L}(t)$, the *energy release rate is not the crack extension force no longer. However, GJ-integral express the crack extension force yet.*

Theorem 14. *For a* $\{\Sigma(t)\} \in \mathrm{SC}_{in}(\Sigma|S)$, *under the varying load* $\mathcal{L}(t)$, *we have the relation*

$$\lim_{t\downarrow 0}\frac{1}{2t}\langle\sigma_{ij}(\boldsymbol{u})\nu_j, [\![u_i(t)-u_i]\!]\rangle_{\Sigma(t)} = J_\omega(\boldsymbol{u}.\boldsymbol{X}), \tag{36}$$

$$\begin{aligned}-\frac{d}{d\tau}\mathcal{E}(\boldsymbol{u}(t);\mathcal{L}(t),\Omega_{\Sigma(t)})\Big|_{t=0} &= J_\omega(\boldsymbol{u};\boldsymbol{X}) \\ &\quad + \int_{\Omega_\Sigma}\dot{\boldsymbol{f}}\cdot\boldsymbol{u}\,dx + \int_{\Gamma_N}\dot{\boldsymbol{g}}\cdot\boldsymbol{u}\,ds\end{aligned} \tag{37}$$

where $\dot{\boldsymbol{f}} = d\boldsymbol{f}/dt|_{t=0}$, $\dot{\boldsymbol{g}} = d\boldsymbol{g}/dt|_{t=0}$.

6 Numerical analysis of 3D fracture

The calculation of GJ-integral is very important in fracture problem. Here we only state the theoretical results for finite element method, because we need huge power of machines and labour in 3D calculations. The method stated below is essentially same in 2D cases (see [21]).

Let Ω be a bounded polyhedral convex domain and let us assume that S is a flat surface. We make a tetrahedral finite element $\mathcal{T}_h = \{T_1, \cdots T_M\}$ of Ω_Σ,

$$\Omega_\Sigma = \cup_{T\in\mathcal{T}_h} T = T_1 \cup T_2 \cup \cdots \cup T_M, \quad h = \max\{\text{size of } T_i : i = 1, \cdots, m\}.$$

We put

$$V_h(\Omega_\Sigma) = \{\boldsymbol{v} : \boldsymbol{v} \in C^0(\Omega_\Sigma)^3, \boldsymbol{v}^+ = \boldsymbol{v}^- \text{ on } \partial\Sigma, \boldsymbol{v}|_T \in \mathrm{P}_1(T)\,\forall T \in \mathcal{T}_h, \boldsymbol{v} = 0 \text{ on } \Gamma_D\},$$

where $\mathrm{P}_1(T)$ is the space of the polynomials less than one defined in T. Here we notice that the discontinuity of $\boldsymbol{v} \in V_h(\Omega_\Sigma)$ is permitted on Σ. So the nodes on plus side and minus side of Σ must be different.

Find $\boldsymbol{u}_h \in V_h(\Omega_\Sigma)$ such that

$$\mathcal{E}(\boldsymbol{u}_h; \mathcal{L}, \Omega_\Sigma) \le \mathcal{E}(\boldsymbol{v}; \mathcal{L}, \Omega_\Sigma) \quad \text{for } \boldsymbol{v} \in V_h(\Omega_\Sigma). \tag{38}$$

Under the assumption stated in Theorem 11, we can prove the following by similar argument in [27, Theorem 12.1].

$$\|\boldsymbol{u} - \boldsymbol{u}_h\|_{1,\Omega_\Sigma} \le C_0 h^\alpha \qquad \text{for } 0 < \alpha < 1/2. \tag{39}$$

Perhaps, $J_{U_\epsilon(\partial\Sigma)}(\boldsymbol{u}_h, \boldsymbol{X})$ gives the approximation of $J_{U_\epsilon(\partial\Sigma)}(\boldsymbol{u}, \boldsymbol{X})$. However, to get the error estimate of $P_{U_\epsilon(\partial\Sigma)}(\boldsymbol{u}_h, \boldsymbol{X}_k)$, we need the error estimation $\|\boldsymbol{u} - \boldsymbol{u}_h\|_{q,U_{\epsilon-\delta,\epsilon+\delta}(\partial\Sigma)}$ for a number $q > 3/2$ where

$$U_{\epsilon_1,\epsilon_2}(\partial\Sigma) = \left\{x = F(\gamma, \xi_1, \xi_2) : \epsilon_1 \le \sqrt{\xi_1^2 + \xi_2^2} \le \epsilon_2, \gamma \in \partial\Sigma\right\}$$

in $0 < \epsilon_1 < \epsilon < \epsilon_2$. Moreover, the constant in error estimation depends on ϵ_1, ϵ_2.

We now prove the important property of GJ-integral in FEM.

Proposition 15. *Let η be arbitrary function $W^{1,\infty}(\mathrm{I\!R}^3)$ such that $\eta \equiv 1$ on $U_{\epsilon/2}(\partial\Sigma)$ and $\eta \equiv 0$ outside $U_{3\epsilon/4}(\partial\Sigma)$.*

$$J_{U_\epsilon(\partial\Sigma)}(\boldsymbol{u}; \boldsymbol{X}) = R_{U_\epsilon(\partial\Sigma)}(\boldsymbol{u}; \eta\boldsymbol{X}). \tag{40}$$

Remark 16. The space $W^{1,\infty}(\mathrm{I\!R}^3)$ is the set of functions whose weak derivatives are essentially bounded on $\mathrm{I\!R}^3$. V_h-interpolation of the cut-off function $\tilde{\eta} \in C_0^\infty(U_\epsilon(\partial\Sigma))$ belongs to $W^{1,\infty}(\mathrm{I\!R}^3)$.

Proof. Let $\boldsymbol{Z} = (Z_1, Z_3, Z_3) = \boldsymbol{X} - \eta\boldsymbol{X}$. Then $J_{U_\epsilon(\partial\Sigma)}(\boldsymbol{u}, \boldsymbol{Z}) = J_{U_\epsilon(\partial\Sigma)}(\boldsymbol{u}, \boldsymbol{X}) - J_{U_\epsilon(\partial\Sigma)}(\boldsymbol{u}, \eta\boldsymbol{X}) = J_{U_{\epsilon/2,\epsilon}(\partial\Sigma)}(\boldsymbol{u}, \boldsymbol{Z})$. Since $\boldsymbol{u}|_{U_{\epsilon/2,\epsilon}(\partial\Sigma)} \in H^2(U_{\epsilon/2,\epsilon}(\partial\Sigma))^3$, we have by the divergence theorem

$$\int_{U_{\epsilon/2,\epsilon}(\partial\Sigma)\cap\Omega_\Sigma} \boldsymbol{Z}\cdot\nabla E(x,\boldsymbol{u})dx = -\int_{U_{\epsilon/2,\epsilon}(\partial\Sigma)\cap\Omega_\Sigma} E(x,\boldsymbol{u})\mathrm{div}\boldsymbol{Z}\,dx$$

$$+\int_{U_{\epsilon/2,\epsilon}(\partial\Sigma)\cap\Omega_\Sigma} [\![E(x,\boldsymbol{u})]\!](\boldsymbol{Z}\cdot\boldsymbol{\nu})\,ds + \int_{\partial(U_{\epsilon/2,\epsilon}(\partial\Sigma)\cap\Omega_\Sigma)} E(x,\boldsymbol{u})(\boldsymbol{Z}\cdot\boldsymbol{n})\,ds.$$

$$\int_{U_{\epsilon/2,\epsilon}(\partial\Sigma)\cap\Omega_\Sigma} \boldsymbol{Z}\cdot\nabla E(x,\boldsymbol{u})dx = \int_{U_{\epsilon/2,\epsilon}(\partial\Sigma)\cap\Omega_\Sigma} \sigma_{ij}(\boldsymbol{u})(\partial_j Z_l\partial_l u_i) - \sigma_{ij}(\boldsymbol{u})(\partial_j Z_l)\partial_l u_i\,dx$$

$$= \int_{\partial(U_{\epsilon/2,\epsilon}(\partial\Sigma)\cap\Omega)} \sigma_{ij}(\boldsymbol{u})n_j(\boldsymbol{Z}\cdot\nabla u_i)ds - \int_{U_{\epsilon/2,\epsilon}(\partial\Sigma)\cap\Sigma} \sigma_{ij}(\boldsymbol{u})^+\nu_j\boldsymbol{Z}\cdot(\nabla\boldsymbol{u})^+ds$$

$$+\int_{U_{\epsilon/2,\epsilon}(\partial\Sigma)\cap\Sigma} \sigma_{ij}(\boldsymbol{u})^-\nu_j\boldsymbol{Z}\cdot(\nabla u_i)^-dl - \int_{U_{\epsilon/2,\epsilon}(\partial\Sigma)\cap\Omega_\Sigma} \partial_j\sigma_{ij}(\boldsymbol{u})(\boldsymbol{X}\cdot\nabla u_i) + \sigma_{ij}(\boldsymbol{u})(\partial_j Z_l)\partial_l\boldsymbol{u}\,dx.$$

By the hypothesis, we have $\boldsymbol{Z}\cdot\nu = 0$ and $\sigma_{ij}(\boldsymbol{u})^{\pm}\nu_j = 0$ on Σ. Collecting these formulas, we obtain $J_{U_{\epsilon/2,\epsilon}(\partial\Sigma)}(\boldsymbol{u};\boldsymbol{Z}) = 0$. Since $\eta\boldsymbol{X}\equiv 0$ on $\partial U_\epsilon(\partial\Sigma)$, we complete the proof of Proposition 15.

For the displacement $\boldsymbol{u}$ and $\boldsymbol{u}_h$ given in (38), we have the estimate by Schwartz inequality

$$|R_{\Omega_\Sigma}(\boldsymbol{u},\eta\boldsymbol{X}) - R_{\Omega_\Sigma}(\boldsymbol{u}_h,\eta\boldsymbol{X})| \le C_1(\eta)\left(\|\boldsymbol{u}\|_{1,\Omega_\Sigma} + \|f\|_{0,\Omega}\right)\|\boldsymbol{u}-\boldsymbol{u}_h\|_{1,\Omega_\Sigma}$$

with a constant $C_1(\eta) > 0$ independent of $\boldsymbol{u}$, $\boldsymbol{f}$, $\boldsymbol{u}_h$. However $C_1(\eta)$ depend on the norm $\|\eta\|_{W^{1,\infty}(\mathbb{R}^3)}$. From this and (39), we obtain

$$|R_{\Omega_\Sigma}(\boldsymbol{u},\eta\boldsymbol{X}) - R_{\Omega_\Sigma}(\boldsymbol{u}_h,\eta\boldsymbol{X})| \le C_2(\eta,\boldsymbol{f},\boldsymbol{u})h^\alpha \quad \alpha < 1/2. \tag{41}$$

where $C_2(\eta_k,\boldsymbol{f},\boldsymbol{u}) = C_0C_1(\eta)\left(\|\boldsymbol{u}\|_{1,\Omega_\Sigma} + \|f\|_{0,\Omega}\right)$.

References

1. Bach, M., Nazarov, S.A.: Smoothness properties of solutions to variational inequalities describing extension of mode-1 cracks. Mathematical aspects of boundary element methods (Palaiseau, 1998), 23–32, Chapman & Hall/CRC Res. Notes Math., 414, Boca Raton, FL, 2000.
2. Cherepanov, G. P.: Crack propagation in continuous media, J. Appl. Math. Mech., 31(1967), 503-512.
3. Duduchava, R., Wendland, W.L.: The wiener–hopf method for systems of pseudodifferential equations with an application to crack problems, Integral Equations Oper. Theory, 23(1995), 294–335.
4. Dunford, N., Schwartz, J.T.: Linear operators Part I, Interscience Publishers Inc., New York, 1967, pp. 265–266.
5. Griffith, A. A.: The phenomena of rupture and flow in solids, Phil. Trans. Roy. Soc. Ser. A 221(1920), 163–198.
6. Griffith, A. A.: The phenomena of rupture and flow in solids, Philosophical Transactions, Royal Society of London, Series A, Vol. 211, 1921, pp. 163–198.
7. Hörmander, L.: Linear partial differential operators, Springer-Verlag, 1969.

8. Irwin, G. R.: Fracture mechanics, in the book "Structural mechanics", Pergamon Press, 1958, 557–594.
9. Kobayashi, S., Nomizu, K.: Foundations of differential geometry, Volume 1, Interscience Publ., 1963.
10. Lang, S.: Differential Manifolds, Addison-Wesley, Massachusetts·Menlo Park·Calfornia·London·Don Mills·Ontario, 1972
11. Leblond, J.-B., Torlai, O.: The stress field near the front of an arbitrarily shaped crack in a three-dimensional elastic body, J. Elasticity, 29(1992), 97–131
12. Nečas, J.: Méthodes Directes en Th' eorie des Équations Elliptiques, Masson Éditeur, Paris, 1967.
13. Ohtsuka, K.: J-integral and two-dimensional fracture mechanics, RIMS Kokyuroku, Kyoto Univ., 386(1980), 231–248.
14. K. Ohtsuka, Generalized J-integral and three dimensional fracture mechanics I, Hiroshima Math. J., 11(1981), 21–52.
15. Ohtsuka, K.: Remark on Griffith's energy balance for three-dimensional fracture problems, Memor. Hiroshima-Denki Institute of Technology and Hiroshima Junior College of Automotive Engineering, 16(1983), 31–36.
16. Ohtsuka, K.: Generalized J-integral and its applications. I. Basic theory. Japan J. Applied Math., 1985, 2, 329–350.
17. Ohtsuka, K.: Sobolev space H^1 on a domain with a surface cut, Memor. Hiroshima-Denki Institute of Technology and Hiroshima Junior College of Automotive Engineering, 18(1985), 51–57.
18. Ohtsuka, K.: Generalized J-integral and three-dimensional fracture mechanics. II. Surface crack problems, Hiroshima Math. J., 16(1986), 327–352.
19. Ohtsuka, K.: Mathematical aspects of fracture mechanics, Lecture Notes in Num. Appl. Anal., 13(1994), 39–59.
20. Ohtsuka, K.: Mathematical analysis of 3-D fracture phenomenon by Griffith's energy balance theory under increasing loads, Theoretical and Applied Mechanics, 45(1996), 99–103.
21. Ohtsuka, K., Pironneau, O., Hecht, F.: Theoretical and numerical analysis of energy release rate in 2D fracture, INFORMATION, 3(2000), 303–315.
22. Ohtsuka, K., Khludnev, A.: Generalized J-integral method for sensitivity analysis of static shape design, Control and Cybernetics, 29(2000), 513–533.
23. Ohtsuka, K.: What's the crack extension force in three-dimensional quasi-static brittle fracture, Free boundary problems, Theory and Applications II, 2000, 344–357.
24. Teman, R.: Navier-Stokes equations, North-Holland, 1979.
25. Ting, T.C.T.: Asymptotic solutions near the apex of an elastic wedge with curved boundaries. Q. Appl. Math. 42(1985), 467–476.
26. Rice, J.R.: A path-independent integral and the approximate analysis of strain concentration by notches and cracks, J. Appl. Mech., 35(1968), 379–386.
27. Wahlbin, Lars B.: Local behavior in finite element methods, Handbook of numerical analysis Vol.2(editors P.G.Ciarlet and J.L.Lions), 353–522, North-Holland, 1991.

Exploiting Partial or Complete Geometrical Symmetry in Boundary Integral Equation Formulations of Elastodynamic Problems

Marc Bonnet

Laboratoire de Mécanique des Solides (UMR CNRS 7649)
Ecole Polytechnique, F-91128 Palaiseau Cedex, FRANCE

Abstract. Procedures based on group representation theory, allowing the exploitation of geometrical symmetry in symmetric Galerkin BEM formulations of 3D elastodynamic problems, are developed. They are applicable for both commutative and noncommutative finite symmetry groups and to partial geometrical symmetry, where the boundary has two disconnected components, one of which is symmetric.

1 Introduction

When a linear boundary-value problem (BVP) exhibits geometrical symmetry, taking full advantage of it yields substantial computational benefits. In Bossavit [4], the linear representation theory for finite groups [8,9] is shown to lead to the correct definition of (i) decomposition of function spaces into orthogonal subspaces of symmetric, skew-symmetric,... functions, and (ii) reconstruction of the global solution from these components; the (domain-based, FEM-oriented) weak formulation is thus recast into a block-diagonal form, each 'subproblem' being defined on a 'symmetry cell' (a subdomain of smallest measure that, under the action of the symmetry group, generates the entire initial domain) and associated to the corresponding projection of the boundary data. The procedure, being essentially an elaborate superposition technique, assumes linear constitutive properties. Similar principles are used by Allgower et al. [1] to block-diagonalize the discretized equations.

The adaptation of the former approach to boundary element methods (BEMs) is not straightforward, mostly because the symmetry cell usually involve the definition of new boundaries, a feature which is unimportant in FEMs but clearly undesirable in BEMs, where subproblems should be stated only on symmetry cells of the boundary. In an earlier work [2], this issue was adressed for collocation BEMs and commutative symmetry groups (see also [6]). Using standard methods to set up and solve the matrix equations, the theoretical computational gains (in relative terms, compared to using the same discretization without symmetry) were found to be $1/n$, $1/n$ and $1/n^2$ for the matrix storage requirement, matrix set-up time and solution time, respectively, where n is the number of elements in the symmetry group (e.g. $n = 8$ for the group of symmetries with respect to three orthogonal planes).

This contribution aims at extending the concepts and results of [2] in three directions. Firstly, the exploitation of geometrical symmetry is considered here in the framework of symmetric Galerkin BEM formulations. Secondly, procedures are developed for both commutative or noncommutative symmetry groups. Thirdly, the approach is generalized to partial geometrical symmetry, where the boundary has two (or more) disconnected components, one (or more) of which being invariant under a symmetry group. For instance, in defect identification problems, bodies with external geometrical symmetry but containing internal cracks, voids, inclusions... of arbitrary shape and location might be encountered. The formulations developed herein are expected to bring significant gains in computational efficiency by exploiting symmetries of the external boundary.

2 Governing equations

In this paper, the use of geometrical symmetry is fully developed for the Neumann BVP of linear elastodynamics in the frequency-domain, using double-layer integral representations. These BVPs are chosen as representative model problems, and the developments to follow are expected to be easily adaptable to other scalar or vector linear BVPs (e.g. from electromagnetics).

The displacement vector $\boldsymbol{u}$, strain tensor $\boldsymbol{\varepsilon}$ and stress tensor $\boldsymbol{\sigma}$ in a three-dimensional isotropic elastic medium are governed by the dynamic equilibrium, constitutive and compatibility field equations:

$$\begin{aligned} \operatorname{div}\boldsymbol{\sigma} + \rho\omega^2\boldsymbol{u} + \boldsymbol{f} &= \boldsymbol{0} \\ \boldsymbol{\sigma} &= \mu\big[\tfrac{2\nu}{1-2\nu}\operatorname{Tr}(\boldsymbol{\varepsilon})\boldsymbol{1} + 2\boldsymbol{\varepsilon}\big] \\ \boldsymbol{\varepsilon} &= (\boldsymbol{\nabla u} + \boldsymbol{\nabla}^T\boldsymbol{u})/2 \end{aligned} \tag{1}$$

(with μ: shear modulus, ν: Poisson ratio, ρ: mass per unit volume, $\boldsymbol{f}$: body force distribution), which, upon elimination of $\boldsymbol{\varepsilon}$ and $\boldsymbol{\sigma}$, yield the well-known Navier equation, an elliptic second-order vector PDE for the primary field $\boldsymbol{u}$.

In particular, a time-harmonic unit point force (i.e. $\boldsymbol{f} = \delta(\boldsymbol{x} - \tilde{\boldsymbol{x}})\boldsymbol{e}_k$) applied in an infinite elastic body at the fixed point $\tilde{\boldsymbol{x}}$ and along the k-direction defines at $\boldsymbol{x} \in \mathbb{R}^3 \setminus \{\tilde{\boldsymbol{x}}\}$ the well-known elastodynamic *fundamental solution.* The fundamental displacement $U_i^k(\tilde{\boldsymbol{x}}, \boldsymbol{x})$, stress tensor $\Sigma_{ij}^k(\tilde{\boldsymbol{x}}, \boldsymbol{x})$ and traction vector $T_i^k(\tilde{\boldsymbol{x}}, \boldsymbol{x})$ are given by:

$$U_i^k(\tilde{\boldsymbol{x}}, \boldsymbol{x}) = 2(1-\nu)[F_{,aa} + k_L^2 F]\delta_{ik} - F_{,ik} \tag{2}$$

$$\Sigma_{ij}^k(\tilde{\boldsymbol{x}}, \boldsymbol{x}) = \mu\Big[\frac{2\nu}{1-2\nu}\delta_{ij}U_{a,a}^k + U_{i,j}^k + U_{j,i}^k\Big] \tag{3}$$

$$T_i^k(\tilde{\boldsymbol{x}}, \boldsymbol{x}) = \Sigma_{ij}^k n_j \tag{4}$$

in terms of the Somigliana potential [5] F:

$$F(\tilde{\boldsymbol{x}}, \boldsymbol{x}) = \frac{1}{4\pi\mu k_T^2}(e^{ik_L r} - e^{ik_T r})\frac{1}{r} \tag{5}$$

($k_T^2 = \rho\omega^2/\mu$ and $k_L^2 = \kappa k_T^2$, with $\kappa = \frac{1-2\nu}{2(1-\nu)}$: transversal and longitudinal wave numbers; $r = |\boldsymbol{x} - \tilde{\boldsymbol{x}}| = [(\boldsymbol{x} - \tilde{\boldsymbol{x}}).(\boldsymbol{x} - \tilde{\boldsymbol{x}})]^{1/2}$: Euclidian distance between $\boldsymbol{x}, \tilde{\boldsymbol{x}}$). F satisfies the equation:

$$F_{,aabb} = \frac{k_T^2}{4\pi\mu}(\kappa^2 e^{ik_L r} - e^{ik_T r})\frac{1}{r} \tag{6}$$

In this paper, solutions to (1) are assumed to be given by a double-layer integral representation formula:

$$u_k(\tilde{\boldsymbol{x}}) = \int_S T_i^k(\tilde{\boldsymbol{x}}, \boldsymbol{x})\phi_i(\boldsymbol{x})\,\mathrm{d}S_x =: [\boldsymbol{A}\boldsymbol{\phi}](\tilde{\boldsymbol{x}}) \tag{7}$$

where S is a bounded surface, either closed or open (or possibly a set of several such surfaces) and the density $\boldsymbol{\phi}$ depends on the boundary conditions; the case of an open surface is usually associated with scattering of elastodynamic waves by cracks. Representations of the form (7) are often used to formulate boundary integral equations (BIEs) for interior or exterior problems on the domain Ω bounded by S with Neumann boundary data $\bar{\boldsymbol{p}}$ over S. In particular, such problems lead to symmetric Galerkin BIE (SGBIE) formulations through a weighted-residual statement of the Neumann boundary condition:

$$\int_S [\boldsymbol{T}^n \boldsymbol{A}\boldsymbol{\phi}](\tilde{\boldsymbol{x}}).\tilde{\boldsymbol{\phi}}^\star(\tilde{\boldsymbol{x}})\,\mathrm{d}S_{\tilde{x}} = \int_S \bar{\boldsymbol{p}}(\tilde{\boldsymbol{x}}).\tilde{\boldsymbol{\phi}}^\star(\tilde{\boldsymbol{x}})\,\mathrm{d}S_{\tilde{x}} \qquad (\forall \tilde{\boldsymbol{\phi}} \in \mathcal{V}) \tag{8}$$

where the traction vector operator $\boldsymbol{T}^n\boldsymbol{u}$ is defined by $\boldsymbol{T}^n\boldsymbol{u} = \boldsymbol{\sigma}(\boldsymbol{u}).\boldsymbol{n}$. The operation $[\boldsymbol{T}^n\boldsymbol{A}\boldsymbol{\phi}](\tilde{\boldsymbol{x}})$ gives rise to hypersingular kernels involving a r^{-3} singularity. After a well-documented regularization process [3,7] involving two integrations by parts over S, the actual SGBIE formulation, which is the basis for the present development, is:

$$\mathcal{A}(\boldsymbol{\phi}, \tilde{\boldsymbol{\phi}}^\star) = \mathcal{L}(\tilde{\boldsymbol{\phi}}^\star) \qquad (\forall \tilde{\boldsymbol{\phi}} \in \mathcal{V} = [H^{1/2}(S)]^3) \tag{9}$$

where the linear form $\mathcal{L}$ and the symmetric bilinear form $\mathcal{A}$ are given by:

$$\mathcal{L}(\tilde{\boldsymbol{\phi}}^\star) = \frac{1}{2}\int_S \bar{\boldsymbol{p}}(\tilde{\boldsymbol{x}}).\tilde{\boldsymbol{\phi}}(\tilde{\boldsymbol{x}})\,\mathrm{d}S_{\tilde{x}} \tag{10}$$

$$\mathcal{A}(\boldsymbol{\phi}, \tilde{\boldsymbol{\phi}}^\star) = \int_S\int_S B(\boldsymbol{x}, \tilde{\boldsymbol{x}}; \boldsymbol{\phi}, \tilde{\boldsymbol{\phi}}^\star)\,\mathrm{d}S_x\,\mathrm{d}S_{\tilde{x}} \tag{11}$$

$$B(\boldsymbol{x}, \tilde{\boldsymbol{x}}; \boldsymbol{\phi}, \tilde{\boldsymbol{\phi}}^\star) = B_{ikqs}(\tilde{\boldsymbol{x}}, \boldsymbol{x})R_s\phi_k(\boldsymbol{x})R_q\tilde{\phi}(\tilde{\boldsymbol{x}}) + k_T^2 A_{ik}(\tilde{\boldsymbol{x}}, \boldsymbol{x})\phi_k(\boldsymbol{x})\tilde{\phi}(\tilde{\boldsymbol{x}})$$

where the two kernel functions B_{ikqs} and A_{ik} are given by:

$$B_{ikqs}(\tilde{\boldsymbol{x}}, \boldsymbol{x}) = -e_{iep}e_{kgr}\mu^2[4\nu\delta_{pq}\delta_{rs} + 2(1-\nu)(\delta_{pr}\delta_{qs} + \delta_{ps}\delta_{qr})]F_{,eg}$$

$$\begin{aligned} A_{ik}(\tilde{\boldsymbol{x}}, \boldsymbol{x}) = \Big[&[2(1-\nu)(\delta_{ik}\delta_{j\ell} + \delta_{jk}\delta_{i\ell}) + \frac{2\nu}{\kappa}\delta_{ij}\delta_{k\ell}]\frac{F_{,aabb}}{k_T^2} \\ &+ (1-2\nu)(\delta_{ik}F_{,j\ell} + \delta_{j\ell}F_{,ik} + \delta_{jk}F_{,i\ell} + \delta_{i\ell}F_{,jk}) \\ &+ \frac{4\nu^2}{1-2\nu}\delta_{ij}\delta_{k\ell}F_{,aa} + 4\nu(\delta_{ij}F_{,k\ell} + \delta_{k\ell}F_{,ij})\Big]n_j(\tilde{\boldsymbol{x}})n_\ell(\boldsymbol{x}) \end{aligned}$$

and where $R_i f$ denotes the i-th component of the *surface curl* of a scalar function f [7] (e_{abc}: permutation tensor):

$$R_i f = e_{ijk} n_j f_{,k} \tag{12}$$

Besides, if the surface S is defined by a mapping $\Delta \subset \mathbb{R}^2 \to S$, $\boldsymbol{\xi} \to \boldsymbol{x}(\boldsymbol{\xi})$, one has

$$[R_i f](\boldsymbol{x}(\boldsymbol{\xi}))\, \mathrm{d}S_x = [\partial_{\xi_1} f \partial_{\xi_2} x_i - \partial_{\xi_2} f \partial_{\xi_1} x_i]\, \mathrm{d}\boldsymbol{\xi} \tag{13}$$

which shows in particular that R_i is a tangential differential operator. Both B_{ikqs} and A_{ik} are weakly singular in view of (5) and (6) and have symmetry properties which ensure the overall symmetry of $\mathcal{A}(\boldsymbol{\phi}, \tilde{\boldsymbol{\phi}}^\star)$ through:

$$B(\boldsymbol{x}, \tilde{\boldsymbol{x}}; \boldsymbol{\phi}, \tilde{\boldsymbol{\phi}}^\star) = B(\tilde{\boldsymbol{x}}, \boldsymbol{x}; \boldsymbol{\phi}, \tilde{\boldsymbol{\phi}}^\star) = B(\boldsymbol{x}, \tilde{\boldsymbol{x}}; \tilde{\boldsymbol{\phi}}^\star, \boldsymbol{\phi}) \tag{14}$$

3 Geometrical symmetry assumptions

The most important assumption for the present purposes is that the boundary S has either full or partial geometrical symmetry. By this, we mean that there exists a finite group $\mathcal{S} = \{s_1, \dots, s_n\}$ of n isometries of $\mathbb{R}^3$ (n is the *order* of $\mathcal{S}$) and a partition of the boundary S into two disconnected components S^1, S^2 such that S^1 is invariant under $\mathcal{S}$ whereas S^2 is not:

$$(\forall s \in \mathcal{S}) \qquad s(S^1) = S^1$$

One can therefore introduce a *symmetry cell* for S^1, i.e. a subset C of S^1 such that

$$\text{Area}(C) = \text{Area}(S^1)/n \qquad \text{and} \quad S^1 = \bigcup_{s \in \mathcal{S}} s(C)$$

For example, S^1 is the (symmetric) external boundary while S^2 is a (collection of) interior hole(s) or crack(s) of arbitrary shape and location. Full symmetry refers to the case where $S^1 = S$ and $S^2 = \emptyset$, i.e. the whole boundary S (and hence also Ω) is invariant under $\mathcal{S}$. Recall that an isometry of $\mathbb{R}^3$ is a linear application $s: \mathbb{R}^3 \to \mathbb{R}^3$ such that $|s\boldsymbol{x}| = |\boldsymbol{x}|$ ($\forall \boldsymbol{x} \in \mathbb{R}^3$), where $|\boldsymbol{x}| \equiv (\boldsymbol{x}.\boldsymbol{x})^{1/2}$ is the usual Euclidean norm in $\mathbb{R}^3$.

Exploiting (partial) symmetry in the SGBEM formulation (9) essentially consists in transforming integrals over S^1 into integrals over C, so that the matrix operators produced by the discretization process are of smaller size than those corresponding to the original integral equations. Note that no symmetry is assumed regarding the Neumann data $\bar{\boldsymbol{p}}$.

In addition to geometrical symmetry, one must assume that the material properties are also invariant under the symmetry group $\mathcal{S}$. Accordingly, the bilinear form $\mathcal{A}$ is said to have the *equivariance property* if:

$$\mathcal{A}(\boldsymbol{u}_s, \boldsymbol{v}_s) = \mathcal{A}(\boldsymbol{u}, \boldsymbol{v}) \qquad (\forall s \in \mathcal{S},\ \forall \boldsymbol{u}, \boldsymbol{v} \in \mathcal{V}^1) \tag{15}$$

(where $\boldsymbol{u}_s(\boldsymbol{x}) \equiv s\boldsymbol{u}(s^{-1}\boldsymbol{x})$) which is a straightforward adaptation of the definition proposed in [4]. Since S^1 is invariant under $\mathcal{S}$, the changes of variables $(\boldsymbol{x}, \tilde{\boldsymbol{x}}) \rightarrow (s\boldsymbol{x}, s\tilde{\boldsymbol{x}})$ imply:

$$B(\boldsymbol{x}, \tilde{\boldsymbol{x}}; \boldsymbol{u}_s, \boldsymbol{v}_s) = B(s\boldsymbol{x}, s\tilde{\boldsymbol{x}}; s\boldsymbol{u}, s\boldsymbol{v}) \qquad (\forall \boldsymbol{x}, \tilde{\boldsymbol{x}} \in S^1)$$

and (15) thus implies:

$$B(s\boldsymbol{x}, s\tilde{\boldsymbol{x}}; s\boldsymbol{u}, s\boldsymbol{v}) = B(\boldsymbol{x}, \tilde{\boldsymbol{x}}; \boldsymbol{u}, \boldsymbol{v}) \qquad (\forall s \in \mathcal{S}, \forall \boldsymbol{x}, \tilde{\boldsymbol{x}} \in S^1) \tag{16}$$

In fact, it is easy to check that:

$$B_{ikqs}(s\tilde{\boldsymbol{x}}, s\boldsymbol{x}) = s_{ij}s_{k\ell}s_{qr}s_{st}B_{j\ell rt}(\tilde{\boldsymbol{x}}, \boldsymbol{x})$$
$$A_{ik}(s\tilde{\boldsymbol{x}}, s\boldsymbol{x}) = s_{ij}s_{k\ell}A_{j\ell}(\tilde{\boldsymbol{x}}, \boldsymbol{x})$$
$$\{R_i[f_s]_k \,\mathrm{d}S\}(s\boldsymbol{x}) = s_{ij}s_{k\ell}\{R_j[f_\ell]_b \,\mathrm{d}S\}(\boldsymbol{x})$$

from (26_2) and the identities:

$$\boldsymbol{r}(s\tilde{\boldsymbol{x}}, s\boldsymbol{x}) = s\boldsymbol{r}(\tilde{\boldsymbol{x}}, \boldsymbol{x}) \qquad r(s\tilde{\boldsymbol{x}}, s\boldsymbol{x}) = r(\tilde{\boldsymbol{x}}, \boldsymbol{x})$$

where $\boldsymbol{r}(\tilde{\boldsymbol{x}}, \boldsymbol{x}) \equiv \boldsymbol{x} - \tilde{\boldsymbol{x}}$ and $r(\tilde{\boldsymbol{x}}, \boldsymbol{x}) = |\tilde{\boldsymbol{x}}, \boldsymbol{x}|$; then (16) follows easily.

An immediate and useful consequence of property (16) is:

$$B(s\boldsymbol{x}, \tilde{s}\tilde{\boldsymbol{x}}; s\boldsymbol{u}, \tilde{s}\boldsymbol{v}) = B(t\boldsymbol{x}, \tilde{\boldsymbol{x}}; t\boldsymbol{u}, \boldsymbol{v}) \qquad (\forall s, \tilde{s} \in \mathcal{S}) \quad \text{with } t = s\tilde{s}^{-1} \tag{17}$$

4 Using geometrical symmetry: the Abelian case

The present approach is based on the exploitation of some basic results from the the theory of linear representations of finite groups. In this respect, Abelian (or commutative, i.e. $\forall s, t \in \mathcal{S}$, $st = ts$) and non-Abelian symmetry groups lead to quite different formulations. In this section, $\mathcal{S}$ is a *commutative* finite group of order n; this includes the common cases of group P_m of symmetries w.r.t. m orthogonal planes (with $n = 2^m$ and $1 \leq m \leq d$) and the group $C_n = \{\mathrm{Id}, r, r^2, \ldots, r^{n-1}\}$ of cyclic symmetry generated by a rotation r of angle $2\pi/n$, with $2 \leq n$). The non-Abelian case is deferred to section 5.

Review of basic definitions [4,8,9]. Any finite Abelian group $\mathcal{S}$ of order n possess n *irreducible linear representations*, i.e. n applications ρ_ν: $\mathcal{S} \rightarrow \mathbb{C}$ which satisfy the following relations:

$$|\rho_\nu(s)| = 1 \qquad \rho_\nu(st) = \rho_\nu(s)\rho_\nu(t) \qquad \rho_\nu(s^{-1}) = \rho_\nu^\star(s) \tag{18}$$

for any $s, t \in \mathcal{S}$, $1 \leq \nu \leq n$ ($z^\star$ denotes the complex conjugate of z), as well as the 'orthogonality relation':

$$\frac{1}{n}\sum_{s \in \mathcal{S}} \rho_\nu(s)\rho_\mu^\star(s) = \delta_{\mu\nu} \tag{19}$$

The ρ_ν are known for all usual groups; they are shown in table 1 for the groups P_1, P_2, P_3 while for the group C_n, one has

$$\rho_\nu(r) = \exp(2i\pi\nu/n) \qquad (\nu = 0, \ldots, n-1) \tag{20}$$

The ρ_ν: $\mathcal{S} \to \mathbb{C}$ can be view as a group isomorphism between $\mathcal{S}$ and GL($\mathbb{C}$), the multiplicative group of linear endomorphisms of $\mathbb{C}$, and are said to be *of degree one.* In contrast, when $\mathcal{S}$ is not commutative, some of the irreducible linear representations are necessarily of degree ≥ 2.

Any vector function $\boldsymbol{v}$ defined on S^1 then admits the decomposition [4]:

$$\boldsymbol{v} = \sum_{\nu=1}^{n} \boldsymbol{P}_\nu \boldsymbol{v} \tag{21}$$

where the linear operators $\boldsymbol{P}_\nu$ defined by:

$$\boldsymbol{v} \to [\boldsymbol{P}_\nu \boldsymbol{v}](\boldsymbol{x}) = \frac{1}{n} \sum_{t \in \mathcal{S}} \rho_\nu(t) t^{-1} \boldsymbol{v}(t\boldsymbol{x}) \tag{22}$$

are readily shown using (19) to be orthogonal projectors for the L^2 scalar product: if $\mathcal{V}$ is a space of functions defined on S^1, then one has

$$\int_{S^1} (\boldsymbol{P}_\mu \boldsymbol{v}).(\boldsymbol{P}_\nu \boldsymbol{v})^\star \, \mathrm{d}S = 0 \qquad \mu \neq \nu$$

and

$$\mathcal{V} = \bigoplus_{\nu=1}^{n} \boldsymbol{P}_\nu \mathcal{V}$$

Let $\boldsymbol{v}_\nu$ denote the restriction on C of $\boldsymbol{P}_\nu v$. Then, from the properties (18), it is easy to show that, for any $\boldsymbol{x} \in C$ and any $s \in \mathcal{S}$:

$$[\boldsymbol{P}_\nu \boldsymbol{v}](s\boldsymbol{x}) = \rho_\nu(s^{-1}) s \boldsymbol{v}_\nu(\boldsymbol{x}) \tag{23}$$

P_1	Id	s_1
ρ_1	+1	+1
ρ_2	+1	-1

P_2	Id	s_1	s_2	s_1s_2
ρ_1	+1	+1	+1	+1
ρ_2	+1	-1	+1	-1
ρ_3	+1	+1	-1	-1
ρ_4	+1	-1	-1	+1

P_3	Id	s_1	s_2	s_3	s_2s_3	s_1s_3	s_1s_2	$s_1s_2s_3$
ρ_1	+1	+1	+1	+1	+1	+1	+1	+1
ρ_2	+1	+1	+1	-1	-1	-1	+1	-1
ρ_3	+1	-1	+1	+1	+1	-1	-1	-1
ρ_4	+1	-1	+1	-1	-1	+1	-1	+1
ρ_5	+1	+1	-1	+1	-1	+1	-1	-1
ρ_6	+1	+1	-1	-1	+1	-1	-1	+1
ρ_7	+1	-1	-1	+1	-1	-1	+1	+1
ρ_8	+1	-1	-1	-1	+1	+1	+1	-1

Table 1. Irreducible representations for plane symmetries with respect to one, two and three coordinate planes (respective orders $n = 2, 4, 8$).

Moreover, let $I_s = \{\boldsymbol{x} \in C, s\boldsymbol{x} \in C\}$, i.e. I_s is the set of points of C whose images under a given s are in C (in fact, such points necessarily belong to ∂C for C to be actually a symmetry cell). Identity (23) then implies that the $\boldsymbol{v}_\nu$ obtained from a given $\boldsymbol{v} \in \mathcal{V}$ must satisfy the constraints:

$$\boldsymbol{v}_\nu(s\boldsymbol{x}) - \rho_\nu(s^{-1})s\boldsymbol{v}_\nu(\boldsymbol{x}) = 0 \qquad (\forall s \in \mathcal{S},\ \forall \boldsymbol{x} \in I_s) \tag{24}$$

Let $\mathcal{V}_\nu$ denote the set of functions defined on C for which (24) holds.

Finally, for any $\boldsymbol{x} \in C$ and any $s \in \mathcal{S}$, the value $\boldsymbol{v}(s\boldsymbol{x})$ of $\boldsymbol{v}$ at the image $s\boldsymbol{x}$ of $\boldsymbol{x}$ can be expressed in terms of the $\boldsymbol{v}_\nu$:

$$\boldsymbol{v}(s\boldsymbol{x}) = \sum_{\nu=1}^{n} \rho_\nu(s^{-1})s\boldsymbol{v}_\nu(\boldsymbol{x}) \tag{25}$$

Exploiting partial symmetry. Under the assumption of partial geometrical symmetry, one can map each $s(C)$ onto C by $\boldsymbol{x} \in C \to \boldsymbol{z} = s\boldsymbol{x} \in s(C)$ and express integrals over S^1 as sums of integrals over C, with the help of the identities

$$\mathrm{d}S(s\boldsymbol{x}) = \mathrm{d}S(\boldsymbol{x}) \qquad \boldsymbol{n}(s\boldsymbol{x}) = s[\boldsymbol{n}(\boldsymbol{x})] \tag{26}$$

which stem from the fact that s is an isometry. In particular, the bilinear form $\mathcal{A}(\boldsymbol{\phi}, \tilde{\boldsymbol{\phi}}^\star)$ and linear form $\mathcal{L}(\tilde{\boldsymbol{\phi}}^\star)$ defined by (11) and (10) take the form:

$$\mathcal{A}(\boldsymbol{\phi}, \tilde{\boldsymbol{\phi}}^\star) = \mathcal{B}(\boldsymbol{\phi}^1, \tilde{\boldsymbol{\phi}}^{1\star}) + \mathcal{C}(\boldsymbol{\phi}^1, \tilde{\boldsymbol{\phi}}^{2\star}) + \mathcal{C}^T(\boldsymbol{\phi}^2, \tilde{\boldsymbol{\phi}}^{1\star}) + \mathcal{D}(\boldsymbol{\phi}^2, \tilde{\boldsymbol{\phi}}^{2\star}) \tag{27}$$

$$\mathcal{L}(\tilde{\boldsymbol{\phi}}^\star) = \mathcal{F}(\tilde{\boldsymbol{\phi}}^{1\star}) + \mathcal{G}(\tilde{\boldsymbol{\phi}}^{2\star}) \tag{28}$$

where $\boldsymbol{\phi}^1, \tilde{\boldsymbol{\phi}}^{1\star} \in \mathcal{V} = [H^{1/2}(S^1)]^3$ and $\boldsymbol{\phi}^2, \tilde{\boldsymbol{\phi}}^{2\star} \in \mathcal{W} = [H^{1/2}(S^2)]^3$, and with

$$\mathcal{B}(\boldsymbol{\phi}^1, \tilde{\boldsymbol{\phi}}^{1\star}) = \sum_{s\in\mathcal{S}} \sum_{\tilde{s}\in\mathcal{S}} \int_C \int_C B(s\boldsymbol{x}, \tilde{s}\tilde{\boldsymbol{x}}; \boldsymbol{\phi}^1 \circ s, \tilde{\boldsymbol{\phi}}^{1\star} \circ \tilde{s})\, \mathrm{d}S_x\, \mathrm{d}S_{\tilde{x}} \tag{29}$$

$$\mathcal{C}(\boldsymbol{\phi}^1, \tilde{\boldsymbol{\phi}}^{2\star}) = \sum_{s\in\mathcal{S}} \int_{S^2} \int_C B(s\boldsymbol{x}, \tilde{\boldsymbol{x}}; \boldsymbol{\phi}^1 \circ s, \tilde{\boldsymbol{\phi}}^{2\star})\, \mathrm{d}S_x\, \mathrm{d}S_{\tilde{x}} \tag{30}$$

$$\mathcal{D}(\boldsymbol{\phi}^2, \tilde{\boldsymbol{\phi}}^{2\star}) = \sum_{s\in\mathcal{S}} \int_{S^2} \int_{S^2} B(\boldsymbol{x}, \tilde{\boldsymbol{x}}; \boldsymbol{\phi}^2, \tilde{\boldsymbol{\phi}}^{2\star})\, \mathrm{d}S_x\, \mathrm{d}S_{\tilde{x}} \tag{31}$$

and

$$\mathcal{F}(\tilde{\boldsymbol{\phi}}^{1\star}) = \frac{1}{2} \sum_{\tilde{s}\in\mathcal{S}} \int_C \bar{\boldsymbol{p}}(\tilde{s}\tilde{\boldsymbol{x}}).\tilde{\boldsymbol{\phi}}^1(\tilde{s}\tilde{\boldsymbol{x}})\, \mathrm{d}S_{\tilde{x}} \qquad \mathcal{G}(\tilde{\boldsymbol{\phi}}^{2\star}) = \frac{1}{2} \int_{S^2} \bar{\boldsymbol{p}}(\tilde{\boldsymbol{x}}).\tilde{\boldsymbol{\phi}}^2(\tilde{\boldsymbol{x}})\, \mathrm{d}S_{\tilde{x}}$$

Now, inserting the decomposition (25) for both $\boldsymbol{\phi}^1$ and $\tilde{\boldsymbol{\phi}}^1$ in $\mathcal{B}(\boldsymbol{\phi}^1, \tilde{\boldsymbol{\phi}}^{1\star})$ defined by (29), one has:

$$\mathcal{B}(\boldsymbol{\phi}^1, \tilde{\boldsymbol{\phi}}^{1\star}) = \sum_{\mu=1}^{n} \sum_{\nu=1}^{n} \sum_{s\in\mathcal{S}} \sum_{\tilde{s}\in\mathcal{S}} \rho^\star_\nu(\tilde{s}^{-1})\rho_\mu(s^{-1})$$

$$\int_C \int_C B(s\boldsymbol{x}, \tilde{s}\tilde{\boldsymbol{x}}; s\boldsymbol{\phi}_\nu, \tilde{s}\tilde{\boldsymbol{\phi}}^\star_\mu)\, \mathrm{d}S_x\, \mathrm{d}S_{\tilde{x}}$$

since $B(\boldsymbol{x},\tilde{\boldsymbol{x}};\boldsymbol{\phi}^1,\tilde{\boldsymbol{\phi}}^{1\star})$ is bilinear in $\boldsymbol{\phi}^1$ and $\tilde{\boldsymbol{\phi}}^{1\star}$. Next, using the change of variable $t=\tilde{s}^{-1}s$ (i.e. $s=\tilde{s}t$) together with property (18), one gets:

$$\mathcal{B}(\boldsymbol{\phi}^1,\tilde{\boldsymbol{\phi}}^{1\star})=\sum_{\mu=1}^{n}\sum_{\nu=1}^{n}\sum_{t\in\mathcal{S}}\sum_{\tilde{s}\in\mathcal{S}}\rho_\nu^\star(\tilde{s}^{-1})\rho_\mu(t^{-1})\rho_\mu(\tilde{s}^{-1}) \int_C\int_C B(\tilde{s}t\boldsymbol{x},\tilde{s}\tilde{\boldsymbol{x}};\tilde{s}t\boldsymbol{\phi}_\nu,\tilde{s}\tilde{\boldsymbol{\phi}}_\mu^\star)\,\mathrm{d}S_x\,\mathrm{d}S_{\tilde{x}}$$

The equivariance property (17) implies that:

$$\int_C\int_C B(\tilde{s}t\boldsymbol{x},\tilde{s}\tilde{\boldsymbol{x}};s\boldsymbol{\phi}_\nu,\tilde{s}\tilde{\boldsymbol{\phi}}_\mu^\star)\,\mathrm{d}S_x\,\mathrm{d}S_{\tilde{x}}=\mathcal{B}_t(\boldsymbol{\phi}_\nu,\tilde{\boldsymbol{\phi}}_\mu^\star) \tag{32}$$

having put

$$\mathcal{B}_t(\boldsymbol{u},\boldsymbol{v})=\int_C\int_C B(t\boldsymbol{x},\tilde{\boldsymbol{x}};t\boldsymbol{u},\boldsymbol{v})\,\mathrm{d}S_x\,\mathrm{d}S_{\tilde{x}} \tag{33}$$

Then, by virtue of the orthogonality property (19):

$$\begin{aligned}\mathcal{B}(\boldsymbol{\phi}^1,\tilde{\boldsymbol{\phi}}^{1\star})&=\sum_{\mu=1}^{n}\sum_{\nu=1}^{n}\sum_{t\in\mathcal{S}}\rho_\mu(t^{-1})\Big\{\sum_{\tilde{s}\in\mathcal{S}}\rho_\nu(\tilde{s})\rho_\mu(\tilde{s}^{-1})\Big\}\mathcal{B}_t(\boldsymbol{\phi}_\nu,\tilde{\boldsymbol{\phi}}_\mu^\star)\\&=\sum_{\nu=1}^{n}\Big\{n\sum_{t\in\mathcal{S}}\rho_\nu(t^{-1})\mathcal{B}_t(\boldsymbol{\phi}_\nu,\tilde{\boldsymbol{\phi}}_\nu^\star)\Big\}\\&\equiv\sum_{\nu=1}^{n}\mathcal{B}_\nu(\boldsymbol{\phi}_\nu,\tilde{\boldsymbol{\phi}}_\nu^\star)\end{aligned} \tag{34}$$

The bilinear form $\mathcal{B}(\boldsymbol{\phi}^1,\tilde{\boldsymbol{\phi}}^{1\star})$ is thus seen to have been reduced in *block-diagonal* form.

One establishes in a similar way the decompositions:

$$\begin{aligned}\mathcal{C}(\boldsymbol{\phi}^2,\tilde{\boldsymbol{\phi}}^{1\star})&=\sum_{\nu=1}^{n}\left\{\sum_{\tilde{s}\in\mathcal{S}}\rho_\nu(\tilde{s})\int_C\int_{S^2}B(\boldsymbol{x},\tilde{s}\tilde{\boldsymbol{x}};\boldsymbol{\phi}^2,\tilde{s}\tilde{\boldsymbol{\phi}}^\star)\,\mathrm{d}S_x\,\mathrm{d}S_{\tilde{x}}\right\}\\&\equiv\sum_{\nu=1}^{n}\mathcal{C}_\nu(\boldsymbol{\phi}^2,\tilde{\boldsymbol{\phi}}_\nu^\star)\end{aligned} \tag{35}$$

$$\mathcal{F}(\tilde{\boldsymbol{\phi}}^{1\star})=\sum_{\nu=1}^{n}\frac{1}{2}\int_C[\boldsymbol{P}_\nu\bar{\boldsymbol{p}}].\tilde{\boldsymbol{\phi}}_\nu^\star\,\mathrm{d}S\equiv\sum_{\nu=1}^{n}\mathcal{F}_\nu(\tilde{\boldsymbol{\phi}}_\nu^\star) \tag{36}$$

Gathering results (34) (35) and (36), the initial integral equation (9) reduces to a set of SGBIE problems of the form:

Find $\boldsymbol{\phi}_\nu\in\mathcal{V}_\nu$, $\boldsymbol{\phi}^2\in\mathcal{W}$; $\forall\tilde{\boldsymbol{\phi}}_\nu\in\mathcal{V}_\nu,\tilde{\boldsymbol{\phi}}^2\in\mathcal{W}$

$$\left\{\begin{aligned}&\mathcal{B}_\nu(\boldsymbol{\phi}_\nu,\tilde{\boldsymbol{\phi}}_\nu^\star)+\mathcal{C}_\nu(\boldsymbol{\phi}^2,\tilde{\boldsymbol{\phi}}_\nu^\star)=\mathcal{F}_\nu(\tilde{\boldsymbol{\phi}}_\nu^\star)\qquad(1\le\nu\le n)\\&\sum_{\nu=1}^{n}\mathcal{C}_\nu^T(\boldsymbol{\phi}_\nu,\tilde{\boldsymbol{\phi}}^{2\star})+\mathcal{D}(\boldsymbol{\phi}^2,\tilde{\boldsymbol{\phi}}^{2\star})=\mathcal{G}(\tilde{\boldsymbol{\phi}}^{2\star})\end{aligned}\right. \tag{37}$$

5 Using geometrical symmetry: the non-Abelian case

In this section, $\mathcal{S}$ is a non-Abelian finite group of order n, i.e. there exist $s, t \in \mathcal{S}$ such that $st \neq ts$. This includes the important practical case of the *dihedral symmetry group* D_m, i.e. the group of order $n = 2m$ of the affine transformations that leave a regular m-gon unchanged.

Review of basic results [4,8,9]. Here, the irreducible representations ρ_ν of $\mathcal{S}$ are of integer degree $d_\nu \geq 1$:

$$\rho_\nu \; : \; s \in \mathcal{S} \to \rho_\nu(s) \in \mathrm{GL}(\mathbb{C}^{d_\nu})$$

i.e. each $\rho_\nu(s)$ is a linear endomorphism of a d_ν-dimensional complex vector space; moreover, the number $R(\mathcal{S})$ of such representations and their degrees d_ν are such that at least one of them is ≥ 2 and:

$$\sum_{\nu=1}^{R(\mathcal{S})} d_\nu^2 = n$$

The properties of the irreducible representations ρ_ν include the preservation of group structure:

$$\rho_\nu^{ij}(st) = \sum_{k=1}^{d_\nu} \rho_\nu^{ik}(s)\rho_\nu^{kj}(t) \quad (\forall s, t \in \mathcal{S}) \qquad \rho_\nu^{ij}(1) = \delta_{ij} \tag{38}$$

which implies in particular, since ρ_ν is unitary, that:

$$\rho_\nu^{ij}(s^{-1}) = [\rho_\nu^{ji}(s)]^\star \tag{39}$$

and the 'orthogonality relation':

$$\frac{d_\nu}{n} \sum_{s \in \mathcal{S}} \rho_\nu^{ij}(s)[\rho_\mu^{k\ell}(s)]^\star = \delta_{ik}\delta_{j\ell}\delta_{\mu\nu} \tag{40}$$

	Id	r	r^2	s	sr	sr^2
ρ_1	+1	+1	+1	+1	+1	+1
ρ_2	+1	+1	+1	-1	-1	-1
ρ_3^{11}	1	j	j^2	0	0	0
ρ_3^{21}	0	0	0	1	j	j^2
ρ_3^{12}	0	0	0	1	j^2	j
ρ_3^{22}	1	j^2	j	0	0	0

(with $j = \exp(2i\pi/3)$)

Table 2. Irreducible representations for dihedral symmetry $\mathcal{S} = D_3$.

A space $\mathcal{V}$ of vector functions defined on S_1 is decomposed into orthogonal subspaces [4]:

$$\mathcal{V} = \bigoplus_{j,\nu} \boldsymbol{P}_{\nu}^{jj} \mathcal{V} \tag{41}$$

with the projectors $\boldsymbol{P}_{\nu}^{ij}$ defined by:

$$[\boldsymbol{P}_{\nu}^{ij} \boldsymbol{v}](\boldsymbol{x}) = \frac{d_\nu}{n} \sum_{t \in \mathcal{S}} \rho_{\nu}^{ji}(t) t^{-1} \boldsymbol{v}(t\boldsymbol{x})$$

From this definition and the properties of the representations, one has for any $s \in \mathcal{S}$ and $\boldsymbol{v} \in \mathcal{V}$:

$$[\boldsymbol{P}_{\nu}^{ij} \boldsymbol{u}](s\boldsymbol{x}) = \sum_{k=1}^{d_\nu} \rho_{\nu}^{ki}(s^{-1}) s [\boldsymbol{P}_{\nu}^{kj} \boldsymbol{u}](\boldsymbol{x}) \tag{42}$$

Hence, for a given $s \in \mathcal{S}$ and a point $\boldsymbol{x} \in C$, the value $\boldsymbol{v}(s\boldsymbol{x})$ of $\boldsymbol{v}$ at the images of $\boldsymbol{x}$ can be expressed by virtue of (41) and (42) in terms of the restriction $\boldsymbol{v}_{\nu}^{kj}$ on C of the projections $\boldsymbol{P}_{\nu}^{kj} \boldsymbol{v}$:

$$\boldsymbol{v}(s\boldsymbol{x}) = \sum_{\mu=1}^{R(\mathcal{S})} \sum_{j,k=1}^{d_\mu} \rho_{\mu}^{kj}(s^{-1}) s \boldsymbol{v}_{\mu}^{kj}(\boldsymbol{x}) \tag{43}$$

Moreover, let again $I_s = \{\boldsymbol{x} \in \partial C, s\boldsymbol{x} \in \partial C\}$. It is then easy to show, from (42), that the d_ν-uple $\{\boldsymbol{v}_{\nu}^{ij}, 1 \leq i \leq d_\nu\}$ of functions defined on C are subject to the following constraints:

$$\boldsymbol{v}_{\nu}^{ij}(\boldsymbol{x}) - \sum_{k=1}^{d_\nu} \rho_{\nu}^{ki}(s^{-1}) s \boldsymbol{v}_{\nu}^{kj}(s^{-1}\boldsymbol{x}) = 0 \qquad (\forall \boldsymbol{x} \in I_s) \tag{44}$$

(note that the constraint does not depend on the rightmost index j). Accordingly, for the non-Abelian case, let $\mathcal{V}_\nu$ denote the set of d_ν-tuples of functions $\boldsymbol{v}^\ell$ $(1 \leq \ell \leq d_\nu)$ defined on C and such that any pair $(\boldsymbol{v}^i, \boldsymbol{v}^k)$ is linked through the constraints (44) (with the index j omitted).

Exploiting partial symmetry. Again, the decomposition (27) holds. Inserting the decomposition (43) for both $\boldsymbol{\phi}^1$ and $\tilde{\boldsymbol{\phi}}^1$ in $\mathcal{B}(\boldsymbol{\phi}^1, \tilde{\boldsymbol{\phi}}^{1\star})$ defined by (27), one obtains:

$$\mathcal{B}(\boldsymbol{\phi}, \tilde{\boldsymbol{\phi}}^\star) = \sum_{\mu=1}^{R(\mathcal{S})} \sum_{\nu=1}^{R(\mathcal{S})} \sum_{j,k=1}^{d_\mu} \sum_{i,\ell=1}^{d_\nu} \sum_{s \in \mathcal{S}} \sum_{\tilde{s} \in \mathcal{S}} \rho_{\nu}^{\ell i \star}(\tilde{s}^{-1}) \rho_{\mu}^{kj}(s^{-1}) \\ \int_C \int_C B(s\boldsymbol{x}, \tilde{s}\tilde{\boldsymbol{x}}; s\boldsymbol{\phi}_{\nu}^{kj}, \tilde{s}\tilde{\boldsymbol{\phi}}_{\mu}^{\star \ell i}) \, \mathrm{d}S_x \, \mathrm{d}S_{\tilde{x}}$$

Then, making the change of variable $s = \tilde{s}t$ and using (38), (39):

$$\mathcal{B}(\phi, \tilde{\phi}^\star) = \sum_{\mu=1}^{R(\mathcal{S})} \sum_{\nu=1}^{R(\mathcal{S})} \sum_{j,k=1}^{d_\mu} \sum_{\ell,i=1}^{d_\nu} \sum_{m=1}^{d_\nu} \sum_{t\in\mathcal{S}} \sum_{\tilde{s}\in\mathcal{S}} \rho_\nu^{i\ell}(\tilde{s}) \rho_\mu^{km}(t^{-1}) \rho_\mu^{mj}(\tilde{s}^{-1}) \int_C \int_C B(\tilde{s}t\boldsymbol{x}, \tilde{s}\tilde{\boldsymbol{x}}; s\phi_\nu^{kj}, \tilde{s}\tilde{\phi}_\mu^{\star\ell i}) \, \mathrm{d}S_x \, \mathrm{d}S_{\tilde{x}}$$

so that, using (40) and the equivariance property (32), one obtains:

$$\begin{aligned}
\mathcal{B}(\phi, \tilde{\phi}^\star) &= \sum_{\mu=1}^{R(\mathcal{S})} \sum_{\nu=1}^{R(\mathcal{S})} \sum_{j,k=1}^{d_\mu} \sum_{\ell,i=1}^{d_\nu} \sum_{m=1}^{d_\nu} \sum_{t\in\mathcal{S}} \Big\{ \sum_{\tilde{s}\in\mathcal{S}} \rho_\nu^{i\ell}(\tilde{s}) \rho_\mu^{mj}(\tilde{s}^{-1}) \Big\} \rho_\mu^{km}(t^{-1}) \mathcal{B}_t(\phi_\nu^{kj}, \tilde{\phi}_\mu^{\star\ell i}) \\
&= \sum_{\mu=1}^{R(\mathcal{S})} \sum_{\nu=1}^{R(\mathcal{S})} \sum_{j,k=1}^{d_\mu} \sum_{\ell,i=1}^{d_\nu} \sum_{m=1}^{d_\nu} \sum_{t\in\mathcal{S}} \frac{n}{d_\nu} \rho_\mu^{km}(t^{-1}) \delta_{ij} \delta_{\ell m} \delta_{\mu\nu} \mathcal{B}_t(\phi_\nu^{kj}, \tilde{\phi}_\mu^{\star\ell i}) \\
&= \sum_{\nu=1}^{R(\mathcal{S})} \sum_{i=1}^{d_\mu} \sum_{k,\ell=1}^{d_\mu} \Big\{ \sum_{t\in\mathcal{S}} \frac{n}{d_\nu} \rho_\nu^{k\ell}(t^{-1}) \Big\} \mathcal{B}_t(\phi_\nu^{kj}, \tilde{\phi}_\nu^{\star\ell i}) \\
&\equiv \sum_{\nu=1}^{R(\mathcal{S})} \sum_{k,\ell=1}^{d_\nu} \sum_{i=1}^{d_\nu} \mathcal{B}_\nu^{k\ell}(\phi_\nu^{ki}, \tilde{\phi}_\nu^{\star\ell i})
\end{aligned} \tag{45}$$

One establishes in a similar way the decompositions:

$$\begin{aligned}
\mathcal{C}(\phi^2, \tilde{\phi}^{1\star}) &= \sum_{\nu=1}^{R(\mathcal{S})} \Big\{ \sum_{\ell,i=1}^{d_\nu} \sum_{\tilde{s}\in\mathcal{S}} \rho_\nu^{i\ell}(\tilde{s}) \int_C \int_{S^2} B(\boldsymbol{x}, \tilde{s}\tilde{\boldsymbol{x}}; \phi^2, \tilde{s}\tilde{\phi}_\nu^{\star\ell i}) \, \mathrm{d}S_x \, \mathrm{d}S_{\tilde{x}} \Big\} \\
&\equiv \sum_{\nu=1}^{R(\mathcal{S})} \sum_{\ell,i=1}^{d_\nu} \mathcal{C}_\nu^{\ell i}(\phi^2, \tilde{\phi}_\nu^{\star\ell i})
\end{aligned} \tag{46}$$

and

$$\begin{aligned}
\mathcal{F}(\tilde{\phi}^{1\star}) &= \frac{1}{2} \sum_{\nu=1}^{R(\mathcal{S})} \sum_{t\in\mathcal{S}} \sum_{i,\ell=1}^{d_\nu} \rho_\nu^{i\ell}(t) \int_C \bar{\mathbf{p}}(t\tilde{\boldsymbol{x}}).t\tilde{\phi}_\nu^{\star\ell i}(\tilde{\boldsymbol{x}}) \, \mathrm{d}S_{\tilde{x}} \\
&= \frac{1}{2} \sum_{\nu=1}^{R(\mathcal{S})} \frac{n}{d_\nu} \sum_{i,\ell=1}^{d_\nu} \int_C [\boldsymbol{P}_\nu^{\ell i} \bar{\mathbf{p}}](\tilde{\boldsymbol{x}}).\tilde{\phi}_\nu^{\star\ell i}(\tilde{\boldsymbol{x}}) \, \mathrm{d}S_{\tilde{x}} \\
&= \sum_{\nu} \sum_{i,\ell=1}^{d_\nu} \mathcal{F}_\nu^{\ell i}(\tilde{\phi}_\nu^{\star\ell i})
\end{aligned} \tag{47}$$

Gathering results (45), (46) and (47), the initial integral equation (9) reduces to a set of SGBIE problems of the form:

$$\text{Find } \{\phi_\nu^{ki}\}_{1\le k\le d_\nu} \in \mathcal{V}_\nu,\ \phi^2 \in \mathcal{W};\ \forall \{\tilde{\phi}_\nu^{\ell}\}_{1\le \ell\le d_\nu} \in \mathcal{V}_\nu, \tilde{\phi}^2 \in \mathcal{W}$$

$$\left\{\begin{array}{l} \displaystyle\sum_{k,\ell=1}^{d_\nu} \mathcal{B}_\nu^{k\ell}(\phi_\nu^{ki}, \tilde{\phi}_\nu^{\star\ell}) + \sum_{\ell=1}^{d_\nu} \mathcal{C}_\nu^{\ell i}(\phi^2, \tilde{\phi}_\nu^{\star\ell}) = \sum_{\ell=1}^{d_\nu} \mathcal{F}_\nu^{\ell i}(\tilde{\phi}_\nu^{\star\ell}) \\ \hfill (1 \le \nu \le R(\mathcal{S}), 1 \le i \le d_\nu) \\ \displaystyle\sum_{\nu=1}^{R(\mathcal{S})} \sum_{\ell,i=1}^{d_\nu} [\mathcal{C}_\nu^{\ell i}]^T(\phi_\nu^{\ell i}, \tilde{\phi}^{2\star}) + \mathcal{D}(\phi^2, \tilde{\phi}^{2\star}) = \mathcal{G}(\tilde{\phi}^2) \end{array}\right. \tag{48}$$

6 Calculation of field values at interior points

Displacement values at selected interior points $\tilde{\boldsymbol{x}}$ can be computed explicitly using the representation formula (7) once the density $\boldsymbol{\phi}$ is known, and related quantities (strains, stresses) at $\tilde{\boldsymbol{x}}$ can be easily obtained as well.

Let $\boldsymbol{u} = \boldsymbol{u}^1 + \boldsymbol{u}^2$ in (7), where $\boldsymbol{u}^I$ is the contribution of the integration over S^I. Exploiting symmetry affects the computation of $\boldsymbol{u}^1$. Inserting the decomposition (43) into (7) and following the now usual pattern, one obtains:

$$u_k^1(\tilde{s}\tilde{\boldsymbol{x}}) = \sum_{s\in\mathcal{S}} \sum_{\nu=1}^{R(\mathcal{S})} \sum_{a,b=1}^{d_\nu} \rho_\nu^{ab}(s^{-1}) \int_C T_i^k(\tilde{s}\tilde{\boldsymbol{x}}, s\boldsymbol{x}) s_{ij} [\phi_\nu^{ab}]_j(\boldsymbol{x})\, \mathrm{d}S_x$$

Then, putting again $s = \tilde{s}t$ and using the equivariance property (16), which holds also for the kernel T_i^k, one obtains:

$$u_k^1(\tilde{s}\tilde{\boldsymbol{x}}) = \sum_{s\in\mathcal{S}} \sum_{\nu=1}^{R(\mathcal{S})} \sum_{a,b,c=1}^{d_\nu} \rho_\nu^{cb}(\tilde{s}^{-1}) \left\{ \rho_\nu^{ac}(t^{-1}) \int_C s_{k\ell} T_i^\ell(t^{-1}\tilde{\boldsymbol{x}}, \boldsymbol{x}) [\phi_\nu^{ab}]_i(\boldsymbol{x})\, \mathrm{d}S_x \right\} \tag{49}$$

(note that $s^t s = s s^t = Id$). A close examination of (49) thus reveals that, for a given interior point $\boldsymbol{x}$, the same numerical quadrature effort is required by (7) and (49). However, the terms within curly brackets in (49) do not depend on $\tilde{s}$, so that the same numerical integrations can be reused (with different weights $\rho_\nu^{cb}(\tilde{s}^{-1})$) to evaluate $\boldsymbol{u}^1$ at all the n images of $\tilde{\boldsymbol{x}}$ under $\mathcal{S}$.

7 Computational implications

7.1 Reduction of numerical quadrature effort

It is obvious from (56) that a reduction of both setup and solution computational efforts results from the block-diagonalization of the operator $\mathcal{B}$. The numerical quadrature effort consists in evaluating discretized versions of

$$\mathcal{B}_t(\boldsymbol{u},\boldsymbol{v}) \equiv \int_C \int_C B(t\boldsymbol{x},\tilde{\boldsymbol{x}};t\boldsymbol{u},\boldsymbol{v})\,\mathrm{d}S_x\,\mathrm{d}S_{\tilde{x}}$$

for all $t \in \mathcal{S}$ instead of

$$\mathcal{B}(\boldsymbol{u},\boldsymbol{v}) \equiv \int_{S^1} \int_{S^1} B(\boldsymbol{x},\tilde{\boldsymbol{x}};\boldsymbol{u},\boldsymbol{v})\,\mathrm{d}S_x\,\mathrm{d}S_{\tilde{x}}$$

Moreover, a useful consequence of equivariance (16) and the symmetry properties (14) of B is:

$$\mathcal{B}_t(\boldsymbol{u},\boldsymbol{v}) = \mathcal{B}_{t^{-1}}(\boldsymbol{v},\boldsymbol{u}) \tag{50}$$

From this identity and the symmetry of the original bilinear form $\mathcal{B}$, the block-diagonalized $\mathcal{B}$ is seen to entail a numerical quadrature effort n times smaller than the original $\mathcal{B}$.

7.2 Symmetry properties of the matrix equations

Abelian case. The irreducible representations ρ_ν are usually complex-valued functions over $\mathcal{S}$ (e.g. (20) for cyclic groups). In that case, it can be shown that the ρ_ν can be associated by conjugate pairs, i.e. that for any ν such that ρ_ν is complex-valued, there exists $\nu^\star$ such that $\rho_{\nu^\star}(\mathcal{S}) = \rho^\star_\nu(\mathcal{S})$. In that case, from (24), $\boldsymbol{v} \in \mathcal{V}_\nu \Rightarrow \boldsymbol{v}^\star \in \mathcal{V}_{\nu^\star}$. Besides, using (50), one can show that:

$$\mathcal{B}_\nu(\boldsymbol{u},\boldsymbol{v}) = [\mathcal{B}_{\nu^\star}]^T(\boldsymbol{u},\boldsymbol{v}) \qquad \text{complex-valued } \rho_\nu \tag{51}$$

i.e. that, although $\mathcal{B}$ is symmetric, the $\mathcal{B}_\nu(\boldsymbol{u},\boldsymbol{v})$ are not individually symmetric, but have a 'reciprocal symmetry'. In some cases, including the very common one of symmetry with respect to coordinate planes, the ρ_ν are real-valued (see table 1); then, $\boldsymbol{v} \in \mathcal{V}_\nu \Rightarrow \boldsymbol{v}^\star \in \mathcal{V}_\nu$ and the $\mathcal{B}_\nu$ are symmetric:

$$\mathcal{B}_\nu(\boldsymbol{u},\boldsymbol{v}) = [\mathcal{B}_\nu]^T(\boldsymbol{u},\boldsymbol{v}) \qquad \text{real-valued } \rho_\nu \tag{52}$$

Non-Abelian case. The symmetry properties of the matrices associated with degree one representations are as in the Abelian case. Otherwise, one has from (45):

$$\mathcal{B}^{k\ell}_\nu(\boldsymbol{u},\boldsymbol{v}) = \frac{n}{d_\nu}\sum_{t\in\mathcal{S}'} \rho^{k\ell}_\nu(t^{-1})\mathcal{B}_t(\boldsymbol{u},\boldsymbol{v}) + \frac{n}{d_\nu}\sum_{t\in\mathcal{S}''}\left\{\rho^{k\ell}_\nu(t)\mathcal{B}_t(\boldsymbol{v},\boldsymbol{u}) + \rho^{\star\ell k}_\nu(t)\mathcal{B}_t(\boldsymbol{u},\boldsymbol{v})\right\} \tag{53}$$

where $\mathcal{S}' = \{t \in \mathcal{S}, t = t^{-1}\}$ and $\mathcal{S}'' \subset \mathcal{S}$ is chosen such that $\mathcal{S}' \cap \mathcal{S}'' = \emptyset$ and $\mathcal{S} = \mathcal{S}' \cup \mathcal{S}'' \cup \{t^{-1}, t \in \mathcal{S}''\}$. First, as a consequence of (50):

$$\mathcal{B}_t(\boldsymbol{u},\boldsymbol{v}) = \mathcal{B}_{t^{-1}}(\boldsymbol{u},\boldsymbol{v}) \qquad (\text{if } t = t^{-1}) \tag{54}$$

Also, whenever the irreducible representations $\rho_\nu^{k\ell}$ and $\rho_\nu^{\ell k}$ are real-valued, one has

$$\rho_\nu^{k\ell}(t)\mathcal{B}_t(\boldsymbol{v},\boldsymbol{u}) + \rho_\nu^{\ell k}(t)\mathcal{B}_t(\boldsymbol{u},\boldsymbol{v}) = [\rho_\nu^{\ell k}(t)\mathcal{B}_t(\boldsymbol{v},\boldsymbol{u}) + \rho_\nu^{k\ell}(t)\mathcal{B}_t(\boldsymbol{u},\boldsymbol{v})]^T$$

Besides, from (44):

$$\{\boldsymbol{v}^\ell\}_{1\le\ell\le d_\nu} \in \mathcal{V}_\nu \Rightarrow \{\boldsymbol{v}^{\ell\star}\}_{1\le\ell\le d_\nu} \in \mathcal{V}_\nu$$

Thus, if all $\rho_\nu^{k\ell}(t)$ are real-valued for a given ν, the bilinear form

$$\sum_{k,\ell=1}^{d_\nu} \mathcal{B}_\nu^{k\ell}(\boldsymbol{u}^k, \boldsymbol{v}^{\star\ell})$$

(where $\{\boldsymbol{u}^k\}_{1\le\ell\le d_\nu} \in \mathcal{V}_\nu$ and $\{\boldsymbol{v}^\ell\}_{1\le\ell\le d_\nu} \in \mathcal{V}_\nu$) is symmetric. On the other hand, if some $\rho_\nu^{k\ell}(t)$ are complex-valued, it is not clear how to establish the symmetry of the above bilinear form from the general properties of the representations.

Besides, it is also important to note that in (48) the same bilinear form $\sum_{k,\ell=1}^{d_\nu} \mathcal{B}_\nu^{k\ell}(\boldsymbol{u}^k, \boldsymbol{v}^{\star\ell})$ appears d_ν times; it should thus be assembled and factored once and then used to solve for all d_ν-uples $\{\boldsymbol{\phi}_\nu^{ki}\}_{1\le k\le d_\nu}$ with $i = 1,\dots,d_\nu$.

Example: the dihedral group $\mathcal{S} = D_3$. Let Σ_r and Σ_s denote two distinct planes in $\mathbb{R}^3$ which intersect along the coordinate line Ox_3 and such that the angle (Σ_r, Σ_s) is $\pi/3$. The dihedral group D_3, which is the simplest non-Abelian one, is generated by the symmetry s w.r.t. Σ_s and the $2\pi/3$ rotation r around Ox_3. Its irreducible representations are shown in Table 2; one has $R(\mathcal{S}) = 3$, $d_1 = d_2 = 1$, $d_3 = 2$.

For the case $\nu = 3$, more explicit expression for the $\mathcal{B}_3^{k\ell}$ are obtained as follows, using Table 2 and (50):

$$\begin{aligned}
\mathcal{B}_3^{11}(\boldsymbol{u}^1,\boldsymbol{v}^{1\star}) &= \mathcal{B}_{Id}(\boldsymbol{u}^1,\boldsymbol{v}^{1\star}) + j\mathcal{B}_r(\boldsymbol{u}^1,\boldsymbol{v}^{1\star}) + j^2\mathcal{B}_r(\boldsymbol{v}^{1\star},\boldsymbol{u}^1)\\
\mathcal{B}_3^{21}(\boldsymbol{u}^2,\boldsymbol{v}^{1\star}) &= \mathcal{B}_s(\boldsymbol{u}^2,\boldsymbol{v}^{1\star}) + j\mathcal{B}_{sr}(\boldsymbol{u}^2,\boldsymbol{v}^{1\star}) + j^2\mathcal{B}_{sr^2}(\boldsymbol{u}^2,\boldsymbol{v}^{1\star})\\
\mathcal{B}_3^{12}(\boldsymbol{u}^1,\boldsymbol{v}^{2\star}) &= \mathcal{B}_s(\boldsymbol{u}^1,\boldsymbol{v}^{2\star}) + j^2\mathcal{B}_{sr}(\boldsymbol{u}^1,\boldsymbol{v}^{2\star}) + j\mathcal{B}_{sr^2}(\boldsymbol{u}^1,\boldsymbol{v}^{2\star})\\
\mathcal{B}_3^{22}(\boldsymbol{u}^2,\boldsymbol{v}^{2\star}) &= \mathcal{B}_{Id}(\boldsymbol{u}^2,\boldsymbol{v}^{2\star}) + j^2\mathcal{B}_r(\boldsymbol{u}^2,\boldsymbol{v}^{2\star}) + j\mathcal{B}_r(\boldsymbol{v}^{2\star},\boldsymbol{u}^2)
\end{aligned}$$

It appears that $\mathcal{B}_3^{11}(\boldsymbol{u},\boldsymbol{v}) = .\mathcal{B}_3^{22}(\boldsymbol{v},\boldsymbol{u})$; besides, since $s = s^{-1}$, $sr = (sr)^{-1}$ and $sr^2 = (sr^2)^{-1}$, (54) implies that $\mathcal{B}_3^{21}(\boldsymbol{u},\boldsymbol{v})$ and $\mathcal{B}_3^{12}(\boldsymbol{u},\boldsymbol{v})$ are symmetric (in both cases disregarding for the moment the constraints (44)).

In addition, the constraints (44) reduce to two independent restrictions, as follows. If $\boldsymbol{x} \in \Sigma_s$, $\boldsymbol{x} = s\boldsymbol{x}$, thus:

$$I_s = \partial C \cup \Sigma_s \qquad \text{with} \quad \boldsymbol{u}^2(\boldsymbol{x}) = s\boldsymbol{u}^1(\boldsymbol{x})$$

whereas if $\boldsymbol{x} \in \Sigma_s$, $\boldsymbol{x} = sr\boldsymbol{x}$, which yields:

$$I_{sr} = \partial C \cup \Sigma_r \qquad \text{with} \quad j^2\boldsymbol{u}^2(\boldsymbol{x}) = sr\boldsymbol{u}^1(\boldsymbol{x})$$

From these, it is easy to infer that

$$\{\boldsymbol{v}^1, \boldsymbol{v}^2\} \in \mathcal{V}_3 \Rightarrow \{\boldsymbol{v}^{2\star}, \boldsymbol{v}^{1\star}\} \in \mathcal{V}_3 \tag{55}$$

Hence, the one-to-one substitution $\{\boldsymbol{v}^{2\star}, \boldsymbol{v}^{1\star}\} \in \mathcal{V}_3 = \{\boldsymbol{w}^1, \boldsymbol{w}^2\} \in \mathcal{V}_3$ can be made, and the contributions for $\nu = 3$ in (48) are recast into a form which is symmetric in $(\{\boldsymbol{\phi}^1, \boldsymbol{\phi}^2\}, \{\tilde{\boldsymbol{\phi}}^1, \tilde{\boldsymbol{\phi}}^2\})$:

$$\sum_{k,\ell=1}^{2} \mathcal{B}_3^{k\bar{\ell}}(\boldsymbol{\phi}_3^{ki}, \tilde{\boldsymbol{\phi}}_3^{\ell}) + \sum_{\ell=1}^{2} \mathcal{C}_3^{i\bar{\ell}}(\boldsymbol{\phi}^2, \tilde{\boldsymbol{\phi}}_3^{\ell}) = \sum_{\ell=1}^{2} \mathcal{F}_3^{\bar{\ell}}(\tilde{\boldsymbol{\phi}}^{\ell}) \qquad (i = 1, 2;\ \bar{\ell} = 3 - \ell)$$

Similar conclusions can be reached for all dihedral symmetry groups D_m.

7.3 Reduction in solution time

Let N and γN denote the number of degrees of freedom supported by the BEM discretization of S^1 and S^2 respectively. The system of equations (37) or (48) takes the general form:

$$\begin{bmatrix} \boldsymbol{B} & \boldsymbol{C} \\ \boldsymbol{C}^T & \boldsymbol{D} \end{bmatrix} \begin{Bmatrix} \boldsymbol{\phi}^1 \\ \boldsymbol{\phi}^2 \end{Bmatrix} = \begin{bmatrix} \boldsymbol{F} \\ \boldsymbol{G} \end{bmatrix} \tag{56}$$

where the matrix $\boldsymbol{B}$ is block-diagonal: $\boldsymbol{B} = \text{Diag}(\boldsymbol{B}_\nu^i)$ $(1 \leq \nu \leq R(\mathcal{S}), 1 \leq i \leq d_\nu)$. Each block $\boldsymbol{B}_\nu^i$ is approximately of size $(d_\nu/n) \times N$ (the constraints (44) causing slight variations in size for the same value of d_ν). Besides, as mentioned before, all blocks $\boldsymbol{B}_\nu^i$ $(1 \leq i \leq d_\nu)$ are the same for a given ν.

Solving the original (symmetric) SGBEM system thus entails a $T = O((1+\gamma)^3 N^3/6)$ solution time. For solving the system (56), one must first solve the block-diagonal part, whereby each $\{\boldsymbol{\phi}_\nu^{k\ell}\}$ is expressed in terms of $\{\boldsymbol{F}_\nu^\ell\}$ and $\{\boldsymbol{\phi}^2\}$, and then substitute these results into the remaining part

γ	0	0.1	0.2	0.5	1	2
R $(\mathcal{S} = P_1)$	0.25	0.3238	0.3924	0.5556	0.7188	0.8611
R $(\mathcal{S} = P_2)$	0.0625	0.1266	0.1971	0.3889	0.6016	0.7986
R $(\mathcal{S} = D_3)$	0.0463	0.1206	0.1973	0.3964	0.6100	0.8042
R $(\mathcal{S} = P_3)$	0.015625	0.06320	0.1265	0.3194	0.5488	0.7691

Table 3. Expected asymptotic ratios R of solution CPU time with and without exploitation of partial symmetry, for some groups and various values of γ (ratio of numbers of DOFs on the surfaces S^1 and S^2)

of the system in order to build and solve a final system with a (symmetric) $\gamma N \times \gamma N$ matrix. The estimated time T_s for solving (56) (retaining only the $O(N^3)$ contributions) is

$$T_s = O\Big(\frac{N^3}{6}[\gamma^3 + 3\gamma^2 + (\frac{3\gamma}{n^2} + \frac{1}{n^3})\sum_{\nu=1}^{R(n)} d_\nu^3]\Big)$$

assuming that all blocks either are symmetric or have reciprocal symmetry. Let $R = T_s/((1+\gamma)^3 N^3/6)$; for instance, with $\gamma = 0$ (i.e. full symmetry), one has $R = (1/n^3)\sum_{\nu=1}^{R(n)} d_\nu^3$. Table 3 displays R for the groups $P_{1,2,3}$ and D_3 and various values of γ. Obviously, the highest gains in solution time occur for n large (i.e. high degrees of symmetry) and γ small. Also, $\sum_{\nu=1}^{R(n)} d_\nu^3 = n$ if $\mathcal{S}$ is Abelian, hence in that case $R = 1/n^2$ with $\gamma = 0$ as expected.

Elastostatic problems, Abelian case. In the limit of zero frequency (i.e. $k_T = 0$), the problem (1) becomes real-valued, as does the kernel function B. However, when the ρ_ν are complex, Eqs. (22), (34), (35), (36) show that the subproblems (37) are in general complex-valued even in that case. In fact, it is easy to show in this case that:

$$\mathcal{B}_{\nu^\star}(\boldsymbol{u},\boldsymbol{v}) = [\mathcal{B}_\nu]^\star(\boldsymbol{u},\boldsymbol{v}) \qquad \mathcal{C}_{\nu^\star}(\boldsymbol{u},\boldsymbol{v}) = [\mathcal{C}_\nu]^\star(\boldsymbol{u},\boldsymbol{v}) \qquad \mathcal{F}_{\nu^\star}(\boldsymbol{v}) = [\mathcal{F}_\nu]^\star(\boldsymbol{v})$$

Thus, the equations for the ν-subproblem and the $\nu^\star$-subproblem, and hence their solutions $(\boldsymbol{\phi}_\nu, \boldsymbol{\phi}_{\nu^\star})$, are conjugate to each other and thus redundant. It is sufficient to solve (say) the ν-subproblem for u_ν. The contribution of the conjugate pair $(\boldsymbol{\phi}_\nu, \boldsymbol{\phi}_{\nu^\star})$ to the reconstruction of the (real) global solution u is then:

$$[\boldsymbol{P}_\nu\boldsymbol{\phi}](s\boldsymbol{x}) + [\boldsymbol{P}_{\nu^\star}\boldsymbol{\phi}](s\boldsymbol{x}) = \rho_\nu^\star(s)\boldsymbol{\phi}_\nu + \rho_\nu(s)\boldsymbol{\phi}_\nu^\star = 2\text{Re}(\rho_\nu^\star(s)\boldsymbol{\phi}_\nu)$$

In the FEM framework, adequate combinations of the two conjugate equations are known to yield two *coupled* real-valued subproblems defined on the (volumic) symmetry cell. Here, a similar approach could be applied to the symmetry-reduced SGBEM. However, contrarily to the FEM case, this would result in one subproblem of size $2N/n$, and hence would not bring any advantage over solving directly the complex-valued subproblem of size N/n.

8 Conclusion

The analysis, conducted here for the simple case of Neumann boundary-value problems, can be extended to the SGBEM formulations of more general boundary-value problems. This strategy is especially interesting when S^2 is 'small' (in terms of the number of degrees of freedom involved). This is for instance the case for externally symmetric bodies containing holes, cracks or other defects of arbitrary shape and location. This work is expected to be highly beneficial to some computationally intensive problems like defect identification in complex bodies exhibiting geometrical symmetry.

References

1. ALLGOWER, E. L., GEORG, K., MIRANDA, R., TAUSCH, J. Numerical Exploitation of Equivariance. *Z. Angew. Math. Mech.*, **78**, 795–806 (1998).
2. BONNET, M. On the use of geometrical symmetry in the boundary element methods for 3D elasticity. In C.A. Brebbia (ed.), *Boundary element technology VI*, pp. 185–201. Comp. Mech. Publ., Southampton / Elsevier, Southampton, Boston (1991).
3. BONNET, M., MAIER, G., POLIZZOTTO, C. On symmetric galerkin boundary element method. *Appl. Mech. Rev.*, **51**, 669–704 (1998).
4. BOSSAVIT, A. Symmetry, groups and boundary value problems : a progressive introduction to noncommutative harmonic analysis of partial differential equations in domains with geometrical symmetry. *Comp. Meth. in Appl. Mech. Engng.*, **56**, 167–215 (1986).
5. ERINGEN, A. C., SUHUBI, E. S. *Elastodynamics (vol II - linear theory).* Academic Press (1975).
6. LOBRY, J., BROCHE, CH. Geometrical symmetry in the boundary element method. *Engng. Anal. with Bound. Elem.*, **14**, 229–238 (1994).
7. NEDELEC, J. C. Integral equations with non integrable kernels. *Integral equations and operator theory*, **5**, 562–572 (1982).
8. SERRE, J. P. *Linear representations of finite groups.* Springer-Verlag (1977).
9. VINBERG, E. B. *Linear representations of groups.* Birkhäuser (1989).

A New Fast Multipole Boundary Integral Equation Method in Elastostatic Crack Problems in 3D

Ken-ichi Yoshida[1], Naoshi Nishimura[1], and Shoichi Kobayashi[2]

[1] Dept. Global Env. Eng., Kyoto University, Kyoto 606-8501, Japan
[2] Dept. Construction Eng., Fukui University of Technology, Fukui 910-8505, Japan

Abstract. This paper discusses a formulation and its applications of the new Fast Multipole Method (FMM) to three-dimensional Boundary Integral Equation Method (BIEM) in elastostatic crack problems. It is shown, through numerical experiments, that the new FMM is more efficient than the original FMM.

1 Introduction

In spite of its apparent advantage of the reduced dimensionality, BIEM has so far been applied to relatively small problems, compared to numerical methods of the domain type such as FEM or FDM. This is because the resulting matrix in BIEM is full, which implies that the memory requirement of BIEM is $O(N^2)$, where N is the number of unknowns. Even more serious is the drawback of BIEM in terms of the computational cost. Indeed, the buildup of the coefficient matrix requires an $O(N^2)$ work, which further increases to $O(N^3)$ if one attempts to solve matrix equations with direct methods based on the LU decomposition. However, the appearance of FMM changed the situation drastically: FMM reduces the computational cost of BIEM to $O(N^{1+\alpha}(\log N)^{\beta})$ and the memory requirements to $O(N)$, where α and β are nonnegative numbers. With the help of FMM, BIEM can now be applied to large scale problems.

FMM was proposed by Rokhlin [1] as a fast solver for integral equations for the two-dimensional Laplace equation. This method was then made famous as Greengard [2] improved the algorithm and applied it to multibody problems. Since then FMM has been developed as a fast solution method for large scale problems. We here cite a few papers related to the use of FMM and BIEM in solid mechanics: Nishimura et al. [3] for crack problems for the three-dimensional Laplace equation, Fu et al. [4], Fukui et al. [5] and Takahashi et al. [6] for ordinary problems for three-dimensional elastostatics, Yoshida et al. [7,8] for crack problems in three-dimensional elastostatics and Fujiwara [9] and Yoshida et al. [10] for three-dimensional elastodynamics.

FMM provides a fast method of computing certain potential functions using the multipole expansion in the evaluation of contributions from remote sources. In this approach the effects of remote sources are aggregated into quantities called multipole moments. The effects of these sources are evaluated not directly with the help of multipole expansion but in the form of a series expansion called 'local expansion'. The process of converting multipole moments into the coefficients of the local expansion is called M2L. In the original FMM by Rokhlin the computational cost for the M2L translation dominates the performance especially in Helmholtz' equation or in three-dimensional problems. In view of this Rokhlin introduced the diagonal form [11,12] so as to reduce the computational cost for the M2L translation. However, Rokhlin's diagonal form is known to have numerical instabilities in Laplace's equation [13] or in Helmholtz' equation [14] with low frequency. In order to overcome these problems Hrycak and Rokhlin [15] proposed a new FMM for the two-dimensional Laplace equation, Greengard and Rokhlin [16] and Cheng et al. [17] for the three-dimensional Laplace equation, and Greengard et al. [18] for the three-dimensional Helmholtz equation. Yoshida et al. [19] investigated the use of the new FMM in crack problems for the Laplace equation in 3D. In this paper we discuss an application of the new FMM to three dimensional elastostatic crack problems. We shall show, via numerical experiments, that the new FMM is more efficient than the original FMM, especially when the cracks are distributed densely.

In this paper we shall use both indicial and direct notations for tensorial quantities. The summation convention is used for repeated indices. Also the position vector of a point x will be denoted by either $\boldsymbol{x}$ or $\overrightarrow{Ox}$, the latter being the preferred notation when one needs to show the origin explicitly.

2 Crack Problems and Integral Equations

Let $S \subset \mathbb{R}^3$, or a 'crack', be a union of smooth non-self-intersecting curved surfaces having smooth edges ∂S. Also let $\boldsymbol{n}$ be the unit normal vector to S. Our problem is to find a solution $\boldsymbol{u}$ of the equation of elastostatics

$$C_{ijkl}u_{k,lj} = 0 \quad \text{in } \mathbb{R}^3 \setminus \overline{S}$$

subject to the boundary condition

$$t_i^{\pm} := C_{ijkl}u_{k,l}^{\pm}n_j = 0 \quad \text{on } S \tag{1}$$

regularity

$$\boldsymbol{\phi}(x) := \boldsymbol{u}^+(x) - \boldsymbol{u}^-(x) = 0 \quad \text{on } \partial S \tag{2}$$

and an asymptotic condition given by

$$\boldsymbol{u}(x) \to \boldsymbol{u}^{\infty}(x) \quad \text{as} \quad |\boldsymbol{x}| \to \infty$$

where $\boldsymbol{u}$, C_{ijkl}, $\boldsymbol{t}$, $\boldsymbol{u}^{\infty}$ and $\boldsymbol{\phi}$ stand for the displacement, elasticity tensor, traction vector, an entire solution of the equation of elastostatics and the crack opening displacement, respectively. Also, the superscript + (−) indicates the limit on S from the positive (negative) side of S where the positive side indicates the one into which the unit normal vector $\boldsymbol{n}$ points. The components of C_{ijkl} are expressed with Lamé's constants (λ, μ) and Kronecker's delta δ_{ij} as

$$C_{ijkl} = \lambda\delta_{ij}\delta_{kl} + \mu(\delta_{ik}\delta_{jl} + \delta_{il}\delta_{jk}) \ .$$

The solution $\boldsymbol{u}$ to this problem has an integral representation given by

$$u_i(x) = u_i^\infty(x) + \int_S \Gamma_{\mathrm{I}ij}(x,y)\phi_j(y)dS_y, \quad x \in \mathrm{IR}^3 \setminus \overline{S} \tag{3}$$

$$\Gamma_{\mathrm{I}ij}(x,y) = C_{jlcd}\frac{\partial}{\partial y_d}\Gamma_{ic}(\boldsymbol{x}-\boldsymbol{y})n_l(y),$$

where $\Gamma_{\mathrm{I}ij}$ is the double layer kernel and Γ_{ij} is the fundamental solution of the equation of elastostatics expressed as

$$\Gamma_{ij}(\boldsymbol{x}-\boldsymbol{y}) = \frac{1}{8\pi\mu}\left(\mathcal{F}_{ij}\frac{1}{|\boldsymbol{x}-\boldsymbol{y}|} + \mathcal{G}_i\frac{1}{|\boldsymbol{x}-\boldsymbol{y}|}\overrightarrow{Oy}_j\right) \ . \tag{4}$$

In this formula the operators $\mathcal{F}$ and $\mathcal{G}$ are defined as

$$\mathcal{F}_{ij} = \frac{\lambda+3\mu}{\lambda+2\mu}\delta_{ij} - \frac{\lambda+\mu}{\lambda+2\mu}(\overrightarrow{Ox})_j\frac{\partial}{\partial x_i}, \quad \mathcal{G}_i = \frac{\lambda+\mu}{\lambda+2\mu}\frac{\partial}{\partial x_i} \ .$$

Using (1) and (3), one obtains the following hypersingular integral equation:

$$t_a^\infty(x) = -\text{p.f.}\int_S n_b(x)C_{abik}\frac{\partial}{\partial x_k}\Gamma_{\mathrm{I}ij}(x,y)\phi_j(y)dS_y, \quad x \in S \tag{5}$$

where $\boldsymbol{t}^\infty(x)$ and p.f. indicate the traction associated with $\boldsymbol{u}^\infty(x)$ and the finite part of a divergent integral. Note that (5) can be 'regularised' into the following form with a less singular kernel function:

$$t_a^\infty(x) = \text{v.p.}\int_S n_b(x)C_{abik}e_{rck}C_{cdjl}\frac{\partial}{\partial y_l}\Gamma_{ij}(\boldsymbol{x}-\boldsymbol{y})e_{rqs}\frac{\partial\phi_d(y)}{\partial y_q}n_s(y)dS_y \tag{6}$$

where v.p. indicates Cauchy's principal value.

3 Original FMM in Elastostatics

In this section we present the formulation and algorithm for the original FMM mainly for the convenience of reference. See Yoshida et al. [7] for further details.

3.1 Formulation

The starting point in the application of FMM to BIEM is to expand the fundamental solution $\Gamma_{ij}(\boldsymbol{x}-\boldsymbol{y})$ into a series of products of functions of x and those of y. For this purpose we use the well-known formula given by

$$\frac{1}{|\boldsymbol{x}-\boldsymbol{y}|}=\sum_{n=0}^{\infty}\sum_{m=-n}^{n} S_{n,m}(\overrightarrow{Ox})\overline{R_{n,m}(\overrightarrow{Oy})}, \quad |\overrightarrow{Oy}|<|\overrightarrow{Ox}| \tag{7}$$

where $R_{n,m}$ and $S_{n,m}$ are the solid harmonics defined as

$$R_{n,m}(\overrightarrow{Ox})=\frac{1}{(n+m)!}P_n^m(\cos\theta)e^{im\phi}r^n,$$
$$S_{n,m}(\overrightarrow{Ox})=(n-m)!P_n^m(\cos\theta)e^{im\phi}\frac{1}{r^{n+1}},$$

(r,θ,ϕ) are the polar coordinates of the point x, P_n^m is the associated Legendre function and a superposed bar indicates the complex conjugate. In passing we point out that the use of solid harmonics in FMM has been suggested by Pérez-Jordá and Yang [21] among others. Using (7), we rewrite (4) as [7]

$$\Gamma_{ij}(\boldsymbol{x}-\boldsymbol{y})=$$
$$\frac{1}{8\pi\mu}\sum_{n=0}^{\infty}\sum_{m=-n}^{n}\left(F_{ij,n,m}^S(\overrightarrow{Ox})\overline{R_{n,m}(\overrightarrow{Oy})}+G_{i,n,m}^S(\overrightarrow{Ox})(\overrightarrow{Oy})_j\overline{R_{n,m}(\overrightarrow{Oy})}\right), \tag{8}$$

where $F_{ij,n,m}^S$ and $G_{i,n,m}^S$ are functions defined as [7]

$$F_{ij,n,m}^S(\overrightarrow{Ox})=\mathcal{F}_{ij}S_{n,m}(\overrightarrow{Ox})$$
$$=\frac{\lambda+3\mu}{\lambda+2\mu}\delta_{ij}S_{n,m}(\overrightarrow{Ox})-\frac{\lambda+\mu}{\lambda+2\mu}(\overrightarrow{Ox})_j\frac{\partial}{\partial x_i}S_{n,m}(\overrightarrow{Ox}), \tag{9}$$

$$G_{i,n,m}^S(\overrightarrow{Ox})=\mathcal{G}_iS_{n,m}(\overrightarrow{Ox})=\frac{\lambda+\mu}{\lambda+2\mu}\frac{\partial}{\partial x_i}S_{n,m}(\overrightarrow{Ox}) . \tag{10}$$

Let S_y be a subset of S, and let x be a point such that $|\overrightarrow{Oy}|<|\overrightarrow{Ox}|$ holds for $\forall y\in S_y$. We now compute the integral on the right hand side of (3) over S_y assuming that ϕ is given. Using (8) we obtain

$$\int_{S_y}\Gamma_{1ij}(x,y)\phi_j(y)dS_y$$
$$=\frac{1}{8\pi\mu}\sum_{n=0}^{\infty}\sum_{m=-n}^{n}\left(F_{ij,n,m}^S(\overrightarrow{Ox})\overline{M_{j,n,m}^1}(O)+G_{i,n,m}^S(\overrightarrow{Ox})\overline{M_{n,m}^2}(O)\right), \tag{11}$$

where $M_{j,n,m}^1$ and $M_{n,m}^2$ are the multipole moments centred at O, expressed as

$$M_{j,n,m}^1(O)=\int_{S_y}C_{cdjl}\frac{\partial}{\partial y_l}R_{n,m}(\overrightarrow{Oy})\phi_d(y)n_c(y)dS_y, \tag{12}$$

$$M^2_{n,m}(O) = \int_{S_y} C_{cdjl} \frac{\partial}{\partial y_l}((\overrightarrow{Oy})_j R_{n,m}(\overrightarrow{Oy}))\phi_d(y) n_c(y) dS_y \ . \tag{13}$$

The multipole moments are translated according to the following formulae as the centre of multipole expansion is shifted from O to O':

$$M^1_{j,n,m}(O') = \sum_{n'=0}^{n} \sum_{m'=-n'}^{n'} R_{n',m'}(\overrightarrow{O'O}) M^1_{j,n-n',m-m'}(O), \tag{14}$$

$$M^2_{n,m}(O') = \sum_{n'=0}^{n} \sum_{m'=-n'}^{n'} R_{n',m'}(\overrightarrow{O'O}) \times \left(M^2_{n-n',m-m'}(O) - (\overrightarrow{OO'})_j M^1_{j,n-n',m-m'}(O)\right), \tag{15}$$

where we have used (12), (13) and the addition theorem for spherical harmonics.

In the evaluation of the integral on the right hand side of (5) for a given ϕ one can use the local expansion given by

$$-\int_{S_y} \frac{\partial}{\partial x_k} \Gamma_{Iij}(\boldsymbol{x}-\boldsymbol{y}) \phi_j(y) dS_y =$$
$$-\frac{1}{8\pi\mu} \sum_{n=0}^{\infty} \sum_{m=-n}^{n} \frac{\partial}{\partial x_k} \left(F^R_{ij,n,m}(\overrightarrow{x_0 x}) L^1_{j,n,m}(x_0) + G^R_{i,n,m}(\overrightarrow{x_0 x}) L^2_{n,m}(x_0) \right), \tag{16}$$

where $L^1_{j,n,m}$ and $L^2_{n,m}$ are the coefficients of the local expansion expressed with $M^1_{j,n,m}$ and $M^2_{n,m}$ by

$$L^1_{j,n',m'}(x_0) = \sum_{n=0}^{\infty} \sum_{m=-n}^{n} (-1)^{n'} \overline{S_{n+n',m+m'}(\overrightarrow{Ox_0})} M^1_{j,n,m}(O), \tag{17}$$

$$L^2_{n',m'}(x_0) = \sum_{n=0}^{\infty} \sum_{m=-n}^{n} (-1)^{n'} \overline{S_{n+n',m+m'}(\overrightarrow{Ox_0})} \times (M^2_{n,m}(O) - (\overrightarrow{Ox_0})_j M^1_{j,n,m}(O)), \tag{18}$$

and $F^R_{ij,n,m}$ and $G^R_{i,n,m}$ are functions obtained by replacing $S_{n,m}$ by $R_{n,m}$ in (9) and (10). In these formulae we have used the addition theorem for spherical harmonics and have assumed that the inequality $|\overrightarrow{Ox_0}| > |\overrightarrow{x_0 x}|$ holds. The procedures given by (17) and (18) are called M2L translation. The coefficients of the local expansion are translated according to the following formulae when the centre of the local expansion is shifted from x_0 to x_1

$$L^1_{j,n'',m''}(x_1) = \sum_{n'=n''}^{\infty} \sum_{m'=-n'}^{n'} R_{n'-n'',m'-m''}(\overrightarrow{x_0 x_1}) L^1_{j,n',m'}(x_0), \tag{19}$$

$$L^2_{n'',m''}(x_1) = \sum_{n'=n''}^{\infty} \sum_{m'=-n'}^{n'} R_{n'-n'',m'-m''}(\overrightarrow{x_0x_1}) \times \left(L^2_{n',m'}(x_0) - (\overrightarrow{x_0x_1})_p L^1_{p,n',m'}(x_0) \right), \quad (20)$$

where we have used (16) and the addition theorem for spherical harmonics.

3.2 Algorithm for original FMM

The algorithm of the original FMM is described as follows:

Step 1. Discretisation:

Discretise S in the same manner as in the conventional BIEM.

Step 2. Determination of octree structure:

Consider a cube which circumscribes S and call this cube the cell of level 0. Now take a cell (a parent cell) of level l ($l \geq 0$) and divide it into 8 equal sub cubes whose edge lengths are half of that of the parent cell and call any of them which contains some boundary elements a cell of level $l+1$. Continue the subdivision of cells until the number of boundary elements in a cell is below a given number. A cell having no children is called a leaf.

Step 3. Computation of the multipole moments:

We first compute the multipole moments associated with leaves via (12) and (13) taking the centre (O) of the multipole moment as the centroid of C. For a non-leaf cell C of level l we compute the associated multipole moments recursively by adding all the multipole moments of C's children after translating them via (14) and (15) from the centroids of C's children (O) to that of C (O'). We repeat this procedure tracing the tree structure of cells upward (decreasing l) until we reach level 2 cells.

Step 4. Computation of the local expansion:

We have to prepare some definitions first. We say that two cells are 'adjacent cells at level l' if these cells are both of the level l and share at least one vertex. Two cells are said to be 'well-separated at level l' if they are not adjacent at level l but their parent cells are adjacent at level $l-1$. The list of all the well-separated cells from a level l cell C is called the interaction list of C.

We now compute the coefficients of local expansion associated with a cell C, which are defined to be the sum of the contributions of the forms in (17) and (18) from cells which are not adjacent to C. Obviously this sum is divided into the contribution from cells in the interaction list of C and the contribution from cells which are not adjacent to C's parent. One computes the former by substituting the multipole moments associated with cells in the interaction list of C into (17) and (18). Also, the latter is computed from the coefficients of the local expansion of C's parent as one shifts the centre of the expansion from the centroid of C's parent (x_0) to that of C (x_1) via (19) and (20). We repeat these procedures starting from $l = 2$ and increasing l along the tree of cells until we reach leaves.

Step 5. Evaluation of the integral in (5):

The integral in (5) is now evaluated in leaves (denoted by C). First we compute the contribution from boundary elements in cells adjacent to C using (6) in the same manner as in the conventional BIEM and then compute the contribution from cells which are not adjacent to C using (16). The sum of these contributions gives the contribution from all boundary elements.

In the implementation of this algorithm one has to truncate the infinite series with respect to n in (11) etc. at $n = p$. For a fixed p one shows that the computational complexity of this algorithm is $O(N)$ where N is the number of boundary elements [1,2]. Also, the M2L operation in the 4th step is seen to require an $O(p^4)$ work.

4 New FMM in Elastostatics

4.1 Formulation

We now consider a source point y located at (y_1, y_2, y_3) and a target point x at (x_1, x_2, x_3). The cell containing the source point y is denoted by C_s and the cell containing the target point x by C_t. Assume that $x_3 > y_3$ holds. We then have the following integral representation:

$$\frac{1}{|\boldsymbol{x}-\boldsymbol{y}|} = \frac{1}{2\pi}\int_0^\infty e^{-\lambda(x_3-y_3)} \int_0^{2\pi} e^{i\lambda((x_1-y_1)\cos\alpha+(x_2-y_2)\sin\alpha)} d\alpha d\lambda \ . \quad (21)$$

Suppose that each of the cells C_s and C_t has an edge length of d. Then the double integral in (21) is evaluated with the following double sum:

$$\frac{1}{|\boldsymbol{x}-\boldsymbol{y}|} = \sum_{k=1}^{s(\varepsilon)} \sum_{j=1}^{M(k)} \frac{\omega_k}{M(k)d} Exp(\overrightarrow{x_0 x}, k, j)\, Exp(\overrightarrow{Ox_0}, k, j)\, Exp(-\overrightarrow{Oy}, k, j) + \varepsilon, \quad (22)$$

where

$$Exp(\boldsymbol{x}, k, j) = e^{-\lambda_k x_3/d} e^{i\lambda_k(x_1\cos\alpha_j(k)+x_2\sin\alpha_j(k))/d},$$

x_0 is a point near x, $\alpha_j(k)$ is given by

$$\alpha_j(k) = \frac{2\pi j}{M(k)},$$

ε is the error term and the numbers $s(\varepsilon)$, $M(k)$, Gaussian weights ω_k and nodes λ_k are given in Yarvin and Rokhlin [20]. One may determine these parameters considering the required accuracy. The expansion in (22) is called the exponential expansion for $1/|\boldsymbol{x}-\boldsymbol{y}|$.

With the exponential expansion in (22) and (4) we derive an exponential expansion of the double layer potential which holds for a point x near x_0:

$$\int_{S_y} \Gamma_{\mathrm{I}ij}(x,y)\phi_j(y)dS_y =$$

$$\frac{1}{8\pi}\sum_{p=1}^{s(\varepsilon)}\sum_{q=1}^{M(p)}\left(V_j^1(p,q;x_0)\mathcal{F}_{ij}(\overrightarrow{x_0x}) + V^2(p,q;x_0)\mathcal{G}_i(\overrightarrow{x_0x})\right) Exp(\overrightarrow{x_0x},p,q), \quad (23)$$

where $V_j^1(p,q;x_0)$ and $V^2(p,q;x_0)$ are the coefficients of the incoming exponential expansion at x_0, which are related to another set of coefficients of exponential expansion $W_j^1(p,q;O)$ and $W^2(p,q;O)$ via

$$V_j^1(p,q;x_0) = W_j^1(p,q;O)Exp(\overrightarrow{Ox_0},p,q), \quad (24)$$

$$V^2(p,q;x_0) = (W^2(p,q;O) - (\overrightarrow{Ox_0})_k W_k^1(p,q;O))Exp(\overrightarrow{Ox_0},p,q) \ . \quad (25)$$

The coefficients $W_j^1(p,q;O)$ and $W^2(p,q;O)$, or the coefficients of the outgoing exponential expansion at O, are defined by

$$W_j^1(p,q;O) = \int_{S_y} C_{cdjl}\frac{\partial}{\partial y_l}Exp(\overrightarrow{Oy},p,q)\phi_d(y)n_c(y)dS_y, \quad (26)$$

$$W^2(p,q;O) = \int_{S_y} C_{cdjl}\frac{\partial}{\partial y_l}((\overrightarrow{Oy})_j Exp(\overrightarrow{Oy},p,q)\phi_d(y)n_c(y)dS_y \ . \quad (27)$$

We note that the relations in (23), (24,25) and (26,27) are conceptually similar to (16), (17,18) and (12,13) in the original FMM, respectively. An important difference between (24,25) and (17,18) is that (24) and (25) do not include summation, while (17) and (18) do. In other words, the operations in (24) and (25) are "diagonal". We shall see the implication of this fact later.

Since the function $Exp(\overrightarrow{Ox},k,j)$ is harmonic, it allows an expansion in terms of $R_{n,m}(\overrightarrow{Ox})$. Indeed, one has

$$Exp(\overrightarrow{Ox},p,q) = \sum_{n=0}^{\infty}\sum_{m=-n}^{n}(-\lambda_p/d)^n(-i)^m e^{-im\alpha_q(p)}R_{n,m}(\overrightarrow{Ox}) \ . \quad (28)$$

This equation enables us to relate coefficients of exponential expansions to multipole moments and coefficients of local expansion via

$$W_j^1(p,q;O) = \frac{\omega_p}{M(p)d}\sum_{m=-\infty}^{\infty}(-i)^m e^{-im\alpha_q(p)}\sum_{n=|m|}^{\infty}(\lambda_p/d)^n M_{j,n,m}^1(O), \quad (29)$$

$$W^2(p,q;O) = \frac{\omega_p}{M(p)d}\sum_{m=-\infty}^{\infty}(-i)^m e^{-im\alpha_q(p)}\sum_{n=|m|}^{\infty}(\lambda_p/d)^n M_{n,m}^2(O), \quad (30)$$

$$L_{j,n,m}^1(x_0) = \sum_{p=1}^{s(\varepsilon)}\sum_{q=1}^{M(p)} V_j^1(p,q;x_0)(-i)^m(-\lambda_p/d)^n e^{-im\alpha_q(p)}, \quad (31)$$

$$L^2_{n,m}(x_0) = \sum_{p=1}^{s(\varepsilon)} \sum_{q=1}^{M(p)} V^2(p,q;O)(-i)^m(-\lambda_p/d)^n e^{-im\alpha_q(p)} \ . \tag{32}$$

4.2 Rotation of Coefficients

In this section we shall remove the assumption that C_t is in $+x_3$ direction of C_s made in the previous section. To this end we divide the interaction list of C_s into 6 lists: uplist, downlist, northlist, southlist, eastlist and westlist. The uplist and downlist contain target cells located in $+x_3$ and $-x_3$ directions of C_s, respectively. The northlist and southlist contain target cells located in $+x_2$ and $-x_2$ directions of C_s except those in the uplist or downlist, respectively. The eastlist and westlist contain remaining target cells located in $+x_1$ and $-x_1$ directions of C_s, respectively. If the target cell is included in lists other than the uplist of C_s we rotate the coordinate system so that the target cell is in the positive $\tilde{x}_3$ direction viewed from the source cell, where $\tilde{x}_i$ denotes the new axis. In general the multipole moments in the new coordinate system are obtained as follows:

$$\widetilde{M}^1_{j,n,m}(O) = \sum_{m'=-n}^{n} \mathcal{R}_{n,m,m'}(\boldsymbol{\nu},\alpha) M^1_{j,n,m'}(O), \tag{33}$$

$$\widetilde{M}^2_{n,m}(O) = \sum_{m'=-n}^{n} \mathcal{R}_{n,m,m'}(\boldsymbol{\nu},\alpha) M^2_{n,m'}(O), \tag{34}$$

where $\mathcal{R}_{n,m,m'}(\boldsymbol{\nu},\alpha)$ is the coefficient of rotation, $\boldsymbol{\nu}$ is a unit vector parallel to the rotation axis and α is the rotation angle. The explicit form of $\mathcal{R}_{n,m,m'}(\boldsymbol{\nu},\alpha)$ is given by (See Biedenharn and Louck [22])

$$\mathcal{R}_{n,m,m'}(\boldsymbol{\nu},\alpha) = (-1)^{m+m'}(n+m')!(n-m')! \\ \sum_k \frac{(\alpha_0 - i\alpha_3)^{n+m-k}(-i\alpha_1-\alpha_2)^{m'-m+k}(-i\alpha_1+\alpha_2)^k(\alpha_0+i\alpha_3)^{n-m'-k}}{(n+m-k)!(m'-m+k)!k!(n-m'-k)!}, \tag{35}$$

where $\alpha_0 = \cos(\alpha/2)$ and $\alpha_i = -\nu_i \sin(\alpha/2)$. The summation in (35) is carried out over such k that the powers in the numerator are all non-negative.

We next describe the M2L translation process in the new FMM.

1. Rotation:
   First we rotate the multipole moments via (33) and (34) so as to make the procedure presented in 4.1 applicable. The specific forms of (33) and (34) depend on the location of C_t and are described as follows:
   (a) $C_t \in$ uplist

$$\widetilde{M}^{1U}_{j,n,m}(O) = M^1_{j,n,m}(O), \quad \widetilde{M}^{2U}_{n,m}(O) = M^2_{n,m}(O), \tag{36}$$

(b) $C_t \in$ downlist
Use (33) and (34) with $(\boldsymbol{\nu}, \alpha) = (\boldsymbol{e}_1, \pi)$ to obtain $\widetilde{M}^D$, where $\widetilde{M}^D$ is short for $\widetilde{M}^{1D}_{i,n,m}$ and $\widetilde{M}^{2D}_{n,m}$. We shall use abbreviations of this type in the rest of this paper.
(c) $C_t \in$ northlist
Use (33) and (34) with $(\boldsymbol{\nu}, \alpha) = (\boldsymbol{e}_1, \pi/2)$ to obtain $\widetilde{M}^N$.
(d) $C_t \in$ southlist
Use (33) and (34) with $(\boldsymbol{\nu}, \alpha) = (\boldsymbol{e}_1, -\pi/2)$ to obtain $\widetilde{M}^S$.
(e) $C_t \in$ eastlist
Use (33) and (34) with $(\boldsymbol{\nu}, \alpha) = (\boldsymbol{e}_2, -\pi/2)$ to obtain $\widetilde{M}^E$.
(f) $C_t \in$ westlist
Use (33) and (34) with $(\boldsymbol{\nu}, \alpha) = (\boldsymbol{e}_2, \pi/2)$ to obtain $\widetilde{M}^W$.

In these statements $\boldsymbol{e}_i$ is the base vector for the cartesian coordinates and superposed indices $\{U,\ D,\ N,\ S,\ E,\ W\}$ correspond to the initial letters of {uplist, downlist, northlist, southlist, eastlist, westlist}, respectively.

2. Computation of the coefficients of the exponential expansion:
Compute the coefficients of the exponential expansion via (29) and (30) with W and M replaced by $W^\diamondsuit$ and $\widetilde{M}^\diamondsuit$, respectively. where $\diamondsuit$ is an element of $\{U, D, N, S, E, W\}$.
3. Translation of the coefficients of the exponential expansion:
As the centre of the exponential expansion is shifted from the centroid of C_s (O) to the centroid of C_t $(\widetilde{x}_0)$, the coefficients of the exponential expansion is translated according to (24) and (25) as follows:

$$V_j^{1\diamondsuit}(p,q;\widetilde{x}_0) = W_j^{1\diamondsuit}(p,q;O) Exp(\overrightarrow{O\widetilde{x}_0}, p, q), \tag{37}$$

$$V^{2\diamondsuit}(p,q;\widetilde{x}_0) = (W^{2\diamondsuit}(p,q;O) - W_k^{1\diamondsuit}(p,q;O)(\overrightarrow{Ox_0})_k) Exp(\overrightarrow{O\widetilde{x}_0}, p, q), \tag{38}$$

where $\diamondsuit$ is an element of $\{U, D, N, S, E, W\}$. Notice that $\overrightarrow{O\widetilde{x}_0}$ in the Exp function is expressed with the ˜ coordinate system while the $(\overrightarrow{Ox_0})_k$ factor in (38) is referred to the original coordinate system.
4. Computation of the coefficients of the local expansion:
Compute the coefficients of the local expansion from those of the exponential expansion according to (31) and (32) with V and L replaced by $V^\diamondsuit$ and $\widetilde{L}^\diamondsuit$, respectively, where $\diamondsuit$ is an element of $\{U, D, N, S, E, W\}$. Then rotate $\widetilde{L}^\diamondsuit_{n,m}$ as follows:
(a) $C_t \in$ uplist

$$L^{1U}_{j,n,m}(x_0) = \widetilde{L}^{1U}_{j,n,m}(\widetilde{x}_0), \quad L^{2U}_{n,m}(x_0) = \widetilde{L}^{2U}_{n,m}(\widetilde{x}_0), \tag{39}$$

(b) $C_t \in$ downlist

$$L^{1D}_{j,n,m}(x_0) = \sum_{m'=-n}^{n} \mathcal{R}_{n,m',m}(\boldsymbol{\nu}, \alpha) \widetilde{L}^{1D}_{j,n,m'}(\widetilde{x}_0), \tag{40}$$

$$L^{2D}_{n,m}(x_0) = \sum_{m'=-n}^{n} \mathcal{R}_{n,m',m}(\boldsymbol{\nu}, \alpha) \widetilde{L}^{2D}_{n,m'}(\widetilde{x}_0), \tag{41}$$

where $(\boldsymbol{\nu}, \alpha) = (\boldsymbol{e}_1, \pi)$.

(c) $C_t \in$ northlist
Use relations similar to (40) and (41) with $(\boldsymbol{\nu}, \alpha) = (\boldsymbol{e}_1, \pi/2)$ to obtain L^N.

(d) $C_t \in$ southlist
Use relations similar to (40) and (41) with $(\boldsymbol{\nu}, \alpha) = (\boldsymbol{e}_1, -\pi/2)$ to obtain L^S.

(e) $C_t \in$ eastlist
Use relations similar to (40) and (41) with $(\boldsymbol{\nu}, \alpha) = (\boldsymbol{e}_2, -\pi/2)$ to obtain L^E.

(f) $C_t \in$ westlist
Use relations similar to (40) and (41) with $(\boldsymbol{\nu}, \alpha) = (\boldsymbol{e}_2, \pi/2)$ to obtain L^W.

Finally add L^U, L^D, L^N, L^S, L^E and L^W together via

$$L^1_{j,n,m}(x_0) = \sum_{\Diamond \in U,D,N,S,E,W} L^{1\Diamond}_{j,n,m}(\tilde{x}_0), \tag{42}$$

$$L^2_{n,m}(x_0) = \sum_{\Diamond \in U,D,N,S,E,W} L^{2\Diamond}_{n,m}(\tilde{x}_0) \tag{43}$$

to obtain the coefficients of the local expansion.

4.3 Algorithm for the new FMM

The algorithm for the new FMM is given as follows:
Steps 1–3. Same as the steps 1–3 in 3.2.
Step 4. Computation of the coefficients of the exponential expansions:
Compute the coefficients of the outgoing exponential expansions in each cell using (29)–(30), taking the origin (O) at the centroid of the cell.
Step 5. Computation of the coefficients of the local expansion:
We compute the coefficients of the local expansion of cells of level l, starting from $l = 2$ and increasing l. Consider a level l cell C and another level l cell C' which is contained in the interaction list of C. Depending on the position of C' relative to C, we translate the coefficients of an appropriate outgoing exponential expansion of C to those of the corresponding incoming exponential expansion via (37) and (38) by shifting the centre of the exponential expansion from the centroid of C (O) to that of C' (x_0). We then use (31), (32) and an appropriate rotation to convert the coefficients of the incoming exponential expansion to the coefficients of the local expansion. After carrying out these conversions for all cells in the interaction list of C, we add them together via (42) and (43) to obtain the contribution from the interaction list of C to the coefficients of the local expansion. To this we add the coefficients of the local expansion of the parent of C after shifting the origin from the centroid of the parent (x_0) to that of C (x_1) via (19) and (20) to

complete the calculation of the coefficients of the local expansion associated with C.

Step 6. Same as the 5th step in 3.2.

With this algorithm one shows that the M2L operation reduces to an $O(p^3)$ work [16,17]. Hence the new FMM is considered to be more efficient than the original FMM, which requires an $O(p^4)$ work for M2L, at least when p is not very small.

5 Numerical Examples

In the FMM accelerated BIEM used in this paper we discretise (5) with collocation and piecewise constant boundary elements. The resulting linear system of equations is solved with the preconditioned GMRES, which requires the product of the discretised matrix and the trial solution in each step of the iteration. The computation of this matrix-vector product is carried out efficiently with the FMM approaches discussed in the previous sections without building up the matrix explicitly.

In our implementation in Fortran 77, the integrals in the multipole moments in (12) and (13) are computed numerically with Gaussian quadrature. The infinite series in (11) and (16) are truncated at 10 terms ($p = 10$) and the sums in (22), (23), (31) and (32) are computed with the 109 point generalised Gaussian quadrature formula in Yarvin and Rokhlin [20]. Also, the maximum number of boundary elements in a leaf is set to be 100. In GMRES we adopt the block diagonal matrix corresponding to the leaves as the preconditioner following Nishida and Hayami [23]. Also, the iteration is stopped when the relative residual norm is below 10^{-5}. The performance of our implemementation has been tested on a desktop computer having a DEC Alpha 21264(500 MHz) as the CPU.

5.1 One Crack

We consider an infinite space which contains one penny-shaped crack having the radius of a_0 and the unit normal vector of $\boldsymbol{n} = (0, 0, 1)$. The function $\boldsymbol{t}^\infty(x)$ is given by $\boldsymbol{t}^\infty(x) = \boldsymbol{\sigma}^\infty \boldsymbol{n}(x)$ where $\boldsymbol{\sigma}^\infty$ is a tensor whose components are zero except for $\sigma_{33}^\infty = p_0$. Hence, one has $\boldsymbol{t}^\infty = (0, 0, p_0)$. This asymptotic condition means that the domain is subjected to a uniform uniaxial tension. Also, Poisson's ratio is set to be 1/4, i.e. $\lambda = \mu$. This problem is solved with the conventional BIEM, the original FM-BIEM (Fast Multipole-BIEM) and the new FM-BIEM. Fig.1 shows the 5736 DOF mesh and Fig.2 plots the non-dimensional crack opening displacement $\mu\phi_3/a_0 p_0$ obtained with this mesh. In Fig.2 the symbols indicated 'conv', 'fmm' and 'newfmm' stand for numerical results computed with the conventional BIEM, the original FM-BIEM and the new FM-BIEM, respectively. Fig.2 shows good agreement in numerical results. Fig.3 plots the total CPU time (sec.) vs the number of unknowns. In Fig.3 the lines marked 'Tdir', 'Tfmm' and 'Tfmmnew' indicate the CPU time required with the conventional BIEM, the original FM-BIEM and the new FM-BIEM, respectively. This figure shows that the new FM-BIEM is only slightly faster than the original FM-BIEM. This is because this example is essentially a two-dimensional one where the computational cost for the M2L translation is not dominant. In order to show the efficiency of the new FMM more clearly we need to consider an example where boundary elements are distributed three-dimensionally. Therefore we consider many crack problems in the next example.

Fig. 1. Mesh for a single crack (5736 DOF)

5.2 Many Cracks

We now consider an infinite space which contains an array of $12 \times 12 \times 12 (= 1728)$ penny-shaped cracks (total DOF=1,285,632), each having the same radius a_0 subjected to the same asymptotic condition as in the previous example. The centroids of these cracks are located at the same interval of $4a_0$ in each coordinate direction, but the orientation of each crack is taken random. Fig.4 plots the non-dimensional crack opening displacement $(\mu\phi/a_0p_0)$ on the non-dimensional mesh x/a_0. The required CPU times with FM-BIEM and the new FM-BIEM are 13954(sec.) and 8290(sec.), respectively. This result shows that the new FM-BIEM is more efficient than the original FM-BIEM when distribution of the boundary elements is dense in the domain. In this

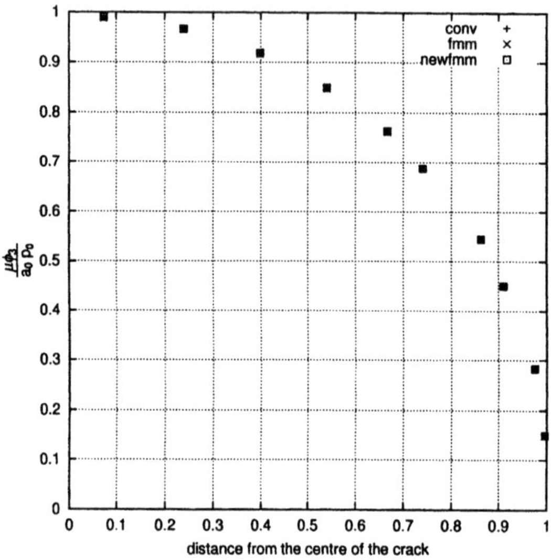

Fig. 2. Crack opening displacement

example the error defined as

$$\text{error} = \frac{||\widetilde{\phi} - \phi||}{||\phi||}$$

is 9.09×10^{-4}, where $\widetilde{\phi}$ is the numerical solution obtained with the new FM-BIEM, ϕ the one obtained with the original FM-BIEM and $||\cdot||$ denotes the L_2-norm.

6 Concluding Remarks

- In this paper we could successfully apply the new FMM to the three-dimensional elastostatic crack problems and could show that the new FMM is faster than the original FMM in sample problems.
- In the future work we plan to use singular elements to take the behaviour of ϕ near the crack tip into account and compute the interior stress field using the FMM techniques proposed in Yoshida et al. [8].
- The proposed techniques can be extended to the Galerkin BIEM which yields highly accurate numerical results for crack problems [8]. Also, the new FMM for the three-dimensional Helmholtz equation proposed by Greengard et al. [18] is expected to be applicable to three-dimensional elastodynamics in the frequency domain.

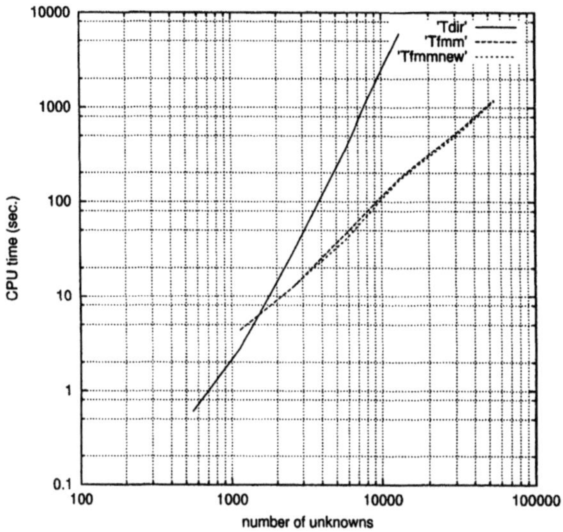

Fig. 3. CPU time (sec.)

References

1. Rokhlin, V.: Rapid solution of integral equations of classical potential theory. J. Comp. Phys. **60** (1985) 187–207
2. Greengard, L.: The Rapid Evaluation of Potential Fields in Particle Systems. The MIT Press, Cambridge (1987)
3. Nishimura, N., Yoshida, K. and Kobayashi, S.: A fast multipole boundary integral equation method for crack problems in 3D. Eng. Anal. Boundary Elements. **23** (1999) 97–105
4. Fu, Y., Klimkowski, K.J., Rodin, G.J., Berger, E., Browne, J.C., Singer, J.K., van de Geijin, R.A. and Vemaganti, K.S.: A fast solution method for three-dimensional many-particle problems of linear elasticity. Int. J. Num. Meth. Eng. **42** (1998) 1215–1229
5. Fukui, T. and Kutsumi, T.: Fast multipole boundary element method in three dimensional elastostatic problems. Proc. 15th Japan Nat. Symp. BEM. **15** (1998) 99–104 (in Japanese)
6. Takahashi, T., Kobayashi, S. and Nishimura, N.: Fast multipole BEM simulation of overcoring in an improved conical-end borehole strain measurement method. Mechanics and Engineering in Honor of Professor Qinghua Du's 80th Anniversary. Tsinghua University Press, Beijing (1999) 120–127
7. Yoshida, K., Nishimura, N. and Kobayashi, S.: Analysis of three dimensional elastostatic crack problems with fast multipole boundary integral equation method. J. Appl. Mech. JSCE. **1** (1998) 365–372 (in Japanese)
8. Yoshida, K., Nishimura, N. and Kobayashi, S.: Application of fast multipole Galerkin boundary integral equation method to elastostatic crack problems in 3D, Int. J. Num. Meth. Eng. **50** (2001) 525–547

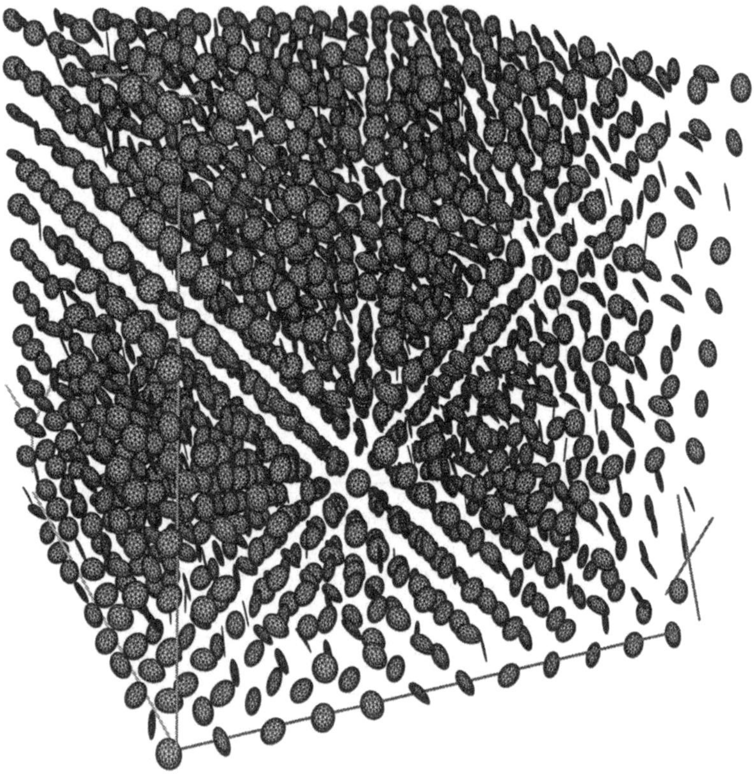

Fig. 4. Crack opening displacement (DOF 1,285,632)

9. Fujiwara, H.: The fast multipole method for solving integral equations of three-dimensional topography and basin problems, Geophys. J. Int. **140** (2000) 198–210
10. Yoshida, K., Nishimura, N. and Kobayashi, S.: Analysis of three dimensional scattering of elastic waves by a crack with fast multipole boundary integral equation method. J. Appl. Mech. JSCE. **3** (2000) 143–150 (in Japanese)
11. Rokhlin, V.: Rapid solution of integral equations of scattering theory in two dimensions. J. Comp. Phys. **86** (1990) 414–439
12. Rokhlin, V.: Diagonal forms of translation operator for the Helmholtz equation in three dimensions. Appl. Comp. Harmon. Anal. **1** (1993) 82–93
13. Elliot, W.D. and Board, J.A. JR.: Fast Fourier transform accelerated fast multipole algorithm. SIAM J. Sci. Comp. **17** (1995) 398–415
14. Dembart, B. and Yip, E.: The accuracy of fast multipole methods for Maxwell's equations. IEEE Comp. Sci. Eng. **5** (1998) 48–56
15. Hrycak, T. and Rokhlin, V.: An improved fast multipole algorithm for potential fields. SIAM J. Sci. Comp. **19** (1998) 1804–1826
16. Greengard, L. and Rokhlin, V.: A new version of the fast multipole method for the Laplace equation in three dimensions. Acta Numerica **6** (1997) 229–270

17. Cheng, H., Greengard, L. and Rokhlin, V.: A fast adaptive multipole algorithm in three dimensions. J. Comp. Phys. **155** (1999) 468–498
18. Greengard, L., Huang, J., Rokhlin, V. and Wandzura, S.: Accelerating fast multipole methods for the Helmholtz equation at low frequencies. IEEE Comp. Sci. Eng. **5** (1998) 32–38
19. Yoshida, K., Nishimura, N. and Kobayashi, S.: Application of new fast multipole boundary integral equation method to crack problems in 3D. Eng. Anal. Boundary Elements (to appear)
20. Yarvin, N. and Rokhlin, V.: Generalized Gaussian quadratures and singular value decomposition of integral operators. SIAM J. Sci. Comp. **20** (1998) 699–718
21. Pérez-Jordá, J.M. and Yang, W.: A concise redefinition of the solid spherical harmonics and its use in fast multipole methods. J. Chem. Phys. **104** (1996) 8003–8006
22. Biedenharn, L.C. and Louck, J.D.: Angular momentum in Quantum Physics: Theory and Application. Addison Wesley, London (1981)
23. Nishida, T. and Hayami, K.: Application of the fast multipole method to the 3D BEM analysis of electron guns. Boundary Elements XIX. Computational Mechanics Publications, Southampton (1997) 613–622

Computational Crack Path Prediction and the Singularities in Elastic-Plastic Stress Fields

Tetsuhiko Miyoshi

Yamaguchi University, Yamaguchi 753-8512, Japan

Abstract. The paper justifies the assumption that the exponent of the first term in asymptotic expansion of two-dimensional stresses at a crack tip of elastic-plastic body is independent of the angle θ in polar coordinates. First we discuss the case of a total deformation theory and then apply the idea used there to an incremental theory. These results can be effectively used to show the validity of a procedure used in computational crack path prediction for elasic-plastic bodies. In Appendix we show that, if the "$\hat{J}$-integral" does not vanish, the exponent is independent of the load parameter t too, and equal to $-\frac{1}{2}$ for stational cracks in the material with hardening, as is seen in elastic stresses.

1 Introduction

This research derives from a practical problem in computational crack path prediction in elastic-plastic bodies. For formulating a crack extension problem we need the criteria to determine the initiation, direction, speed and arrest of the extension, at least. In this paper, however, we limit consideration to only the direction. Several criteria to determine the direction of crack extension in *elastic* media have been proposed. (see, for example,[2],[5],[6],[12]). Some of them are applicable, at least formally, to plastic cracks too. The one which is usually called the maximum stress criterion or max- σ_θ criterion is one of them. This criterion conjectures that the crack will grow in a direction perpendicular to the maximum principal stress. This criterion is simple and has wide applicability.

In a certain procedure applying the max-σ_θ criterion, we meet a problem related to the order of singularity of plasic solutions, which has been simply assumed to be constant in the engineering literatures. The order of singularity of the stresses in 2-dimensional elastic crack is $O(r^{-\frac{1}{2}})$ in polar coordinates. This fact is well known and is used widely in the practical application of the theory of fracture. For plastic deformation there is a singular solution known as HRR solution. This solution was found by Hutchinson[7] and Rice and Rosengren[11]. In seeking this solution they assume that the material is governed by a deformation theory and that the exponent to express the singularity is constant. However this constant assumption must be rigorously examined, especially in applying the incremental plasticity theory, since the exponent might be dependent of the degree of the hardening. In fact the results in [1] for steadily growing cracks show the existence of such dependence even for bilinear hardening materials.

In this paper we examine the validity of the constant exponent assumption for in-plane plastic deformation. We discuss first the case based on a deformation theory and then proceed to the case of an incremental theory for materials with hardening. The starting assumption of the present paper is, like many other literatures [1],[7] and [11], for example, the existence of the stress function, and that the leading term of the stress function is of the form

$$\phi = r^{s(\theta)}(\log r)^j \, \varphi(\theta) \quad \text{or} \quad \phi = r^{s(t,\theta)}(\log r)^j \, \varphi(t,\theta)$$

in polar coordinates (r, θ), where t denotes the loading parameter in the flow theory of plasticity and j is an integer.

The paper shows that in both formulations the exponent s is independent of θ. In Appendix we show that, if an integral which is known as $\hat{J}$ -integral does not vanish as r tends to 0, then s is independent of t too and is $-\frac{1}{2}$ for stresses in the incremental formulation of stationary cracks as is seen in the elastic deformation. Since our approach is based on the use of a stress function, the equation to represent the compatibility condition of strains plays a key role. In treating this equation we use a formal manipulation program. The present study is motivated by the work of Kaminishi [8].

2 The max- σ_θ criterion

Throughout this paper we assume the following for simplicity, unless otherwise stated. See [10], for example, for the details on the general theory of plasticity.

(1) Plane-stress problem.

(2) Straight or curved crack with one end point (crack tip), stationary or steadily extending to the x direction in (x, y) plane. The origin of the (x, y) coordinates is taken at the crack tip, being the x coordinate the tangent to the crack at the crack tip.

(3) Nonlineality appears only in the stress-strain relations.

(4) The von Mises yield function is employed.

Let $(\sigma_{11}, \sigma_{22}, \sigma_{12})$ be the stresses in (x, y) coordinates (suffix 1 and 2 correspond to the coordinates x and y, respectively). Let (r, θ) be the polar coordinates taken at the crack tip, being $\theta = 0$ the plus part of the x coordinate. Let S be an arbitrary surface with direction θ. Then the normal and shear stresses on S are expressed as follows, respectively.

$$\sigma = \sigma_{11}\cos^2\theta + 2\sigma_{12}\sin\theta\cos\theta + \sigma_{22}\sin^2\theta$$
$$\tau = (\sigma_{22} - \sigma_{11})\sin\theta\cos\theta + \sigma_{12}(\cos^2\theta - \sin^2\theta).$$

The stresses should be divergent at the crack tip. However if the leading part of σ can be written as

$$\sigma = r^s\varphi(\theta) + \cdots, \quad s < 0$$

in the vicinity of the crack tip, it is natural to think that, in such a small region, the direction of the maximum tensile stress is, approximately, the direction θ at which the maximum of $\varphi(\theta)$ is attained. The max- σ_θ criterion conjectures, therefore, that the crack will grow in a direction perpendicular to such θ. Now it is easy to see that

$$\frac{d\sigma}{d\theta} = 2\tau,$$

so that the maximum of σ is attained at such θ that $\tau = 0$ is satisfied. Therefore the problem of predicting the crack path reduces to find the direction along which the shear stress $\tau = \tau(r, \theta)$ vanishes. Following experimental and theoretical results, the variation of this stress is rather gentle even near the crack tip and it is not so difficult to find the zero of the shear stress. In fact if the kinking of the crack is small, we can effectively apply the Newton's method as follows. Let (r, θ_0) be an approximate zero of $\tau(r, \theta)$. If

$$0 = \tau(r, \theta_0 + \delta) \approx \tau(r, \theta_0) + \tau_\theta(r, \theta_0)\delta, \qquad \tau(r, \theta) = r^s \psi(\theta) \tag{1}$$

we have the Newton correction

$$\delta = -\frac{\tau(r, \theta_0)}{\tau_\theta(r, \theta_0)} = -\frac{\psi(\theta_0)}{\psi'(\theta_0)}.$$

Therefore a precise approximation of the zero of τ will be obtained (in practical computation some modification is necessary, since the computation ψ' is difficult generally). This procedure, however, may break down if the exponent s in (1) is dependent on θ. In fact if $s = s(\theta)$ we have

$$\delta = -\frac{\psi(\theta_0)}{s'(\theta_0)\psi(\theta_0)\log r + \psi'(\theta_0)},$$

that is, the correction δ may not effectively work for small r. This situation can be clearly illustrated by a simple example. Assume, for example, that

$$\tau(r, \theta) = K r^{-\frac{1}{2}+\theta^2} \sin\theta.$$

Then the Newton correction δ takes the following values for different r.

r	1	0.5	0.1	0.01
δ	−0.199	−0.210	−0.242	−0.310

$$\left(\theta_0 = \frac{\pi}{16} = 0.196\cdots\right)$$

Therefore, it is necessary to examine whether or not the exponent s is really independent of θ. This is the motivation of the present study.

3 Deformation theory

In this section we consider the material with the nonlinearity of the so called Ramberg-Osgood type :

$$\epsilon = \sigma + \alpha\sigma^n, \tag{2}$$

and check the validity of the constant exponent assumption by a method which is applicable to other formulations of plasticity.
We follow the formulation in [7]. The stresses and strains are normalized by $\bar{\sigma_e}, \bar{\sigma_e}/E$, respectively, where $\bar{\sigma_e}$ is the initial yield stress and E the initial tangent of the stress - strain relation. We introduce the deviatonic stresses

$$s_{ij} = \sigma_{ij} - \frac{1}{3}\sigma_{kk}\delta_{ij}, \tag{3}$$

and the equivalent stress σ_e by

$$\sigma_e^2 = \frac{3}{2} s_{ij} s_{ij}. \tag{4}$$

Then the general stress-strain relation which is equivalent to (2) is written as follows. Let ν be the Poisson's ratio:

$$\epsilon_{ij} = (1+\nu)s_{ij} + \frac{1-2\nu}{3}\sigma_{pp}\delta_{ij} + \frac{3}{2}\alpha {\sigma_e}^{n-1} s_{ij}. \tag{5}$$

Let $\sigma = (\sigma_r, \sigma_\theta, \tau_{r\theta})$ be the stresses in polar coordinates. If there exists a stress function ϕ satisfying the following equation, then the equilibrium equations are automatically satisfied.

$$\begin{aligned} \sigma_r &= r^{-1}\phi_r + r^{-2}\phi_{\theta\theta} \\ \sigma_\theta &= \phi_{rr} \\ \tau_{r\theta} &= -(r^{-1}\phi_\theta)_r. \end{aligned}$$

The von Mises yield function is written as follows.

$$\sigma_e = \sigma_r^2 + \sigma_\theta^2 - \sigma_r\sigma_\theta + 3\tau_{r\theta}^2.$$

The stress-strain relations (5) are then written in polar coordinates as

$$\begin{aligned} \epsilon_r &= \sigma_r - \nu\sigma_\theta + \alpha\sigma_e^{n-1}(\sigma_r - \frac{1}{2}\sigma_\theta) \\ \epsilon_\theta &= \sigma_\theta - \nu\sigma_r + \alpha\sigma_e^{n-1}(\sigma_\theta - \frac{1}{2}\sigma_r) \\ \epsilon_{r\theta} &= (1+\nu)\sigma_{r\theta} + \frac{3}{2}\alpha\sigma_e^{n-1}\tau_{r\theta}. \end{aligned}$$

Since the compatibility condition of the strains

$$r^{-1}(r\epsilon_\theta)_{rr} + r^{-2}(\epsilon_r)_{\theta\theta} - r^{-1}(\epsilon_r)_r - 2r^{-2}(r(\epsilon_{r\theta})_\theta)_r = 0 \tag{6}$$

must be satisfied, the governing equation of the stress function is written as follows.

$$\begin{aligned} \Delta^2\phi + \frac{\alpha}{2}\Big\{ & r^{-1}\left(\sigma_e^{n-1}(2r\phi_{rr} - \phi_r - r^{-1}\phi_{\theta\theta})\right)_{rr} \\ & + 6r^{-1}\left(\sigma_e^{n-1} r(r^{-1}\phi_\theta)_r\right)_{r\theta} \\ & + r^{-1}\left(\sigma_e^{n-1}(-2r^{-1}\phi_r - 2r^{-2}\phi_{\theta\theta} + \phi_{rr})\right)_r \\ & + r^{-2}\left(\sigma_e^{n-1}(-\phi_{rr} + 2r^{-1}\phi_r + 2r^{-2}\phi_{\theta\theta})\right)_{\theta\theta}\Big\} = 0. \end{aligned} \tag{7}$$

The HRR solution of this equation is sought in the form

$$\phi = r^s \varphi(\theta), \tag{8}$$

where s is assumed to be constant. By using the J-integral twice or by solving (7) numerically, this constant has been determined in [7] and [11] to be

$$s = \frac{2n+1}{n+1}. \tag{9}$$

We want to examine whether the constant assumption on s is valid. We hence assume that the leading term of the stress function ϕ will be written as follows.

$$\phi = r^{s(\theta)} \varphi(\theta). \tag{10}$$

Note that the stresses in the polar coordinates are then written as follows.

$$\begin{aligned} \sigma_r &= r^{s(\theta)-2}\Bigg[\Big(s(\theta) + (s'(\theta)\log r)^2 + s''(\theta)\log r \Big)\varphi(\theta) \\ &\qquad + 2s'(\theta)\varphi'(\theta)\log r + \varphi''(\theta) \Bigg] \\ \sigma_\theta &= s(\theta)(s(\theta)-1) r^{s(\theta)-2}\varphi(\theta) \\ \tau_{r\theta} &= -r^{s(\theta)-2}\Bigg[\Big(s'(\theta) + (s(\theta)-1)s'(\theta)\log r \Big)\varphi(\theta) \\ &\qquad + (s(\theta)-1)\varphi'(\theta) \Bigg]. \end{aligned} \tag{11}$$

We examine the case where n is odd, so that the governing equation has a polynomial nonlinearity. Substitution (11) into the stress - strain relation leads the equation (7) to a nonlinear equation beginning from the leading terms of the following form.

$$r^{-i+js(\theta)}\Big(F_1(\theta)(\log r)^{k_1} + F_2(\theta)(\log r)^{k_1-1} + \ldots. \Big),$$

where $F_i(\theta)$ is a function of $s = s(\theta)$ and φ, and of their derivatives. The calculation of these coefficients is enormously complicated. We hence employed a formal manipulation program. We cite below the coefficients of $(\log r)^m$ of the terms of highest singularity in r of this equation for $n = 3$ and 5. The nonzero term begins from $m = 8$ and 12 for $n = 3$ and 5, respectively.

For $n = 3$:

$m = 8$: $9\alpha r^{-8+3s(\theta)}\varphi^3(\theta)\{s'(\theta)\}^8$

$m = 7, 6, 5$: 0 (if $s'(\theta) = 0$)

$m = 4$: $9\alpha r^{-8+3s(\theta)}\varphi^3(\theta)\{s''(\theta)\}^4$ (if $s'(\theta) = 0$)

$m = 3$: 0 (if $s'(\theta) = s''(\theta) = 0$)

$m = 2$: $\alpha r^{-8+3s(\theta)}\varphi^2(\theta)\Big(9s(\theta)\varphi(\theta) - 3s^2(\theta)\varphi(\theta) + 6\varphi''(\theta)\Big)$

$\{s'''(\theta)\}^2$ (if $s'(\theta) = s''(\theta) = 0$)

$m = 1$: 0 (if $s'(\theta) = s''(\theta) = s'''(\theta) = s''''(\theta) = 0$).

For $n = 5$:

$m = 12$: $25\alpha r^{-12+5s(\theta)}\varphi^5(\theta)\{s'(\theta)\}^{12}$

$m = 11, 10, 9, 8, 7$: 0 (if $s'(\theta) = 0$)

$m = 6$: $15\alpha r^{-12+5s(\theta)}\varphi^5(\theta)\{s''(\theta)\}^6$ (if $s'(\theta) = 0$)

$m = 5, 4, 3$: 0 (if $s'(\theta) = s''(\theta) = 0$)

$m = 2$: 0 (if $s'(\theta) = s''(\theta) = s'''(\theta) = 0$)

$m = 1$: 0 (if $s'(\theta) = s''(\theta) = s'''(\theta) = s''''(\theta) = 0$).

These coefficients must vanish near the crack tip, as far as the equation (7) holds. Therefore in both cases we have

$$\varphi(\theta)s'(\theta) = 0,$$

from the coefficients for $m = 8$(for $n = 3$) and for $m = 12$(for $n = 5$). It is clear by (11) that $\varphi(\theta) \equiv 0$ does not hold on any interval of positive length, since the equivalent stress σ_e should diverge at the crack tip and therefore the leading term of σ_e should not vanish. $s'(\theta) \neq 0$ implies that it must hold on a certain interval I including θ if $s'(\theta)$ is continuous, and this implies $\varphi(\theta) \equiv 0$ on I. Hence we have

$$s'(\theta) = 0 \qquad \text{for all } \theta,$$

that is, $s(\theta)$ is constant. Other coefficients therefore vanish automatically as is seen from the above result.

Remark. In the above argument we assumed (10) as the leading term of the stress function. Even if we assume the more general form

$$\phi = r^{s(\theta)}(\log r)^j\, \varphi(\theta),$$

the result is the same. In this case the non-zero term begins from $m = 3j+8$, for $n = 3$ for example, and the coefficient of $(\log r)^m$ is again

$$9\alpha r^{-8+3s(\theta)}\varphi^3(\theta)\{s'(\theta)\}^8.$$

4 Incremental theory - stationary cracks

In this section we apply the above method to the solution based on the incremental theory of plasticity. Consider a stationary crack subjected to increasing load. We assume that a sufficiently wide neighborhood of the crack tip under consideration is in plastic state . Also we assume that the material is governed by the so called Prandtl-Reuss' flow rule with kinematic hardning condition.
The equilibrium equations in the Cartecian coordinates are given by

$$\sum_{j=1}^{2} \frac{\partial}{\partial x_j}\sigma_{ij} = 0, \quad i = 1, 2.$$

Let $\alpha = (\alpha_r, \alpha_\theta, \alpha_{r\theta})$ be the parameter to express the center of the yield surface in the polar coordinates and

$$f^2(\sigma - \alpha) = (\sigma_r - \alpha_r)^2 + (\sigma_\theta - \alpha_\theta)^2 - (\sigma_r - \alpha_r)(\sigma_\theta - \alpha_\theta) + 3(\tau_{r\theta} - \alpha_{r\theta})^2.$$

We introduce the vector ∂f by

$$\partial f = \left(\frac{\partial f}{\partial \sigma_r}, \frac{\partial f}{\partial \sigma_\theta}, \frac{\partial f}{\partial \tau_{r\theta}}\right).$$

Then $\dot{\epsilon}$ and $\dot{\alpha}$ are given by the following equation.

$$\dot{\epsilon} = C\dot{\sigma} + \dot{\epsilon}_p \quad (C = D^{-1}), \tag{12}$$

$$\dot{\epsilon}_p = \frac{1}{\eta}\partial f < \partial f, \dot{\sigma} >,$$

$$\dot{\alpha} = (\sigma - \alpha)\frac{< \partial f, \dot{\sigma} >}{\bar{\sigma_e}}, \tag{13}$$

where D is the matrix to denote the modulus of elasticity, $<, >$ denotes the inner product of vectors and $(\dot{\ })$ denotes the differentiation on a loading parameter t. η denotes the rate of hardening which is assumed to be constant in what follows. Also $\bar{\sigma_e}$ is the initial yield stress and assumed to be unity for simplicity. The plastic state is characterized by the conditions

$$f(\sigma - \alpha) = \bar{\sigma_e}, \qquad < \partial f, \dot{\sigma} > \geq 0.$$

As in the preceding section we introduce the stress function. Since the stress should be dependent of the loading parameter we assume that, in the vicinity of the crack tip, the stress function will take the form

$$\phi = r^{s(t,\theta)}\varphi(t, \theta). \tag{14}$$

If this ϕ satisfies the relations

$$\begin{cases} \dot{\sigma}_r &= r^{-1}\phi_r + r^{-2}\phi_{\theta\theta} \\ \dot{\sigma}_\theta &= \phi_{rr} \\ \dot{\tau}_{r\theta} &= -(r^{-1}\phi_\theta)_r, \end{cases} \tag{15}$$

then the equilibrium equations in incremental form are automatically satisfied. Let us introduce a bounded function $\xi = (\xi_r, \xi_\theta, \xi_{r\theta})$ by

$$\xi = \sigma - \alpha. \tag{16}$$

This function is well defined and bounded at any point of the material since the stress point σ is not able to go out of the yield surface. Substituting (14) and (15) into (12) and using the compatibility condition of incremental form

$$r^{-1}(r\dot{\epsilon}_\theta)_{rr} + r^{-2}(\dot{\epsilon}_r)_{\theta\theta} - r^{-1}(\dot{\epsilon}_r)_r - 2r^{-2}(r(\dot{\epsilon}_{r\theta})_\theta)_r = 0, \tag{17}$$

we have an equation to be satisfied by the stress function.

We cite below the coefficients of $r^{-4+s(\theta)}(\log r)^m$, the term of the highest singularity of equation (17). A formal manipulation shows that the non-zero coefficients begin from $m = 4$ and

for $m = 4$: $\left(\dfrac{1}{E} + \dfrac{(\xi_r - 0.5\xi_\theta)^2}{\eta}\right)\varphi(t,\theta)\{s_\theta(t,\theta)\}^4$

for $m = 3$: 0 (if $s_\theta(t,\theta) = 0$)

for $m = 2$: $3\left(\dfrac{1}{E} + \dfrac{(\xi_r - 0.5\xi_\theta)^2}{\eta}\right)\varphi(t,\theta)\{s_{\theta\theta}(t,\theta)\}^2$ (if $s_\theta(t,\theta) = 0$)

for $m = 1$: 0 (if $s_\theta(t,\theta) = s_{\theta\theta}(t,\theta) = s_{\theta\theta\theta}(t,\theta) = s_{\theta\theta\theta\theta}(t,\theta) = 0$)

These results imply again that

$$s_\theta(t,\theta) = 0, \tag{18}$$

as is expected.

Remark 1. Even if we assume the more general form, instead of (13),

$$\phi = r^{s(t,\theta)}(\log r)^j\, \varphi(t,\theta) \tag{19}$$

as the stress function, where j is plus or minus integer, we have the same result. In this case the non-zero term begins from $m = j+4$ and the coefficient of the highest term is again

$$\left(\frac{1}{E} + \frac{(\xi_r - 0.5\xi_\theta)^2}{\eta}\right)\varphi(t,\theta)\{s_\theta(t,\theta)\}^4.$$

Remark 2. The independence of t, that is,

$$s_t(t,\theta) = 0$$

is not proved by the present method. This independence will be shown in Appendix.

5 Incremental theory - steadily growing cracks

The above argument is valid for steadily extending cracks. Consider a crack growing steadily and quasi-statically along the x axis. In this case the differentiation $\dot{(\ )}$ must be understood as

$$\dot{(\ )} = -\frac{\partial}{\partial x}. \tag{20}$$

We assumed for simplicty that the velocity of the crack extension is unity. The form of the stress function is (see [3],[4], for instance)

$$\phi = r^{s(\theta)} \, (\log r)^j \, \varphi(\theta).$$

We require that this ϕ satisfies the relations

$$\begin{cases} \sigma_r &= r^{-1}\phi_r + r^{-2}\phi_{\theta\theta} \\ \sigma_\theta &= \phi_{rr} \\ \tau_{r\theta} &= -(r^{-1}\phi_\theta)_r, \end{cases}$$

so that the equilibrium equations are automatically satisfied. To define the stress rates in polar coordinates we introduce the well known matrix M to transform the stresses represented by polar coordinates into those in Cartesian coordinates:

$$M = \begin{pmatrix} \cos^2\theta & \sin^2\theta & -\sin 2\theta \\ \sin^2\theta & \cos^2\theta & \sin 2\theta \\ \sin\theta\cos\theta & -\sin\theta\cos\theta & \cos 2\theta \end{pmatrix}.$$

The stress rates in Cartesian coordinates are obtained by differentiating

$$\begin{pmatrix} \sigma_{11} \\ \sigma_{22} \\ \sigma_{12} \end{pmatrix} = M \begin{pmatrix} \sigma_r \\ \sigma_\theta \\ \tau_{r\theta} \end{pmatrix}$$

in the sense of (20). The stress rates in the polar coordinates are obtained by inverting this relation, that is,

$$\begin{pmatrix} \dot\sigma_r \\ \dot\sigma_\theta \\ \dot\tau_{r\theta} \end{pmatrix} = M^{-1} \begin{pmatrix} \dot\sigma_{11} \\ \dot\sigma_{22} \\ \dot\sigma_{12} \end{pmatrix}$$

The other treatments are the same as in the stationary case. In this case the non-zero term begins from $m = 5 + j$ and the coefficient of the highest term $r^{-5+s(\theta)}(\log r)^{5+j}$ reads as follows.

$$-\left(\frac{1}{E} + \frac{(\xi_r - 0.5\xi_\theta)^2}{\eta}\right)\varphi(\theta)\sin\theta\{s'(\theta)\}^5. \tag{21}$$

Hence the situation is exactly the same as in the stationary cracks and the exponent $s(\theta)$ is independent of θ in this case too.

A Appendix : Estimation of the exponent

It is still open if the exponent is independent of the loading parameter t. In this appendix we will show that this independence is assured if the so called $\hat{J}$-integral does not vanish. At the same time it will be shown that the exponent is $-\frac{1}{2}$ for stationary cracks.

In estimating the exponent of the singularity of stresses, Hutchinson[7] and Rice and Rosengren[11] have employed the so called J-integral which is valid for the deformation theory. The key role of this integral is the path-independence, or for the present purpose that it does not vanish as r tends to 0. In this section we discuss the case of the incremental formulation for stationary cracks. For steadily growing cracks the present approach will not be valid since the integral of the J-integral type will vanish due to the low singularity of the stresses and strains[1]. We employ an integral known as $\hat{J}$-integral [9], which is originally derived so as to apply to various problems including plasticity analysis. We will show that the exponent of singularity is $-\frac{1}{2}$, as far as this integral does not vanish. In the following argument we assume that

(1) The whole region that we consider is in plastic state.

(2) The singularity of the stresses and strains occurs only at the crack tip.

(3) The order of singularity in stresses at the crack tip is estimated as

$$\sigma = O(r^{-\frac{1}{2}+\delta}),\ (\frac{1}{2} > \delta \geq 0). \tag{22}$$

The assumption (22) is motivated by the singularity of the HRR solution.

Before introducing the $\hat{J}$-integral we derive some preliminary estimates of the stresses, strains and the displacements. Let S be defined by

$$S = \begin{pmatrix} 1 & -0.5 & 0 \\ -0.5 & 1 & 0 \\ 0 & 0 & 3 \end{pmatrix}.$$

Then, since

$$\partial f = \frac{S(\sigma - \alpha)}{f}, \tag{23}$$

it is easy to see that

$$\dot{\epsilon_p} = \frac{1}{\eta} S\dot{\alpha}, \tag{24}$$

and that, since η is assumed to be constant,

$$\epsilon = C\sigma + \frac{1}{\eta} S\alpha. \tag{25}$$

To get the estimates of the singularities, we introduce a stress function ϕ as in Section 3. We have already known that the exponent s of the stress function is independent of the angle θ in polar coordinates (see the Remark of Section 4). Therefore we take a function of the form

$$\phi = r^{s(t)}\varphi(t,\theta), \quad (\frac{3}{2} \leq s(t) < 2)$$

as the stress function corresponding to the assumption (22). The stresses are then expressed as follows for certain functions φ_1.

$$\sigma = r^{s(t)-2}\varphi_1(t,\theta). \tag{26}$$

Since the condition

$$f(\sigma - \alpha) \leq \sigma_e$$

must holds always, we have

$$\alpha = r^{s(t)-2}\varphi_1(t,\theta) + (\text{bounded term}), \tag{27}$$

and by (25)

$$\epsilon = r^{s(t)-2}\varphi_2(t,\theta) + (\text{term of lower singularity}). \tag{28}$$

Using the strain - displacement relation in polar coordinates

$$\begin{cases} \epsilon_r &= u_{r,r} \\ \epsilon_\theta &= \frac{1}{r}(u_r + u_{\theta,\theta}) \\ \gamma_{r\theta} &= \frac{1}{r}(u_{r,\theta} - u_\theta) + u_{\theta,r}, \end{cases}$$

where $u_{,r}$ denotes the differentiation on r, we integrate the strains to get the displacements. We then have

$$u = r^{s(t)-1}\varphi_3(t,\theta) + (\text{higher order term}). \tag{29}$$

It is also clear by (24) that

$$\epsilon_p = r^{s(t)-2}\varphi_4(t,\theta) + (\text{term of lower singularity}). \tag{30}$$

Now the $\hat{J}$-integral is derived as follows. Let Γ_a and Γ_δ be the circles centered at a crack tip under consideration and of radious a and δ $(a > \delta)$, respectively. (Remark : As the following discussion shows, Γ_a and Γ_δ need not to be circles. Also, in [9], the integration to define the $\hat{J}$-integral is carried outside the *process region* in which the actual fracture should occur and the application of continuum mechanics breaks down. We here assume, however, that the plasticity theory introduced before is valid in all the material and neglect such an exceptional region). By $\Omega_{a,\delta}$ we denote the region bounded by Γ_a, Γ_δ and by the upper and lower sides of the crack. We start from the following identity.

$$\int_{\Omega_{a,\delta}} < \sigma, C\sigma_{,x} - \epsilon_{,x} + \frac{1}{\eta} S\alpha_{,x} > dxdy = 0. \tag{31}$$

Integration by parts in (31) leads to the following equation.

$$\int_{\Gamma_a - \Gamma_\delta} \{\frac{1}{2}[< \sigma, C\sigma > + \frac{1}{\eta} < \alpha, S\alpha >]\nu_x - < T(\sigma), u_{,x} >\}ds \tag{32}$$
$$+ \int_{\Omega_{a,\delta}} < \sigma - \alpha, \epsilon_{p,x} > dxdy = 0,$$

where $T(\sigma)$ denotes the traction vector on the circles and ν_x the directional cosine to x axis. The $\hat{J}$ integral is therefore defined by

$$\hat{J} = \int_{\Gamma_a} \{\frac{1}{2}[< \sigma, C\sigma > + \frac{1}{\eta} < \alpha, S\alpha >]\nu_x - < T(\sigma), u_{,x} >\}ds \tag{33}$$
$$+ \int_{\Omega_a} < \sigma - \alpha, \epsilon_{p,x} > dxdy,$$

where Ω_a is the inner zone of Γ_a. Note that the last term in the right-hand side of (33), the integral over Ω_a, is finite and hence vanishes as a tends to 0. In fact, by (30) we have

$$\epsilon_{p,x} = O(r^{s(t)-3}), \tag{34}$$

which implies the integrability of $< \sigma - \alpha, \epsilon_{p,x} >$. This integral is slightly different from the original $\hat{J}$-integral. In fact, in the original derivation the $\hat{J}$-integral includes the term of the form

$$\int_{\Omega_a} < \sigma, \epsilon_{p,x} > dxdy \tag{35}$$

as the integral over the region Ω_a and does not include the term for the back stress α. Note also that it is not clear if the integral in (35) is well defined. $\hat{J}$ stands for the so called energy release rate when this integration is applied to purely elastic problems. This is the outline to derive the $\hat{J}$ -integral. It is clear that this integral is path independent under the previous assumptions.

Now we substitute (26),(27),(29) and (34) into (33). We then have

$$|\hat{J}| \leq \text{constant} \cdot (r^{2s(t)-3} + r^{s(t)-1}). \tag{36}$$

Therefore, if $\hat{J}$ does not vanish as r tends to zero, we have

$$2s(t) - 3 \leq 0.$$

However, $2s(t) - 3 < 0$ implies that the stresses have the singularity stronger than $O(r^{-\frac{1}{2}})$, which contradicts to the assumption (22). Hence the only possible case is

$$s(t) = \frac{3}{2},$$

that is, the exponent of singularity in the stresses is independent of the parameter t and equal to $-\frac{1}{2}$ as far as the $\hat{J}$-integral does not vanish as r tends to zero.

Acknowledgment. The author expresses his sincere thanks to Professor Shigeru Aoki, Tokyo Institute of Technology for kind comment on this investigation. He thanks Professor Yoichi Sumi, Yokohama National University and Professor B.Andersson, Aeronautical Institute of Sweden for referring him to several papers.

References

1. Amazigo,J.C. and Hutchinson,J.W., Crack-tip fields in steady crack-growth with linear strain-hardening, J.Mech.Phys. Solids(1977)81-97.
2. Cotterell,B. and Rice,J.R., Slightly curved or kinked cracks, Int. Journ. of Fracture 16(1980)155-169.
3. Drugan,W.J., Rice,J.R. and Sham,T.L., Asymptotic analysis of growing plane strain tensile cracks in elastic-ideally plastic solids, J.Mech.Phys.Solids,Vol.30-6(1982)447-473.
4. Drugan,W.J., Limitations to leading-order asymptotic solutions for elastic-plastic crack growth, J.Mech.Phys.Solids,Vol.46-12(1998)2361-2386.
5. Erdogan,F. and Sih,G.C., On the crack extension in plates under plane loading and transverse shear, J.Basic Engng, 85,519-527(1963).
6. Hussain,M.A., Pu,S.L. and Underwood,J., Strain energy release for a crack under combined mode I and mode II, Fracture Analysis,ASTM STP 560,2-28(1974).
7. Hutchinson,J.W., Singular behaviour at the end of a tensile crack in a hardening material, J.Mech.Phys.Solids,Vol.16(1968)13-31.
8. Kaminishi,K., Prediction of the fatigue crack growth life in microelectronic solder joints, (in this Proceedings)
9. Kishimoto,K., Aoki,S. and Sakata, M., On the path-independent integral $-\hat{J}$, Engineering Fracture Mechanics 13(1980)841-850.
10. Miyoshi,T., Foundations of the Numerical Analysis of Plasticity, North-Holland Mathematical Studies 107,Chapter 1 (1984).
11. Rice,J.R. and Rosengren,G.F., Plane strain deformation near a crack tip in a power-law hardening material, J.Mech.Phys.Solids,Vol.16(1968)1-12.
12. Sih, G.C., Strain energy density factor applied to mixed mode crack problems, Int.J. Fracture 10,305-321(1974).

Editorial Policy

§1. Volumes in the following four categories will be published in LNCSE:

i) Research monographs
ii) Lecture and seminar notes
iii) Conference proceedings
iv) Textbooks

Those considering a book which might be suitable for the series are strongly advised to contact the publisher or the series editors at an early stage.

§2. Categories i) and ii). These categories will be emphasized by Lecture Notes in Computational Science and Engineering. **Submissions by interdisciplinary teams of authors are encouraged.** The goal is to report new developments – quickly, informally, and in a way that will make them accessible to non-specialists. In the evaluation of submissions timeliness of the work is an important criterion. Texts should be well-rounded, well-written and reasonably self-contained. In most cases the work will contain results of others as well as those of the author(s). In each case the author(s) should provide sufficient motivation, examples, and applications. In this respect, Ph.D. theses will usually be deemed unsuitable for the Lecture Notes series. Proposals for volumes in these categories should be submitted either to one of the series editors or to Springer-Verlag, Heidelberg, and will be refereed. A provisional judgment on the acceptability of a project can be based on partial information about the work: a detailed outline describing the contents of each chapter, the estimated length, a bibliography, and one or two sample chapters – or a first draft. A final decision whether to accept will rest on an evaluation of the completed work which should include

- at least 100 pages of text;
- a table of contents;
- an informative introduction perhaps with some historical remarks which should be accessible to readers unfamiliar with the topic treated;
- a subject index.

§3. Category iii). Conference proceedings will be considered for publication provided that they are both of exceptional interest and devoted to a single topic. One (or more) expert participants will act as the scientific editor(s) of the volume. They select the papers which are suitable for inclusion and have them individually refereed as for a journal. Papers not closely related to the central topic are to be excluded. Organizers should contact Lecture Notes in Computational Science and Engineering at the planning stage.

In exceptional cases some other multi-author-volumes may be considered in this category.

§4. Category iv) Textbooks on topics in the field of computational science and engineering will be considered. They should be written for courses in CSE education. Both graduate and undergraduate level are appropriate. Multidisciplinary topics are especially welcome.

§5. Format. Only works in English are considered. They should be submitted in camera-ready form according to Springer-Verlag's specifications. Electronic material can be included if appropriate. Please contact the publisher. Technical instructions and/or TeX macros are available via http://www.springer.de/author/tex/help-tex.html; the name of the macro package is "LNCSE – LaTeX2e class for Lecture Notes in Computational Science and Engineering". The macros can also be sent on request.

General Remarks

Lecture Notes are printed by photo-offset from the master-copy delivered in camera-ready form by the authors. For this purpose Springer-Verlag provides technical instructions for the preparation of manuscripts. See also *Editorial Policy.*

Careful preparation of manuscripts will help keep production time short and ensure a satisfactory appearance of the finished book. The actual production of a Lecture Notes volume normally takes approximately 12 weeks.

The following terms and conditions hold:

Categories i), ii), and iii):
Authors receive 50 free copies of their book. No royalty is paid. Commitment to publish is made by letter of intent rather than by signing a formal contract. Springer-Verlag secures the copyright for each volume.

For conference proceedings, editors receive a total of 50 free copies of their volume for distribution to the contributing authors.

Category iv):
Regarding free copies and royalties, the standard terms for Springer mathematics monographs and textbooks hold. Please write to Peters@springer.de for details. The standard contracts are used for publishing agreements.

All categories:
Authors are entitled to purchase further copies of their book and other Springer mathematics books for their personal use, at a discount of 33,3 % directly from Springer-Verlag.

Addresses:

Professor M. Griebel
Institut für Angewandte Mathematik
der Universität Bonn
Wegelerstr. 6
D-53115 Bonn, Germany
e-mail: griebel@iam.uni-bonn.de

Professor D. E. Keyes
Computer Science Department
Old Dominion University
Norfolk, VA 23529-0162, USA
e-mail: keyes@cs.odu.edu

Professor R. M. Nieminen
Laboratory of Physics
Helsinki University of Technology
02150 Espoo, Finland
e-mail: rni@fyslab.hut.fi

Professor D. Roose
Department of Computer Science
Katholieke Universiteit Leuven
Celestijnenlaan 200A
3001 Leuven-Heverlee, Belgium
e-mail: dirk.roose@cs.kuleuven.ac.be

Professor T. Schlick
Department of Chemistry and
Courant Institute of Mathematical
Sciences
New York University
and Howard Hughes Medical Institute
251 Mercer Street, Rm 509
New York, NY 10012-1548, USA
e-mail: schlick@nyu.edu

Springer-Verlag, Mathematics Editorial
Tiergartenstrasse 17
D-69121 Heidelberg, Germany
Tel.: *49 (6221) 487-185
e-mail: peters@springer.de
http://www.springer.de/math/
peters.html

Lecture Notes in Computational Science and Engineering

Vol. 1 D. Funaro, *Spectral Elements for Transport-Dominated Equations.* 1997. X, 211 pp. Softcover. ISBN 3-540-62649-2

Vol. 2 H. P. Langtangen, *Computational Partial Differential Equations.* Numerical Methods and Diffpack Programming. 1999. XXIII, 682 pp. Hardcover. ISBN 3-540-65274-4

Vol. 3 W. Hackbusch, G. Wittum (eds.), *Multigrid Methods V.* Proceedings of the Fifth European Multigrid Conference held in Stuttgart, Germany, October 1-4, 1996. 1998. VIII, 334 pp. Softcover. ISBN 3-540-63133-X

Vol. 4 P. Deuflhard, J. Hermans, B. Leimkuhler, A. E. Mark, S. Reich, R. D. Skeel (eds.), *Computational Molecular Dynamics: Challenges, Methods, Ideas.* Proceedings of the 2nd International Symposium on Algorithms for Macromolecular Modelling, Berlin, May 21-24, 1997. 1998. XI, 489 pp. Softcover. ISBN 3-540-63242-5

Vol. 5 D. Kröner, M. Ohlberger, C. Rohde (eds.), *An Introduction to Recent Developments in Theory and Numerics for Conservation Laws.* Proceedings of the International School on Theory and Numerics for Conservation Laws, Freiburg / Littenweiler, October 20-24, 1997. 1998. VII, 285 pp. Softcover. ISBN 3-540-65081-4

Vol. 6 S. Turek, *Efficient Solvers for Incompressible Flow Problems.* An Algorithmic and Computational Approach. 1999. XVII, 352 pp, with CD-ROM. Hardcover. ISBN 3-540-65433-X

Vol. 7 R. von Schwerin, *Multi Body System SIMulation.* Numerical Methods, Algorithms, and Software. 1999. XX, 338 pp. Softcover. ISBN 3-540-65662-6

Vol. 8 H.-J. Bungartz, F. Durst, C. Zenger (eds.), *High Performance Scientific and Engineering Computing.* Proceedings of the International FORTWIHR Conference on HPSEC, Munich, March 16-18, 1998. 1999. X, 471 pp. Softcover. 3-540-65730-4

Vol. 9 T. J. Barth, H. Deconinck (eds.), *High-Order Methods for Computational Physics.* 1999. VII, 582 pp. Hardcover. 3-540-65893-9

Vol. 10 H. P. Langtangen, A. M. Bruaset, E. Quak (eds.), *Advances in Software Tools for Scientific Computing.* 2000. X, 357 pp. Softcover. 3-540-66557-9

Vol. 11 B. Cockburn, G. E. Karniadakis, C.-W. Shu (eds.), *Discontinuous Galerkin Methods.* Theory, Computation and Applications. 2000. XI, 470 pp. Hardcover. 3-540-66787-3

Vol. 12 U. van Rienen, *Numerical Methods in Computational Electrodynamics.* Linear Systems in Practical Applications. 2000. XIII, 375 pp. Softcover. 3-540-67629-5

Vol. 13 B. Engquist, L. Johnsson, M. Hammill, F. Short (eds.), *Simulation and Visualization on the Grid.* Parallelldatorcentrum Seventh Annual Conference, Stockholm, December 1999, Proceedings. 2000. XIII, 301 pp. Softcover. 3-540-67264-8

Vol. 14 E. Dick, K. Riemslagh, J. Vierendeels (eds.), *Multigrid Methods VI.* Proceedings of the Sixth European Multigrid Conference Held in Gent, Belgium, September 27-30, 1999. 2000. IX, 293 pp. Softcover. 3-540-67157-9

Vol. 15 A. Frommer, T. Lippert, B. Medeke, K. Schilling (eds.), *Numerical Challenges in Lattice Quantum Chromodynamics.* Joint Interdisciplinary Workshop of John von Neumann Institute for Computing, Jülich and Institute of Applied Computer Science, Wuppertal University, August 1999. 2000. VIII, 184 pp. Softcover. 3-540-67732-1

Vol. 16 J. Lang, *Adaptive Multilevel Solution of Nonlinear Parabolic PDE Systems.* Theory, Algorithm, and Applications. 2001. XII, 157 pp. Softcover. 3-540-67900-6

Vol. 17 B. I. Wohlmuth, *Discretization Methods and Iterative Solvers Based on Domain Decomposition.* 2001. X, 197 pp. Softcover. 3-540-41083-X

Vol. 18 U. van Rienen, M. Günther, D. Hecht (eds.), *Scientific Computing in Electrical Engineering.* Proceedings of the 3rd International Workshop, August 20-23, 2000, Warnemünde, Germany. 2001. XII, 428 pp. Softcover. 3-540-42173-4

Vol. 19 I. Babuška, P. G. Ciarlet, T. Miyoshi (eds.), *Mathematical Modeling and Numerical Simulation in Continuum Mechanics.* Proceedings of the International Symposium on Mathematical Modeling and Numerical Simulation in Continuum Mechanics, September 29 - October 3, 2000, Yamaguchi, Japan. 2002. VIII, 301 pp. Softcover. 3-540-42399-0

Vol. 20 T. J. Barth, T. Chan, R. Haimes (eds.), *Multiscale and Multiresolution Methods.* Theory and Applications. 2002. X, 389 pp. Softcover. 3-540-42420-2

For further information on these books please have a look at our mathematics catalogue at the following URL: `http://www.springer.de/math/index.html`

GPSR Compliance

The European Union's (EU) General Product Safety Regulation (GPSR) is a set of rules that requires consumer products to be safe and our obligations to ensure this.

If you have any concerns about our products, you can contact us on ProductSafety@springernature.com

In case Publisher is established outside the EU, the EU authorized representative is:

Springer Nature Customer Service Center GmbH
Europaplatz 3
69115 Heidelberg, Germany

Batch number: 10704711

Printed by Printforce, the Netherlands